Technological Solutions for Water Sustainability: Challenges and Prospects

Technological Solutions for Water Sustainability: Challenges and Prospects

Towards a Water-secure India

Edited by

Ligy Philip, Thalappil Pradeep and S. Murty Bhallamudi

IWA PUBLISHING

Published by IWA Publishing
Unit 104–105, Export Building
1 Clove Crescent
London E14 2BA, UK
Telephone: +44 (0)20 7654 5500
Fax: +44 (0)20 7654 5555
Email: publications@iwap.co.uk
Web: www.iwapublishing.com

First published 2023

British Library Cataloguing in Publication Data
A CIP catalogue record for this book is available from the British Library

ISBN: 9781789063707 (Paperback)
ISBN: 9781789063714 (eBook)

This eBook was made Open Access in November 2023.

Doi: 10.2166/9781789063714

Contents

Section 5: Urban Water Management

Preface

India is soon to become the most populous nation in the world. Also, it is projected that by 2050, more than 800 million people will live in urban areas in India. As the resources are limited, expanding cities have been facing severe crises concerning the pollution of water bodies, the increasing gap between the demand and supply of domestic water, and the widening chasm between the amount of wastewater generated and the installed capacity of wastewater treatment plants. Challenges are also posed by lack of space, monitoring and data acquisition on water quality and quantity, appropriate and sustainable technologies, and vulnerability to climate change. Moreover, the existing infrastructure is aged and dilapidated, urging immediate attention.

Linkages between the right amount of water supply, sanitation, wastewater management, hygiene, and public health have been well established. Water supply and management of wastewater go hand in hand. One has to protect the water sources from getting contaminated and becoming unfit as drinking water sources. In many emerging economies, the absence of appropriate measures for treating wastewater, sullage, and septage from burgeoning septic tanks has contaminated many of the surface water and groundwater bodies. Conventionally, cities have been planned and operated on a linear flow of resources to feed, water, and shelter the growing urban population. However, recently, it has been recognized that the reuse and recycling of treated wastewater are economically viable and are attractive options for bridging the gap between the demand and supply of water. The development of tertiary treatment technologies makes it possible to treat wastewater to a level wherein it can be safely reused and recycled. But there are several concerns regarding the implication of large-scale recycling of treated wastewater from the perspective of emerging contaminants. It has also been recognized in recent times that flood risk and water scarcity can be tackled simultaneously through engineered groundwater recharge. There is an immediate need to develop new materials, processes, and technologies for efficient water and wastewater treatment, wastewater recycling, removal of residual pollution like emerging contaminants, nutrients, etc. Accordingly, appropriate remedial measures should be developed for reclaiming the polluted water bodies and providing safe water supply. Also, there is an immediate need for efficient monitoring and surveillance of water quality to safeguard public health and prevent environmental degradation. The technologies of supply should go hand in hand with the planning, designing, and operation of water infrastructure systems to optimally minimize water consumption, reduce water leakage, and improve water use efficiency.

The present edited volume provides comprehensive information about water and wastewater management, with a major focus on urban areas. Several technologies have been developed in the recent past in the Global North for advanced water and wastewater treatment, as well as for resource recovery and reuse. However, these technologies and processes from Global North may not be applicable directly in India and in other countries in Global South. Planning and installation of expensive water infrastructure in rapidly expanding urban agglomerations in India are complex because of uncertainties associated with the prediction of urbanization, such as lack of land area, transforming governance structures, limited finances, etc. The intermittent water supply systems have to transition to 24 × 7 systems. There is a necessity for large-scale monitoring of water quality and water quantity parameters and the employment of easy-to-use and economical sensors to ensure safety. Some of the other key factors are the availability of skilled manpower for operations, maintenance, and socio-economic conditions. The unique feature of the proposed edited volume is that it addresses all the above issues originating from the special perspective of India. It provides information about the adaption of technologies and the development of new technologies and management practices, which are context-driven and region-specific. It also deals with economical and easy-to-use sensors for large-scale monitoring of water quality and water quantity parameters.

The purpose of this book is to provide comprehensive information about water and wastewater management which will help to achieve water circularity in growing urban agglomerations in India. It is expected that this book will be highly valuable to all those who are concerned with technologies, planning concepts, and management practices. It is also expected that the material presented in this book will be of relevance to other emerging economies in Global South. The purpose is achieved by putting together edited chapters authored by academics from India based on studies conducted by them in the last five years. Most of the contributing authors are a part of Water Technology Research and Innovation Centre 'SUTRAM of EASY WATER' (Centre for Sustainable Treatment, Reuse and Management for Efficient, Affordable and Synergistic solutions for Water) funded by Water Technology Initiative (WTI) of the Department of Science and Technology (DST), Government of India (GoI). This is a consortium of nine institutes across India. Authors from other countries who have researched water issues in India and in other nations belonging to the Global South have also contributed.

The book is divided into five sections. Section 1 consists of five chapters and they provide information about the status and challenges for sustainable water management in India, from the perspective of water quality, industrial and domestic waste water treatment, urban water infrastructure and policy and governance towards water security. Section 2 deals with new age materials for water and wastewater treatment. Five chapters in this section discuss new framework solids for water purification; new materials for arsenic and fluoride removal, nano-composites for water and wastewater treatment and removal of hazardous materials, and toxicity of these materials. Section 3 presents the new technologies developed for water and wastewater treatment. It has seven chapters dealing with pulsed power technology, constructed wetlands, nutrient recovery, low-cost filters, and pollution abatement using waste-derived materials. It also has a chapter especially dealing with technology evaluation for sustainability. Section 4 focuses on sensors: the four chapters of this part present the development of low-cost colorimetric sensors for eutrophying ions, sensors for conductivity and flow parameters, and multi-analyte assessment for water quality. The last section of the book has five chapters, which address the issues related to urban water infrastructure, sustainable urban drainage, and integrated flood and water scarcity management. This section also discusses the virtual water and challenges faced in the implementation of new projects and technologies.

This book is intended as a reference book for graduate students, working professionals, and policymakers.

Acknowledgements

This book would not have materialized without immense help from various funding agencies and individuals. The team thanks WTI of the DST, GoI for setting up the Centre of Excellence SUTRAM for EASY Water, which is instrumental in bringing together many researchers to carry out focused work in the area of water and sustainability. Most of the contributing authors of this book are part of this centre. The team acknowledges the Institute of Eminence funding from the University Grants Commission, GoI through Indian Institute of Technology Madras for setting up the Research Centre on Water and Sustainability, which immensely helped the research activities of the group. The team also places on record the financial support received from the DST, GoI in the form of extramural research grants to several contributors of the book.

The help received from the IWA team was immense. We thank the entire team, especially Mr. Mark Hammond. We would like to thank the entire WTI team for their constant encouragement and support. We place on record our grateful thanks to all the contributors and reviewers of the book for their time and efforts, without which this book would not have seen the light. Our gratitude goes to the researchers and students involved in the water and sustainability team, and to our partnering Centre of Excellence, SUTRAM for EASY Water, as well as our international collaborators for their commitment and support to this endeavour. Their active involvement and dedication were the key to the success of this initiative.

A special thanks to Mr. D. Kumaran for coordinating the activities and to the team from M/S Skillskapes for their editorial support. We thank the Centre for continuing education, Indian Institute of Technology for partially funding the book writing project.

Ligy Philip, Thalappil Pradeep and S. Murty Bhallamudi

Section 1

The Status and Challenges for Sustainable Water Management in India

INTRODUCTION

Water covers two-thirds of the earth's surface. However, freshwater accounts for only 0.5% of the total volume. In addition to this, the demand for water has been ever-increasing and is expected to reach a whopping 5890 km^3/year by the middle of the present century. While the per capita demand for water has been increasing, the availability of the same has been decreasing due to various anthropogenic activities. Presently, more than 50% of the world's population lives under the threat of water-stress conditions. The challenge pertains to the unavailability of an adequate quantity of water that is suitable for use. Thus, water is now an integral focus of several sustainable development goals: SDG-2 (zero hunger); SDG-6 (clean water and sanitation); SDG–11 (sustainable cities and communities); and SDG-12 (responsible consumption and production).

In this context, it is critical to examine the status of water resources from both perspectives of availability and quality. The five chapters in Section 1 focus on this aspect. Much of the discussion in this section concentrates on India, with some additional references to other countries in the Global South. The chapter on **Sustainable management of water** discusses the impending water crisis in India and other nations in the Global South and how sustainability in domestic water supply, both in urban and rural areas, can be achieved. The chapter also discusses the challenges to achieving water sustainability and the way forward. The chapter on **Water quality status and challenges in India and its neighbour Nepal** reviews the water quality status. It suggests some solutions to fulfil water quality objectives sustainably. The focus is on the quality of water in the sources and how this quality is affected by mismanagement, poor sanitation facilities, improper handling of liquid and solid wastes, and uncontrolled population growth. A prerequisite for maintaining quality in water sources is treating the wastewater generated in homes and factories. The chapter on **Status and challenges in domestic and industrial wastewater treatment in India** discusses the role of domestic sewage and industrial effluents in water pollution and the challenges encountered in preventing this pollution. The importance of effective wastewater reuse practices and zero liquid discharge is illustrated through important example case studies. The chapter on the **Status and challenges of urban water infrastructure in India** highlights the impending water crisis in urban areas as the population in these locations continues to rise. The overview in this chapter includes domestic water supply, sewerage, and stormwater drainage systems. The challenges on the way to attaining water-sensitive cities are identified, and the main issues to be considered for overcoming the difficulties are discussed.

The chapter on **Designing water policy in India for sustainability** takes up the essential role water policy and governance play in achieving water sustainability. Effective and legitimate implementation of governance regimes for water has become critical due to climate change, leading to widespread interest in 'adaptive governance'. This chapter provides historical context, the current situation of governance, and existing challenges at national and subnational levels. Possible pathways to sustainability that could constitute adaptive governance in an existing peri-urban landscape are presented.

doi: 10.2166/9781789063714_0003

Chapter 1

Sustainable management of water

*S. Murty Bhallamudi and Ligy Philip**

Department of Civil Engineering, Indian Institute of Technology Madras, Chennai, India
**Corresponding author: ligy@iitm.ac.in*

ABSTRACT

Water is an integral component of several sustainable development goals: SDG-2 (zero hunger); SDG-6 (clean water and sanitation); SDG-11 (sustainable cities and communities) and SDG-12 (responsible consumption and production). It is used in four major sectors: agricultural; domestic; industrial and ecological, and these needs could often be conflicting with each other. Further, there is a significant temporal and spatial variation in the availability of fresh water in India and in many countries in the Global South. Many countries in the Global South are already facing a major water crisis due to stressors such as increasing population and climate change, among others. Therefore, it is important to manage the scarcely available water in a sustainable way to not only meet the present needs, but also the needs of future generations. In this chapter, we briefly discuss the impending water crisis in India and other nations in the Global South and how to achieve water sustainability, both in urban and rural areas. We discuss index-based methods for measuring sustainability. We also discuss the challenges to achieve water sustainability and the way forward.

Keywords: water sustainability, Global South, water scarcity, water management, climate change

1.1 INTRODUCTION

Water scarcity is one of the main problems faced by many countries. The gap between the demand for water and its availability is ever increasing due to stressors such as rapid development, increasing population, climate change, and dwindling water sources due to anthropogenic contamination. On the one hand, the water available in surface sources such as rivers, reservoirs, and lakes is inadequate to meet the increasing demands of domestic water supply, irrigated agriculture, industry, and water needed for ecosystem services. On the other, most of the water resources are getting contaminated due to the indiscriminate discharge of untreated and partially treated wastewater from domestic and industrial sectors. Even as there is a significant decrease in groundwater levels due to over exploitation (pumping), and reduced recharge due to urbanization and changes in land-use land cover (LULC), existing aquifers are getting contaminated by pollutants of natural and anthropogenic origin.

Rural and peri-urban India is mostly dependent on unprotected surface or groundwater for drinking purpose, but emerging pollutants such as pharmaceuticals, personal care products, and pesticides are finding their way into drinking water sources. Added to the woes of the non-availability of adequate amount of water, there is a significant loss of water in distribution systems and the water use efficiency in agriculture sector is abysmal. Water-intensive industries are facing severe problems with respect to water availability for the processes and treatment of complex wastewaters generated from these industries. Flood and drought frequencies are increasing due to the climate change effects. However, appropriate management strategies for using storm water and treated wastewater to augment water sources are lacking. Thus, there is an urgent need for sustainable management of the water resources to provide adequate quantity of good quality of water for all purposes, to all sections of society, while preserving this precious resource for the needs of future generations.

1.2 THE IMPENDING WATER CRISIS

Water covers more than two-thirds of the Earth's surface. But fresh water represents less than 0.5% of the total water on Earth. In 2017, the global freshwater consumption was 3800 km^3/year and is projected to increase to 5890 km^3/year by 2050 (Islam & Karim, 2019). Half of the world's population lives under high-to-low-water stress conditions, with one-fourth lacking access to safe and treated drinking water. The current status on sustainable development goal 6 on water and sanitation (SDG-6) is as follows: 26% and 46% of the world's population lacks a safely managed drinking water source and sanitation facility, respectively, and 19% of the world's renewable water resources are being withdrawn. Forty per cent of the world's monitored water bodies do not have good ambient water quality (United Nations, 2023a). Figure 1.1 shows the predicted water-stressed regions in the world in 2040.

It is predicted that many regions such as South Asia, Southeast Asia, Middle East, Southern Africa, Saharan Africa, Chile and Argentina will face high-to-extremely high-water stress in 2040. Several urban agglomerations in Global South are already facing water crisis. Beard and Mitlin (2021) analysed data from 15 cities in sub-Saharan Africa, South Asia, and Latin America and found that

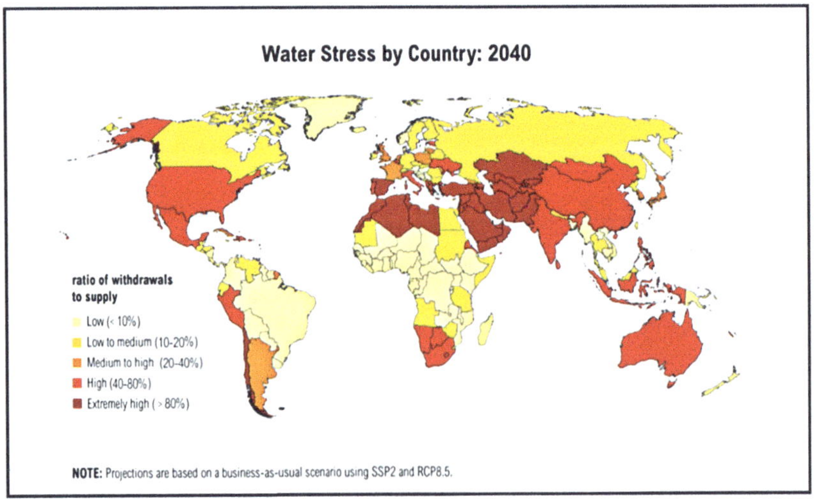

Figure 1.1 Map showing water stress regions in 2040. (*Source*: World Resource Institute: https://reliefweb.int/map/world/water-stress-country-2040).

municipal water supply was intermittent in 12 out of 15 cities. Most importantly, families that do not have access to public water supply obtain water from alternative sources at 52 times the normal cost. Importance of sustainable water management in cities in Global South became obvious after 'Day Zero' in Cape Town, South Africa on 23 May 2017 and in Chennai, India on 19 June 2019.

1.3 THE GAP BETWEEN AVAILABILITY AND NEED FOR WATER IN INDIA

India accounts for more than 17% of the world's population. However, its share of fresh water sources is only 4%. Presently, India has 1137 BCM (billion cubic metres) of water in surface and groundwater sources put together. As per the recent report prepared by the National Institution for Transforming India (NITI Aayog), the per capita water availability in 12 river basins including Ganga River basin is around 1000 m^3 (NITI Aayog, 2019). Approximately 820 million people living in these river basins are facing high-to-extremely high water stress. Due to inadequate wastewater treatment facilities, almost 70% of surface water sources are contaminated. The approximate generation of wastewater in 2020 was around 72,368 MLD against a functional treatment capacity of 26,869 MLD (CPCB, 2021). This has resulted in 163 million people not having clean water access in close proximity to their houses. Out of more than 80% of rural population, only 18% has access to piped water supply. It has been reported that the drought-prone areas in India have increased by 57% from the base year of 1997 (Down to Earth, 2022). More than four droughts occur in almost 30% of districts, affecting 50 million people. The estimates for water demand in different sectors in 2025 and 2050 are provided in Table 1.1. The problem of water challenge in India in future becomes obvious when the demand is compared with the estimated average per capita water availability of 1140 m^3 in 2050 (NITI Aayog, 2019).

1.4 SUSTAINABLE WATER MANAGEMENT

1.4.1 Water sustainability

Resolution 64/292 of the United Nations General Assembly, passed on 28 July 2010, established the human right to water and sanitation (United Nations, 2023b). It has explicitly recognized that clean drinking water and sanitation are essential to the realization of all other human rights. Based on this, sustainable water availability on a global scale will be achieved only when every person in the world gets affordable access to 20–50 litres of water per day.

As per the vision statement of the Water Services Regulation Authority (Ofwat) in England and Wales, sustainable water is described as '*a sustainable water cycle in which we are able to meet our needs for water and sewerage services while enabling future generations to meet their own needs*'. In general, water infrastructure systems have a long planning horizon and have to be operated and maintained over several decades. Therefore, the sustainability policy of the Environmental Protection Agency (EPA) of the United States of America mandates that drinking water and wastewater infrastructure projects be planned to ensure: (1) cost-effectiveness over the lifecycle; (2) maximization of resource efficiency, and (3) consistency with other relevant community goals. In this context, it should be recognized that there are multiple demands on water, chief among them being domestic, agriculture, industrial, energy, and ecological. Therefore, water sustainability is achieved only when

Table 1.1 Future water demand in India in BCM.

Year	Irrigation	Drinking	Industry	Energy	Others	Total
2025	910	73	23	15	72	1093
2050	1072	102	63	130	80	1447

https://www.statista.com/statistics/1111839/india-water-demand-by-sector/ downloaded on 7 January 2023.

there is a multidisciplinary, integrated, and holistic management of water resources, which focuses equally on economical, technical, ecological, aesthetic, social, and cultural issues.

1.4.2 Sustainability indices

In the last two decades, there has been a significant emphasis on evolving water resources management practices which reduce the vulnerability of these systems, that is, the magnitude of an adverse impact, in future scenarios. A sustainability index (SI) can be used to evaluate and compare the performances of various alternative water resources projects, subject to different actions and policies under different scenarios in the future. Juwana *et al.* (2012) made a comparative study of six SIs by exploring: (a) the selection of performance indicators, (b) methodology for obtaining sub-index values, (c) weighting schemes, (d) aggregation methodology; (e) robustness and (f) interpretation of the final index value. Recently, Vieira and Sandoval-Solis (2018) have used the following SI based on four performance criteria: volumetric reliability; resilience; vulnerability, and maximum deficit. Each of these criteria is measured using deficit in water supplied, D_t^i where i indicates the ith water user (irrigated area, livestock, urban water supply, and rural water supply) and t indicates the time period.

$$D_t^i = X_{\text{target},t}^i - X_{\text{supplied},t}^i \quad \text{if} \quad X_{\text{target},t}^i > X_{\text{supplied},t}^i; \quad \text{else}\, D_t^i = 0 \tag{1.1}$$

where X_{target} is the demand and X_{supplied} is the actual amount supplied. The four performance criteria are evaluated for each water user as follows:

$$\begin{aligned}
&\text{Rel}^i = \frac{\sum_{t=1}^{t=n} X_{\text{Supplied},t}^i}{\sum_{t=1}^{t=n} X_{\text{target},t}^i};\ \text{Res}^i = \frac{\text{No. of times } D_t^i = 0 \text{ follows } D_t^i > 0}{\text{No. of times } D_t^i > 0 \text{ occurred}} \\
&\text{Vul}^i = \frac{\left(\dfrac{\sum_{t=1}^{t=n} D_t^i}{\text{No. of times } D_t^i > 0 \text{ occurred}}\right)}{X_{\text{target}}^i}\ \text{Max Def}_i = \frac{\text{Max}\left(D_{\text{annual}}^i\right)}{\text{Water demand}^i}
\end{aligned} \tag{1.2}$$

The SI for a water user is then calculated as

$$SI^i = \left[\text{Rel}^i \times \text{Res}^i \times \left(1 - \text{Vul}^i\right) \times \left(1 - \text{Max def}^i\right)\right]^{1/4} \tag{1.3}$$

Finally, sustainability for the group is obtained as weighted sum of *SI*. Vieira and Sandoval-Solis (2018) have used this index for studying the sustainability of a water-stressed basin in Brazil.

1.4.3 Urban water sustainability

It is projected that by 2025, more than 50% of the population in the Global South will be living in urban areas accounting for a population of 3.75 billion. Urbanizations in different countries belonging to the Global South have the common characteristics such as: (1) strong urban–rural connect, (2) growing urban spread, and (3) increasing inequalities within urban areas. As the resources are limited, expanding cities in the Global South have been facing severe crisis with regard to water pollution, resource depletion, lack of space, equitable access to critical infrastructure and vulnerability to climate change (Bandauko *et al.*, 2022). Moreover, the existing infrastructure is aged and is in a dilapidated condition.

In the context of water, there is a conflict between urban domestic water demand and pre-urban agriculture demand. The growing cities should ensure equitable access to all stakeholders, increase the reliability of water supply to both the domestic and agriculture sector, and provide the maximum protection to the existing water bodies from contamination. Our urban agglomerations should provide necessary characteristics of resilience. To achieve a sustainable balance, practices and policies which create virtuous loops should be adopted in order to move towards circular systems. Bell (2020) has identified five distinct but overlapping frameworks for providing a deeper understanding of different conceptions of urban water sustainability. These include sustainable development, ecological modernization, socio-technical framings, urban political ecology, and radical ecology, which can be strategically deployed for high impact. Urban water infrastructure should be regenerative so that they minimize resource (water, energy, and materials) depletion and protect the water quality in sources. They should: (a) reduce the water consumption; (b) reuse the treated water for different purposes with suitably fit-for-purpose quality; (c) upcycle nutrients and organic matter and (d) recover the energy. The cities should also adopt nature-based solutions for simultaneously achieving flood protection and increasing the availability of fresh water.

Use of treated wastewater for potable and non-potable uses has been in practice for several decades in many water-scarce areas around the world, such as in Windhoek, Namibia since 1968. Forty per cent of water demand in Singapore is met by its NEWater scheme. Several cities in Brazil, Mexico, Kuwait, and India are planning projects for potable water reuse (Tortajada, 2020). In 2019, the city of Chennai in India has implemented two Tertiary Treatment Reverse Osmosis (TTRO) plants at Kodungaiyur and Koyambedu to treat 90 MLD of sewage and supply treated wastewater to industries in Chennai's northern belt. To improve climate resilience of water infrastructure, the city has also commissioned a tertiary treatment ultrafiltration plant at Nesapakkam, which provides treated wastewater for augmenting the drinking water supply in the city. Plans are being made to utilize abandoned stone quarries and rejuvenated ponds for storm water management as well as augmenting water supply sources.

1.4.4 Rural water sustainability

In the context of rural water sustainability, it is important to note that the major sector which uses water is agriculture. As mentioned earlier, the present water use efficiency in agriculture in India is significantly low. The situation is further worsened by adaptation of water-intensive cash crops. Sustainable water management in rural areas should address this issue. However, in this chapter, we restrict our discussion to domestic water supply. Readers are referred to the publication by OECD (2010).

UNICEF has estimated that access to safely managed water services in rural areas world over has increased from 39% to 53% between 2000 and 2017. As per the current status report of Jal Jeevan Mission (JJM) of India, 108,336,930 out of 192,657,780 households have been provided with tap water connection as on date. The percentage of connections has increased from 16.8% in 2019 to 56.3% in 2022. Several recent studies (Molinos-Senante *et al.*, 2019) concluded that sustainability of rural water supply is affected by deficiencies in maintenance, operation, and financial management, which mainly arise due to: (a) low population density; (b) low incomes; (c) unregulated services, and (d) non-availability of skilled work force and competent managers. Recently, Mvongo *et al.* (2021) conducted an extensive review of sustainability indices and concluded that most of the existing sustainability assessment frameworks are for water resources management and have not focused on integrative and contextualized analysis of the sustainability of water services in rural areas.

In many rural areas, drinking water is provided, but adequate sanitation and wastewater management systems are absent, leading to significant increase in the contamination of surface and groundwater sources. The Jal Shakti Ministry of the Government of India has identified the source sustainability as a priority area. It should be noted that source sustainability should not be limited to only protection of drinking water sources but also consider the water use for agriculture, climate change adaptation and community involvement. Capacity building of implementing agencies is a prime requirement for achieving sustainability.

A national wastewater reuse policy has been in existence in Tunisia since the early 1980s. Several states in India have also formulated policies for treated wastewater reuse in the last decade. A national framework for safe reuse of treated water was made in November 2022. In South Africa, 43% of treated wastewater reuse is in agriculture sector while it is around 58.4% in Greece and 78% in India (Kesari *et al.*, 2021). Recently, authors have implemented an integrated water management system in a rural area in Southern Indian state of Tamil Nadu. All the storm water from the village is collected in drains and is carried to a nearby rejuvenated surface water body. This pond recharges the groundwater as well as provides agricultural water. Simultaneously, the grey water from the households and the septic tank effluent are carried by a small-bore system, laid within the storm drainage system, to a constructed wetland for nature-based treatment. The treated wastewater is then used for agriculture purpose. Simultaneously, a municipal solid waste management system is put in place in order to protect the water bodies. Biodegradable waste is composted and is sold in the local market. Non-biodegradable waste is further sorted for recovering reusable material. Finances generated cover 30% of the operation and maintenance cost. Local people are trained to take care of operation and maintenance.

1.5 CHALLENGES TO ACHIEVING SUSTAINABILITY

1.5.1 Climate change

Projections for 2050, without the climate change effect, indicate that most of the river basins in India, except Godavari, Brahmani-Baitarani, Mahanadi and Narmada will be facing water scarcity. Climate change will further exacerbate the problem. It has been well recognized that climate change is going to pose a significant challenge to sustainable water management world over (Mujumdar, 2008). India already faces a significant uneven spatial distribution of water resources, for instance, the wide gap between the 11,430 mm of annual rainfall in Cherrapunji and less than 100 mm in Jaisalmer. The country has experienced several extreme climate events in the last decade; while the 2018 floods in Kerala caused an economic loss of about 4.3 billion USD, the 2016 drought affected around 330 million people and caused an economic loss of 100 billion USD. India supplies food to more than 17% of the world population, with rain-fed farming covering more than 56% the net sown area. Samantaray *et al.* (2022) predicted a likely increase in drought risks in the northwest, eastern and south regions of India in the near future, even under moderate climate change scenarios. Ali *et al.* (2019) have predicted an increase in flood risk in the Indian sub-continent in the future due to a significant increase in the multi-day precipitation extremes. The IPCC Report of 2021 has also indicated that heatwaves and humid heat stress will be more intense and occur more frequently in South Asian region in the future. It has also predicted heavy monsoon rainfalls in the future although there is a decline in southwest monsoon presently. Increased occurrence of flooding and drought events in India in future will significantly impact agriculture, water resources, and reservoir operations, and hence the sustainable management of water resources. The impacts on water resources due to climate change will be further exacerbated by the anthropogenic changes to land use and land cover and increased demand for water. It is expected that climate change may substantially affect irrigation withdrawals although it may not be the primary driver for increase in domestic and industrial water demand (Mujumdar, 2008).

1.5.2 Urbanization

The rapid urbanization in India and other cities in Global South will pose significant challenges to sustainable water management. The major drivers for increase in water demand in urban areas are: (a) a direct increase in demand due to increase in population; (b) demand aggregation due to migration of population from surrounding rural areas, (c) change in lifestyles due to increase in per capita income, and (d) an increase in industrial water demand. Many times, urbanization may occur in places which do not have water sources nearby and it may become necessary to transport water from far away locations. The city of Chennai in India transported around 80 million m^3 of water from Srisailam Reservoir, over a distance of 406 km in 2022. Bengaluru in India sources water from the Cauvery

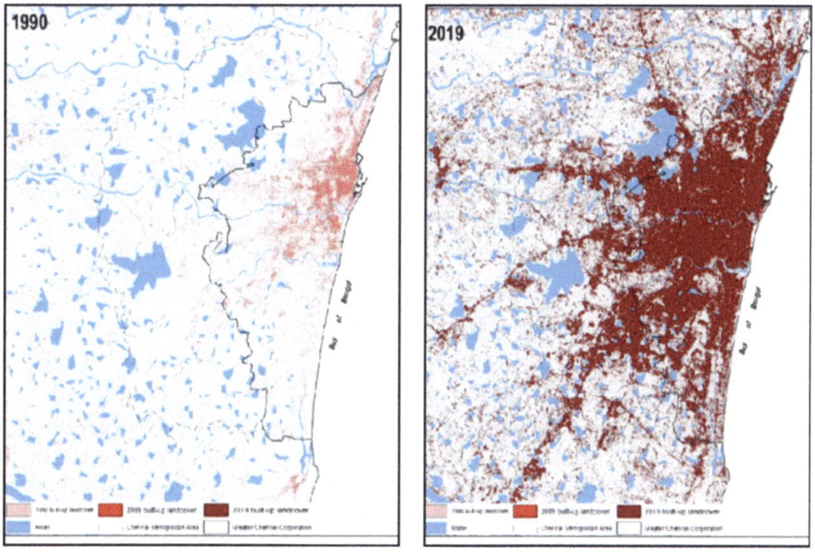

Figure 1.2 Growth of Chennai city from 1990 to 2019. (Indo-German Center for Sustainability, IIT Madras).

River which is 86 km away from the city. Transporting water over long distances to meet the water demand may not be economical and may also require significant energy. Also, a large amount of water may be lost during the transport.

Urbanization alters the local hydrological cycle significantly because it alters the land use and land cover, as shown for Chennai in Figure 1.2. The increased urbanization reduces the infiltration of rainwater during storms, causing a substantial increase in the runoff and frequency of flooding. Urbanization also leads to disappearance of small water bodies which could act as detention storages and mitigate floods (Devi *et al.*, 2019). Reduction in rainfall infiltration leads to reduction in groundwater recharge and thus the rapid lowering of groundwater table due to continued withdrawals. This in turn increases the sea water intrusion into aquifers in coastal cities such as Chennai. Thus, urbanization poses challenges to solving both the water scarcity and water excess problems.

1.5.3 Other challenges

Adaptation of sustainable water management practices and new technologies for water and wastewater treatment require an appropriate policy framework. For example, controlling sea water intrusion into aquifers through optimal groundwater pumping will require a policy on installation and operation of pumping wells by individual households for sourcing drinking water, by farmers for irrigation water and by the industry. Similarly, a policy framework may be needed for rainwater harvesting for not only decreasing the domestic water demand but also increasing the groundwater recharge. If the city is planning a sewerage system for collecting and treating domestic wastewater, it may have to pass a policy on household connections so that all households are connected to the system. Often, inappropriate policies may pose a challenge to the implementation of sustainable water management. For example, subsidies provided to farmers for power usage in many states in India has led to indiscriminate use of groundwater for irrigation. In many instances, there may not be political will to frame appropriate policies due to various extraneous reasons. Even if the policies are framed and laws are passed, there may not be political will to monitor and implement the laws. Challenges to sustainable water management also stem from social acceptance, availability of finances for implementation and the technical capacity for planning, operation and maintenance of infrastructure, and capacity for management.

India's population is poised to increase to 1.67 billion by 2050 from the current 1.40 billion people. Industrial production in India has increased by 7.1% in 2022 compared to that in 2021 and it is expected to achieve $5 trillion gross domestic product by 2024–2025. Thus, an increased demand for domestic as well as industrial water needs in the future, along with increasing generation of wastewater will pose challenges to sustainable water management. Although India is making significant strides in providing potable water to all in recent years, there is a significant gap between the amount of wastewater generated and the present infrastructure capacity for proper wastewater management.

1.6 THE WAY FORWARD FOR ACHIEVING SUSTAINABILITY

1.6.1 Circular economy

The gap between water availability and water demand has been widening in many places. This has forced many cities to source water either from faraway places or to go for costly desalination. The reuse of treated wastewater is an attractive option for bridging the gap between the demand and supply of water. In 2017, IIT Madras has implemented a water infrastructure system for 100% reuse of the treated wastewater for: (a) toilet flushing; (b) horticulture; (c) maintenance of sports facilities; (d) centralized air conditioning plants; (e) selling the treated water to external users, and (f) for storing in a lake for future use. The system costed 100 million INR for installation, and this investment was recovered within four years of installation. The efficient and safe operation of this 4 MLD capacity system has convinced Chennai city for scaling it to 260 MLD system to bridge the gap between the demand and the supply in the city. Wastewater is a source of energy, carbon, and nutrients. Technologies and management strategies need to be developed to make the wastewater/organic waste management a completely closed system with respect to water, nutrients, carbon, and energy.

It has also been recognized in recent times that the flood risk and water scarcity can be tackled simultaneously through local tank storage along with engineered groundwater recharge (Vanegas-Espinosa *et al.*, 2022). However, there are several concerns regarding the implication for large-scale engineered recharge and recycling of treated wastewater from the perspective of water quality and presence of emerging contaminants (Philip & Bhallamudi, 2019). The close monitoring of the system for water quality is necessary to prevent outbreak of diseases. Appropriate treatment units for the removal of residual pollutants such as nutrients, heavy metals, pharmaceuticals, personnel care products and so on should be included. Excess nutrients can cause eutrophication of the receiving water bodies, which deteriorate the water quality. In addition, algal toxins are a concern when such water is used for drinking purpose. It is important to study the presence of antimicrobial-resistant genes (ARGs) and antimicrobial-resistant bacteria (ARBs), as they can pose a serious threat to human health and wellbeing.

1.6.2 Integral management for increased resilience

The impact of increased imperviousness in causing urban floods is severe in the case of unplanned development in many countries such as India. Most of the present-day drainage systems in India and elsewhere do not pay much attention to flood control, water quality or biodiversity requirements in an integrated way. Sustainable urban drainage measures attempt to reduce surface runoff generation and increase infiltration of water into the soil. The long-term stability of water resources and increasing the resilience of water supply in urban areas can be achieved by adopting sustainable drainage and linking it to water supply. Implementing such measures should induce minimal impact on the watershed flow processes. Bioretention cells, rainwater harvesting and green roofs as low impact development techniques (LIDs) were studied by Palanisamy and Chui (2015) for reducing runoff generated from urban catchments. These drainage systems achieve sustainability by (a) controlling surface run-off, (b) reducing the impact of urbanization, (c) protecting water quality, (d) considering needs of the natural environment and community, (e) creating new wildlife habitats among the watercourses, and (f) promoting natural groundwater recharge.

A sustainable water supply system, with increased resilience, can be achieved by incorporating the measures of recycling of treated wastewater, rainwater harvesting at household level and managed aquifer recharge for increasing the groundwater storage. This necessarily means one has to plan all the three infrastructures, that is, the water supply system, the wastewater management system, and the storm water drainage system in an integrated way, unlike the present practice of planning the three components individually. It also makes it necessary to integrate the operation of all these three components of a city water infrastructure. Software tools need to be developed for optimal design and operation of all water infrastructures as a single integrated system.

1.6.3 Adaptive planning

Planning of expensive water infrastructure in rapidly expanding urban agglomerations in the Global South is complex because of uncertainties associated with prediction of urbanization, lack of land area, transforming governance structures, limited finances and so on. This requires the appropriate framing of water allocation studies based on changing land-use patterns, uncertainties in future demands, impact of climate change on water availability, spatial distribution of sources, and types of sources in the future.

The planning should be flexible to uncertainties in future urbanization. Innovative frameworks for multi-period and staged infrastructure planning and design need to be developed since traditional flexible design formulations are not equipped to handle deep uncertainties arising from climate change, population growth and migration, availability of innovative and cost-effective technologies in future, where a single probability distribution cannot be specified. New decision-making frameworks need to be developed for considering these deep uncertainties while planning the water infrastructure. Recently, Cunha *et al.* (2019) developed a novel conceptual framework for designing a water distribution network which adapts to changing conditions over a long planning horizon, divided into phases. Their methodology takes into account the uncertainty in the demands by considering many demand scenarios. They also consider four different criteria for decision making, including cost, greenhouse gas emissions, generalized resilience/failure index and loop diameter uniformity. A multicriteria decision analysis (MCDA), embracing an optimization model is used to size flexible alternatives.

1.7 CONCLUSIONS

Water scarcity is a major problem in most of the countries in Global South. Climate change, urbanization, increasing population and lack of appropriate policies will aggravate the water stress in these regions. Indiscriminate discharge of partially treated or untreated wastewater has been deteriorating the water quality in many surfaces as well as groundwater sources. The widening gap between the existing and required infrastructure for the wastewater treatment is a cause for concern. It is essential to adopt the circular economy concepts with respect to used water to handle the grievous situation. There is a need for integrated planning and design of water infrastructure for increasing the resilience of the system. It is also advisable to have adaptive/flexible planning to address the uncertainties especially in the ever-growing urban agglomerations. It is the right time to have appropriate policies towards sustainable water management and generate awareness among the citizens about the need to protect, preserve, and augment existing water resources.

REFERENCES

Ali H., Modi P. and Mishra V. (2019). Increased flood risk in Indian sub-continent under the warming climate. *Weather and Climate Extremes*, **25**, 100212, https://doi.org/10.1016/j.wace.2019.100212

Bandauko E., Arku G. and Nyantakyi-Frimpong H. (2022). A systematic review of gated communities and the challenge of urban transformation in African cities. *Journal of Housing and the Built Environment*, **37**(1), 339–368, https://doi.org/10.1007/s10901-021-09840-1

Beard V. A. and Mitlin D. (2021). Water access in Global South cities: the challenges of intermittency and affordability. *World Development*, **147**, 105625, https://doi.org/10.1016/j.worlddev.2021.105625

Bell S. J. (2020). Frameworks for urban water sustainability. *Wiley Interdisciplinary Reviews: Water*, **7**(2), 1411.

CPCB (2021). National inventory of sewage treatment plants. Report prepared by *Central Pollution Control Board (CPCB)*, Government of India.

Cunha M., Marques J., Creaco E. and Savić D. (2019). A dynamic adaptive approach for water distribution network design. *Journal of Water Resources Planning and Management*, **145**(7), 04019026, https://doi.org/10.1061/(ASCE)WR.1943-5452.0001085

Devi N. N., Sridharan B. and Kuiry S. N. (2019). Impact of urban sprawl on future flooding in Chennai city, India. *Journal of Hydrology*, **574**, 486–496, https://doi.org/10.1016/j.jhydrol.2019.04.041

Down to Earth (2022). https://www.downtoearth.org.in/news/climate-change/desertification-droughts-reduced-india-s-gdp-by-up-to-5-in-20-years--82794 (downloaded on 24 January 2023).

Islam S. M. F. and Karim Z. (2019). World's demand for food and water: The consequences of climate change. In: Desalination: Challenges and Opportunities, M. H. D. A. Farahani, V. Vatanpour and A. H. Taheri (eds.), pp. 57–84, Intech Open, London.

Juwana I., Muttil N. and Perera B. J. C. (2012). Indicator-based water sustainability assessment – a review. *Science of the Total Environment*, **438**, 357–371, https://doi.org/10.1016/j.scitotenv.2012.08.093

Kesari K. K., Soni R., Jamal Q. M. S., Tripathi P., Lal J. A., Jha N. K., Siddiqui M. H., Kumar P., Tripathi V. and Ruokolainen J. (2021). Wastewater treatment and reuse: a review of its applications and health implications. *Water, Air, & Soil Pollution*, **232**(5), 1–, https://doi.org/10.1007/s11270-021-05154-8

Molinos-Senante M., Muñoz S. and Chamorro A. (2019). Assessing the quality of service for drinking water supplies in rural settings: a synthetic index approach. *Journal of Environmental Management*, **247**, 613–623, https://doi.org/10.1016/j.jenvman.2019.06.112

Mujumdar P. P. (2008). Implications of climate change for sustainable water resources management in India. *Physics and Chemistry of the Earth, Parts A/B/C*, **33**(5), 354–358, https://doi.org/10.1016/j.pce.2008.02.014

Mvongo V. D., Defo C. and Tchoffo M. (2021). Sustainability of rural water services in rural sub-Saharan Africa environments: developing a Water Service Sustainability Index. *Sustainable Water Resources Management*, **7**(3), 1–17, https://doi.org/10.1007/s40899-021-00526-8

NITI Aayog (2019). Composite water management index. Report prepared by *NITI Aayog*.

OECD (2010). Sustainable management of water resources in agriculture.

Palanisamy B. and Chui T. F. M. (2015). Rehabilitation of concrete canals in urban catchments using low impact development techniques. *Journal of Hydrology*, **523**, 309–319, https://doi.org/10.1016/j.jhydrol.2015.01.034

Philip L. and Bhallamudi S. M. (2019). Editorial perspectives: innovation needs for the water sector in India to achieve sustainable development goals. *Environmental Science: Water Research & Technology*, **5**(7), 1200–1201, https://doi.org/10.1039/C9EW90026H

Samantaray A. K., Ramdas M. and Panda R. K. (2022). Changes in drought characteristics based on rainfall pattern drought index and the CMIP6 multi-model ensemble. *Agricultural Water Management*, **266**, 107568, https://doi.org/10.1016/j.agwat.2022.107568

Tortajada C. (2020). Contributions of recycled wastewater to clean water and sanitation: Sustainable development goals. *NPJ Clean Water*, **3**(1), 1–6, https://doi.org/10.1038/s41545-020-0069-3

United Nations (2023a). https://www.sdg6data.org/en/node/1 (downloaded on 9 May 2023).

United Nations (2023b). https://www.un.org/waterforlifedecade/human_right_to_water.shtml#:~:text=On%2028%20July%202010%2C%20through,realisation%20of20all%20human%20rights.

Vanegas-Espinosa L. I., Vargas-del-Río D., Ochoa-Covarrubias G. and Grindlay A. L. (2022). Flood mitigation in urban areas through deep aquifer recharge: the case of the metropolitan area of Guadalajara. *International Journal of Environmental Research and Public Health*, **19**(6), 3160, https://doi.org/10.3390/ijerph19063160

Vieira E. D. O. and Sandoval-Solis S. (2018). Water resources sustainability index for a water-stressed basin in Brazil. *Journal of Hydrology: Regional Studies*, **19**, 97–109, https://doi.org/10.1016/j.ejrh.2018.08.003

doi: 10.2166/9781789063714_0013

Chapter 2

Water quality status and challenges in India and Nepal

Syed Arwa A. Balkhi[1], *Bhesh Kumar Karki*[2], *Ligy Philip*[2]* *and Shihabudheen M. Maliyekkal*[1]

[1]*Department of Civil and Environmental Engineering, Indian Institute of Technology Tirupati, Tirupati, Andhra Pradesh, India*
[2]*Department of Civil Engineering, Indian Institute of Technology Madras, Chennai, India*
**Corresponding author: ligy@iitm.ac.in*

ABSTRACT

Water is a life-sustaining sector, and adequate access to safe water is fundamental to a healthy life and driving countries' economic growth. It is an endowment from nature that must be protected and defended. However, the gross negligence and mismanagement of the water bodies, poor sanitation facilities, improper handling of liquid and solid wastes, and uncontrolled population growth pose a severe threat to the freshwater sector in developing economies like India. The widespread presence of healthcare products, pharmaceutically active compounds, and similar unregulated compounds in freshwater is an emerging concern due to their increasing societal dependence. These pollutants are causing additional threats to the dwindling freshwater reserves and posing significant challenges to achieving water quality objectives. The frequency of occurrence of geogenic pollutants, including radioactive pollutants, is also increased due to anthropogenic factors. Though various measures have been taken to address the problem, lack of planning, regulatory guidelines, poor institutional framework, and corruption have hampered the beneficial outcome. As a result, many underprivileged people from rural and urban areas in India and other nations in the Global South still lack access to safe drinking water. Considering the complexity of the problem, a comprehensive approach is needed to improve the existing condition and provide safe water to everyone. This chapter reviews various challenges pertaining to water quality in India and Nepal and solutions to obtain water quality objectives in a sustainable way.

Keywords: challenges in the drinking water sector, emerging contaminants, monitoring of pollutants, water quality

2.1 INTRODUCTION

Water is a precious gift by nature that plays a crucial role in the health and economic development of individuals and communities. The environment also relies on water for various purposes, such as regulating temperature and supporting the growth of plants and animals. Additionally, safe water is essential for public health, as it is necessary for drinking and cooking, apart from industrial and agricultural activities. Water pollution, caused by anthropogenic activities and natural hazards such as agricultural runoff and industrial waste, mishandling of solid and liquid waste, and flooding, can have severe consequences for public health, socioeconomic structure, and the environment. It is more concerning that global water shortage is exacerbated by population growth, urbanization, and climate change, which is more vulnerable in developing countries like India.

India is dealing with severe water quality and availability issues, affecting human well-being, economic development, and ecological sustainability. India supports and sustains 17.7% of the world's population. However, it only possesses 4% of the world's water resources, making it challenging to manage limited and heterogeneously distributed supplies for a large population (NITI Aayog, 2019). Access to safe drinking water in the country is limited to 25% of dwellings, and more than 800 million people in India live with per capita water availability of nearly 1000 m^3/year. This indicates that more than half the population of the country lives in water-stressed areas (NITI Aayog, 2018). India recently ranked 139th out of 180 countries in the environmental performance index regarding drinking water and sanitation (Wolf *et al.*, 2022). The data is worrisome, and if adequate measures are not taken, it can pose a significant challenge to the country's economic growth.

Unlike India, neighbouring Nepal's average annual per capita water availability is significantly high (7173 m^3/year). However, more than 80% of the population in Nepal lacks access to safe drinking water (JICA, 2019). This is mainly due to a fourfold urban population rise (2021 census). Such dense urban population are forced to rely on inadequate and ageing water infrastructure due to limited resources along with the procrastination of bureaucratic systems, forcing them to consume contaminated water. On the other hand, scattered rural communities and the depletion of spring sources in hilly regions are the major causes of poor water availability. Apart from that, major natural catastrophes, such as the frequent occurrence of large earthquakes, frequently devastate water facilities. These constraints severely impede the country's economic progress. As discussed above, in order to alleviate water quality concerns in the Global South zone, particularly in Indian territories, it is vital to analyse the current and future drinking water challenges. Therefore, this chapter comprehensively discusses major aspects of water quality concerns and potential solutions to improve the poor public health, environment, and socio-economic aspects sustainably.

2.2 CURRENT AND FUTURE WATER QUALITY CHALLENGES IN THE INDIAN AND NEPALESE WATER SECTORS

The water quality challenges in the Indian and Nepalese water sectors are complex and multifaceted. Some of the current and future challenges are discussed below.

2.2.1 Water pollution

Most of the surface and groundwater bodies in the world are contaminated with diverse pollutants, including geogenic, microbiological, emerging contaminants, and radioactive compounds (Allabakshi *et al.*, 2022). About 1.96 million Indian homes have water that has chemical contamination, primarily from arsenic and fluoride (UNICEF, 2022). The South Asian Belt, including India, and Nepal, is reported to have the highest arsenic pollution levels (Shaji *et al.*, 2021). It is projected that 31% of the population in Nepal is exposed to arsenic contamination. According to the Central Groundwater Pollution Board, 24.2% of wells in Punjab, followed by 19.2% of wells in Haryana, are contaminated with uranium concentrations above the World Health Organization (WHO) limit (CGWB, 2020b). Salinity in water is also a common issue in India that has adverse effects on soil, and ecosystems. In India, salinity in water with an electrical conductivity of more than 4000 μS/cm is reported in a region of about 2×10^5 km^2 (CGWB, 2020a). Similarly, India and Nepal face issues with high nitrate concentrations, and heavy metals, including lead, cadmium, iron, and chromium in surface and groundwaters (CGWB, 2019). Figure 2.1 shows a state-wise depiction of various inorganic pollutants detected in water bodies in India and Nepal beyond the permissible limit.

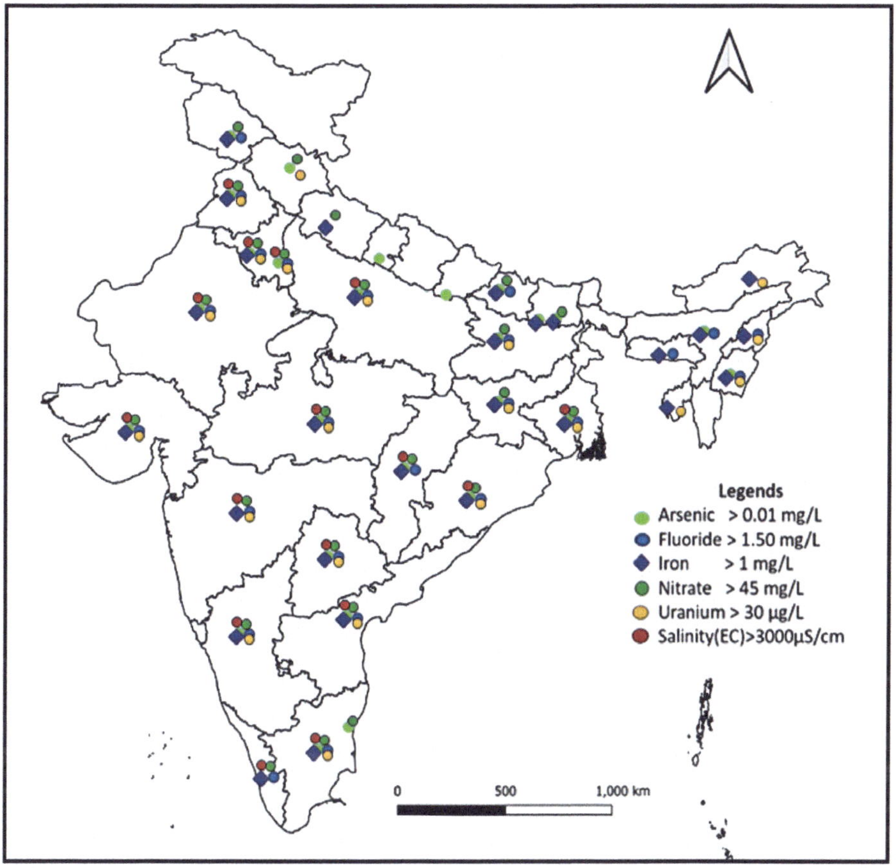

Figure 2.1 State-wise distribution of various inorganic pollutants in water bodies of India and Nepal. (*Data-sources*: CGWB, 2019, 2020b; Shaji *et al.*, 2021).

Besides inorganic pollutants, most water bodies in India and Nepal are contaminated with pathogenic organisms. McGuinness *et al.* (2020) reported that 69% of stored water samples collected from households in rural parts of India had been contaminated with *Escherichia coli* (*E. coli*). The primary reason for the contamination is the disposal of untreated or partially treated wastewater and poor sanitation and hygienic conditions. The same is true for Nepal, where a study found widespread faecal pollution of many water sources and a total disease burden of 0.32 disability-adjusted life years (DALYs) per person per year (PPPY), which is significantly higher than the WHO-recommended threshold of 0.0001 DALYs PPPY (Uprety *et al.*, 2020). Additionally, pharmaceutically active compounds, personal care products, endocrine-disrupting chemicals, disinfection by-products, pesticides, and microplastics are increasingly being detected in water bodies. These pollutants can come from various sources, including hospital effluents, industrial and agricultural practices, waste disposal, and the use of certain products. India has giant hitches as a result of having one of the highest rates of antibiotic use per capita, which is the main reason behind a serious contemporary healthcare and antimicrobial resistance gene in the water system. Moreover, it is not well understood how emerging contaminants will affect the environment, aquatic habitats, and human health in the long haul. The presence of many contaminants in this region's water system demonstrate the water system's complicated behaviour. Many geological formations with considerable lithological and temporal variances, a complex tectonic structure, climatological variables, unique hydro-chemical conditions, and uncontrolled anthropogenic activities all contribute to this. Knowing the distribution and prevalence of these contaminants in the aquatic system and creating management approaches are critical to addressing the growing problems.

2.2.2 Overuse and groundwater depletion

Most of the population in India and Nepal depends on groundwater for their water requirement, economic prosperity, and food security. However, it is reported that the groundwater level is dropping by 1–2 m/year in various regions of India and 2.5 m/year in Kathmandu, Nepal due to overexploitation (Dahal *et al.*, 2019; Dangar *et al.*, 2021). Extreme water depletion affects two-thirds of India's entire districts, which is a major worrisome condition for such a hefty population (World Bank, 2023). This leads to higher pumping expenses or wells drying up, land desertification, reduced agricultural products, saltwater intrusions, land subsidence, substantially reduced storage, and possibly infrastructure damage. These repercussions can be minimized by managing groundwater resources sustainably through an integrated strategy with proper water security planning incorporating more efficient water use, crop rotations, rainwater harvesting, and encouraging regulated aquifer recharging during the monsoon.

2.2.3 Intermittent supply and aged water infrastructure

India and Nepal use intermittent water distribution systems in water-stressed areas, mainly with outdated water infrastructures. More than 70% of water supply schemes in Nepal are dysfunctional (Sharma. *et al.*, 2021). These systems have poor hydraulic integrity and are prone to contamination during periods of non-supply. The withdrawal of water from the disruption system using illegal pumps can lead to leaky joints and water contamination. In the core city regions, water supply lines also frequently pass through or close to sanitary sewer systems, making the water supply more vulnerable to contamination. To tackle such issues, regular monitoring and periodic maintenance should be done.

2.2.4 Lack of wastewater treatment facilities

Insufficient wastewater treatment is the main driver of water quality degradation worldwide. In India, the amount of urban sewage generated is estimated to be 72,368 million litres per day (MLD) (CPCB, 2021). However, out of the existing treatment capacity of 26,869 MLD, only around 45% of treated effluents meet the national standards, indicating poor operational conditions. Approximately 27% of the country's wastewater is treated, which is more than the global average of 20%. Unfortunately, considering the magnitude of the problem, it is far from adequate. About 60% of industrial wastewater generated from manufacturing clusters is believed to be treated in India, but over 6200 MLD remains untreated, primarily because of the numerous constraints faced by small- and medium-sized industries (India Infrastructure, 2021). Due to limited treatment facilities, it is reported that more than 279 Indian rivers are found to be severely polluted which might reduce the downstream regions' GDP growth by up to a third (Desbureaux *et al.*, 2019). Similar to other low-income countries, about 30% of Nepal's generated urban wastewater is collected through sewer lines, out of which only about 3% is treated (JICA, 2019). These worst conditions also reveal a need for considerable immediate input in the wastewater management sectors in both countries so that surface and groundwater quality can be restored and protected. A comparison of wastewater generated and treated in the South Asian countries is presented in Table 2.1.

Table 2.1 General information regarding wastewater generation and treatment in South Asia.

Country	Urban Population (%)	Water Availability per Capita (m^3/year)	Population Covered by Basic Drinking Water (%)	Population Access to Safe Drinking Water (%)	Public Sewerage Coverage to Urban Population (%)	Municipal Sewage Generated (MLD)	Treatment to Wastewater Generated (%)	Source
India	34.93	1651	90	–	37	72368.00	27.96	World Bank data, CPCB (2021), FAO and Aquastat
Nepal	21.4	7347	90	18	30	1560.00	3.00	
Pakistan	37.17	1050	90	36	50	8384.00	1.00	
Bangladesh	38.81	7451	98	59	–	1986.30	6.04	
Bhutan	37.8	94,500	97	37	20	26.10	38.32	
Sri Lanka	18.8	2437	92	–	2.5	323.56	14.54	
Maldives	40.67	56.5	100	–	48	10.14	0.00	

2.2.5 Climate change

Climate change poses a significant threat to water security in regions such as India and Nepal. Global warming can cause changes in precipitation which can lead to droughts or floods. It can also lead to groundwater contamination and the melting of glaciers, posing a risk to freshwater resources. In addition, hotter temperatures and severe weather can also cause contaminants to be released from different sources such as mineral rocks, and increasing water pollution. Rising global temperatures may lead to saline water intrusion into groundwater aquifers, negatively affecting communities dependent on them. Implementing good water, sanitation, and hygiene systems can help create a stable and productive livelihood and increase a nation's capacity to withstand shocks caused by climate change shocks.

According to US EPA, various adaptation strategies can be adapted to mitigate the effect of climate change on water quality, including stormwater runoff management, controlling erosion and sedimentation by promoting green infrastructures, and algal bloom control using various mechanisms such as creating models to analyse potential changes in water quality, controlling the water quality in reservoirs and using effluent cooling systems (US EPA, 2023). This can incur higher costs to low- and middle-income countries, which might exceed 50% of the gross domestic product (Paudel & Pant, 2020), which is a worrisome situation for the future.

2.2.6 Transboundary water issues

Transboundary water issues can be complex, involving multiple stakeholders, including governments, organizations, and local communities. In India, several water resources are shared with neighbouring countries and states, such as the Indus River. Pollution and other variables at the cross-border stage can impact the quality of these resources, posing a risk to human health and the environment. Collaborative efforts between stakeholders are essential to managing transboundary water resources sustainably and equitably, considering the needs and concerns of all affected parties. This can promote economic and social development while ensuring the fair use of resources by all users.

2.3 WATER QUALITY CRITERIA AND REGULATIONS

Water quality criteria and regulations are established to ensure water quality objectives and protect public health and the environment. These standards are set based on guidelines from the WHO and other international organizations, and national and local regulations. The Central Pollution Control

Table 2.2 Comparative table for common basic drinking water quality standards.

S. No.	Parameters	India	Nepal	Remarks
		(BIS 10500: 2012)	(NDWQS–2022)	
1	Turbidity (NTU)	1 (5)	5	NHBV
2	pH	6.5–8.5	6.5–8.5	NHBV
3	Colour (TCU)	5 (15)	5	NHBV
4	Taste and odour	Agreeable	Agreeable	NHBV
5	EC (μs/cm)	–	1500	NHBV
6	Iron (mg/L)	0.3	0.30 (3)	NHBV
7	Manganese (mg/L)	0.1	0.2	NHBV
8	Arsenic (mg/L)	0.01 (0.05)	0.05	HBV
9	Fluoride (mg/L)	1.0 (1.5)	0.50–1.50	HBV
10	Ammonia-N (mg/L)	0.5	1.5	NHBV
11	Chloride (mg/L)	250	250	NHBV
12	Sulphate (mg/L)	200 (400)	250	NHBV
13	Nitrate (mg/L) (as NO_3)	45	50	HBV
14	Copper (mg/L)	0.05 (1.50)	1	NHBV
15	Zinc (mg/L)	5 (15)	3	NHBV
16	Aluminium (mg/L)	0.03 (0.2)	0.2	NHBV
17	Total hardness (mg/L)	200 (600)	500	NHBV
18	Residual chlorine (mg/L)	0.2 (1)	0.10–0.50	HBV
19	*E. coli* (CFU/100 mL)	Nil	Nil	HBV

Note: EC: electrical conductivity; HBV: health-based value; NHBV: non-health-based value; values within parentheses applicable only for no alternative availability of water sources.

Board (CPCB) is the national regulatory agency that monitors and controls pollution in the country, including setting standards for water quality in rivers, lakes, and other bodies of water. The water quality criteria set by CPCB divide water into different classes such as A, B, C, D, and E as per the 'Designated Best Use'. India and Nepal have their own drinking water quality standards, which are the Indian Standards for Drinking Water BIS 10500: 2012 and 'Nepal Drinking Water Quality Standards (NDWQS) – 2022', respectively. Both water quality standards establish permissible limits for numerous pollutants, including physical, chemical, and biological contaminants, to guarantee that the water is safe for human consumption. These standards are well accepted by academics, medical experts, and public health authorities. Table 2.2 shows the typical common water quality metrics from both standards. To make these standards more trustworthy, they must be revised on time, with permitted limits for specific contaminants including emerging contaminants.

2.4 PUBLIC HEALTH AND ENVIRONMENTAL IMPACTS OF WATER POLLUTION

According to the WHO, 2 billion people use unsafe water worldwide, resulting in approximately 829,000 deaths annually (Osseiran, 2017). In India, diseases related to water and sanitation account for 60% of the environmental health burden (Prüss-Ustün *et al.*, 2019). The World Bank estimates that water-related illnesses like cholera, dysentery, jaundice, and diarrhoea account for 21% of illnesses in India. Annually, approximately 37.7 million Indians are affected by waterborne diseases, resulting in 1.5 million deaths among children due to diarrhoea and a loss of 73 million working days. Children under five are particularly vulnerable, with an estimated 44,000 deaths annually in Nepal alone (TWP, 2021). The health impact of various pollutants is shown in Figure 2.2(a).

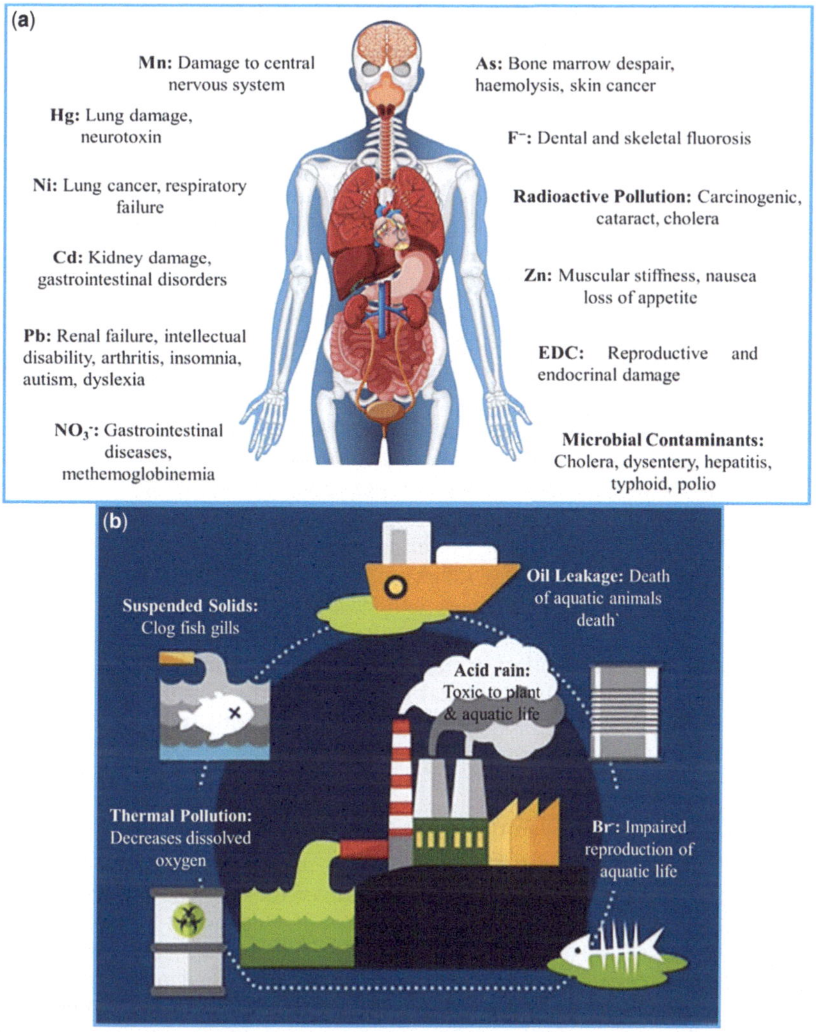

Figure 2.2 Effects of various pollutants such as manganese (Mn), mercury (Hg), nickel (Ni), cadmium (Cd), lead (Pb), nitrate (NO^{3-}), arsenic (As), fluoride (F^-), radioactive pollutants, zinc (Zn), endocrine disruptive chemicals (EDCs), bromide (Br^-), persistent organic pollutants (POPs), etc. on: (a) human health and (b) environment.

The source of water pollution such as acid rains, oil leakage, the presence of suspended solids and toxins, and thermal pollution has severe environmental impacts. These effects are summarized in Figure 2.2(b). In India, the River Ganga is one of the most polluted water bodies, with industrial waste and untreated sewage being discharged into the river. In Nepal, the Bagmati River, which is considered a holy river, is heavily polluted with untreated sewage and waste from households and industries. The pollution has caused damage to the ecosystem, with fish and other aquatic life being severely affected. The government and various organizations in both countries have taken steps to improve the situation, including the construction of sewage treatment plants and awareness campaigns to educate the public about the importance of clean water.

2.5 SOCIO-ECONOMIC IMPLICATION OF WATER POLLUTION

Water pollution has numerous economic repercussions. The economic repercussions manifest as property damage, land damage, decreased crop output, plant loss, tourism industry loss, the decline in waterfront property value, and other forms of economic loss. According to a World Bank analysis, heavily polluted water reduced economic growth by up to a third in some nations (World Bank, 2019). Various economic and societal problems arise due to the degradation of water quality, including reduced fish catches, navigational challenges, decreased water availability, and loss of recreational opportunities. In addition, industries may experience financial losses as a result of polluted water sources, which can lead to increased expenses for water treatment and cleaning efforts. These expenses can be significant and impact both businesses and individuals. Consequently, there is often a need for substantial investments in efforts such as river cleaning, chlorination, and drinking water treatment to mitigate these negative effects. Industries may also be required to spend more on treating contaminated water before discharging it into water bodies. Overall, the financial impact of water pollution is significant and has wide-ranging effects on society.

Each year, 200 million working days are lost due to water-related illnesses worldwide, killing 1.5 million children under the age of five. In India, water pollution-related health costs are estimated to be around INR 470–610 billion ($6.7–8.7 billion per year). The survival of 400,000 people is lost annually in India due to a lack of water, sanitation, and hygiene, in addition to the financial cost (Priyank & Vikas, 2019). In Nepal, water pollution is also a significant issue, with many people relying on contaminated water sources for drinking, cooking, and other household activities. This leads to a high incidence of waterborne diseases, which can result in significant medical costs and lost productivity. The pollution also affects fishing and other water-based industries, leading to economic losses for many communities. In addition, Nepal is a popular destination for ecotourism, but water pollution can lead to the degradation of natural habitats and the loss of tourism revenue. Overall, water pollution has significant socio-economic implications in both India and Nepal, affecting the health and livelihoods of millions of people.

2.6 CHALLENGES IN WATER RECLAMATION AND REUSE

Wastewater reclamation is the process of treating or processing wastewater to make it reusable, while wastewater reuse refers to the use of wastewater for various beneficial purposes. Water reuse is based on three principles: (1) providing reliable treatment of wastewater to meet strict water quality requirements for the intended reuse application, (2) protecting public health, and (3) gaining public acceptance. In India, the Central Pollution Control Board has developed guidelines for reusing treated wastewater. These guidelines provide recommendations on the treatment processes and quality standards for reclaimed water, as well as the types of reuse applications that are suitable based on the quality of the reclaimed water. However, the suitability of water reuse for a specific location is determined by various factors, including economic considerations, the potential uses for reclaimed water, and the level of strictness of waste discharge regulations. Public policies can be implemented that promote water conservation and reuse rather than the costly development of additional water resources with considerable environmental expenditures.

As a part of the reclamation, the Panipat refinery in India was able to expand its operations, while also increasing economic activity. This move resulted in a cost saving of EUR 5000–12,400 per year by utilizing reclaimed water, and a decrease in water demand by 16,000 m^3 per day. Moreover, it had a beneficial environmental impact by reducing freshwater withdrawal by 7000 m^3 per day and preventing sewage discharge in the Yamuna Canal (Jehan Gulamussen *et al.*, 2019).

However, gaining public acceptance of water reuse is a challenge, as there may be concerns about the safety and quality of reclaimed water. Despite the limited examples of wastewater reclamation and water reuse in Indian cities, there remains a hesitancy among people to utilize reclaimed water for potable purposes (Panwar *et al.*, 2015). Therefore, it is necessary to make a concerted effort to educate the public on the advantages of water reuse. Economic considerations are also a challenge in implementing water reuse, as there may be significant costs associated with the treatment and distribution of reclaimed water. In addition to this, technical challenges may be associated with the distribution and storage of reclaimed water. For example, it may be necessary to construct new pipelines or other infrastructure to transport the reclaimed water to its intended use.

Overall, the reclamation and reuse of wastewater present several challenges that need to be carefully considered and addressed to effectively and safely implement this approach to water management.

2.7 FUTURE PERSPECTIVE AND THE WAY FORWARD

Several factors contribute to poor water quality. Improving the water quality status will require a holistic approach through sustainable management addressing the underlying causes of water pollution and scarcity, as shown in Figure 2.3. This will involve a combination of government policies and programmes, investment in infrastructure and technology, timely updating of water quality criteria and standards, effective regulations to control pollution, community engagement, and raising awareness. In India, the 'Jal Jeevan Mission' programme was launched to provide functional household tap connections to every rural household by 2024 and aims to improve the overall water quality status in the country (Ministry of Water Resources, 2021). However, these efforts have been hindered by a lack of resources and capacity, and a lack of political will to implement the necessary transformations. Nepal's government has also implemented policies and programmes aimed at improving water quality, such as the National Water Plan and the Integrated Watershed Management Program (JICA, 2019). Because of a shortage of resources, these initiatives are only partially successful and take more time to implement. Some preliminary outputs and outcomes of these initiatives can be noted like the National Water Plan has made significant progress in the development of water policy and institutional frameworks in Nepal. These help in building the capacity of communities in watershed management, promoting community-based forest management, and improving soil and

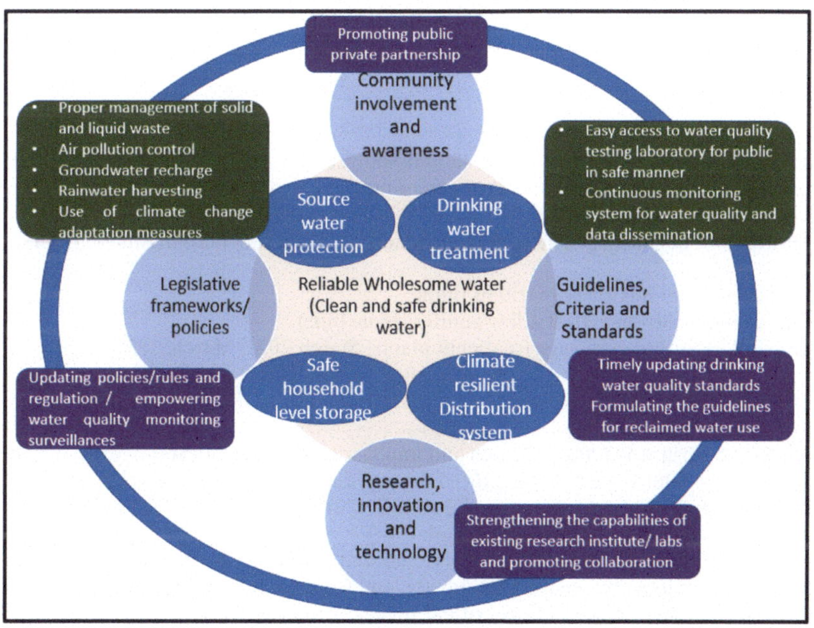

Figure 2.3 Proposed sustainable way forward for improving water quality.

water conservation practices. The programme has also supported the development of community-led micro-irrigation schemes, which have increased access to water for irrigation and improved agricultural productivity (Panwar *et al.*, 2015).

In addition, climate change adaptation measures are essential to mitigate the impact of climate change on water resources. Adaptation measures such as water harvesting, climate-resilient water infrastructure, recharge and storage, water conservation, use of reclaimed water, and more efficient water use are required to mitigate the impact of climate change on water resources. Overall, there is a need for greater cooperation and collaboration among the concerned stakeholders to address the current and future water quality challenges in both India and Nepal. The government must take a more proactive role in addressing water quality issues. The private sector and civil society should be engaged in implementing policies and programmes that help to improve water quality.

2.8 SUMMARY

The growing urbanization, industrialization, and climate change are accelerating the many problems in the drinking water sectors, such as increased water demand, restricted access to good-quality water, depletion of groundwater level, and increased waste generation. The widespread presence of numerous geogenic and other contaminants, together with insufficient wastewater treatment facilities, play critical roles in deteriorating the water quality and resulting in poor public health. To avoid such problems in the future, it is necessary to take a few more key areas into consideration, which include the implementation of low-cost sanitation systems with fewer subsidies, efficient water quality monitoring including emerging contaminants, planning, rural drainage systems, awareness-raising, involving NGOs and local groups, development of efficient treatment technologies, and human resource development. Government policies on quality and legal action such as penalties against polluters mostly industries should also be strictly implemented. Furthermore, the government should take care of socio-economic issues and implementation of reclamation activities to address water scarcity. Additionally, the strategy on sustainable approaches for treatment and their public unacceptance should be implemented and resolved. Developing effective policies, emphasizing sustainability with political commitment, and using the proper public–private partnerships are requirements for change.

REFERENCES

Allabakshi S. M., Srikar P. S. N. S. R., Gangwar R. K. and Maliyekkal S. M. (2022). Application of plasma-assisted advanced oxidation processes for removal of emerging contaminants in water. *Energy, Environment, and Sustainability*, 333–370, https://doi.org/10.1007/978-981-16-8367-1_15/COVER

CGWB (2019). States wise details of partly affected districts with select contaminants in ground water of India *. pp. 1–17.

CGWB (2020a). Ground water quality scenario in India.

CGWB (2020b). Uranium occurrence in shallow aquifer (Issue June).

CPCB (2021). National inventory of sewage treatment plants, March 2021. March, 183.

Dahal A., Khanal R. and Mishra B. K. (2019). Identification of critical location for enhancing groundwater recharge in Kathmandu Valley, Nepal. *Groundwater for Sustainable Development*, **9**(November 2018), 100253, https://doi.org/10.1016/j.gsd.2019.100253

Dangar S., Asoka A. and Mishra V. (2021). Causes and implications of groundwater depletion in India: A review. *Journal of Hydrology*, **596**(August 2020), 126103, https://doi.org/10.1016/j.jhydrol.2021.126103

Desbureaux S., Damania R., Rodella A., Russ J. and Zaveri E. (2019). The impact of water quality on GDP growth: evidence from around the world. The World Bank.

India Infrastructure (2021). Working with waste: Key trends and developments in industrial effluent management.

Jehan Gulamussen N., Arsénio A. M., Pedro Matsinhe N. and Cornelis Rietveld L. (2019). Water reclamation for industrial use in sub-Saharan Africa – A critical review. *Drinking Water Engineering and Science*, **12**(2), 45–58, https://doi.org/10.5194/DWES-12-45-2019

JICA (2019). Federal Democratic Republic of Nepal data collection survey on water supply and waste water sector in Nepal.

McGuinness S. L., O'Toole J., Barker S. F., Forbes A. B., Boving T. B., Giriyan A., Patil K., D'Souza F., Vhaval R., Cheng A. C. and Leder K. (2020). Household water storage management, hygiene practices, and associated drinking water quality in rural India. *Environmental Science & Technology*, **54**(8), 4963–4973, https://doi.org/10.1021/acs.est.9b04818

Ministry of Water Resources, G. (2021). Drinking water quality monitoring & surveillance framework.

NITI Aayog (2018). Comprehensive water management index (CWMI).

NITI Aayog (2019). Composite water management index. pp. 11–214.

Osseiran N. (2017). 2.1 Billion People Lack Safe Drinking Water at Home, More Than Twice as Many Lack Safe Sanitation. World Health Organization, pp. 1–6, https://www.who.int/news/item/12-07-2017-2-1-billion-people-lack-safe-drinking-water-at-home-more-than-twice-as-many-lack-safe-sanitation

Panwar M., Manoj Panwar A. and Sunil Antil M. (2015). Issues, challenges and prospects of water supply in urban India. *IOSR Journal of Humanities and Social Science (IOSR-JHSS)*, **20**(5), 68, https://doi.org/10.9790/0837-20526873

Paudel U. and Pant K. P. (2020). Estimation of household health cost and climate adaptation cost with its health related determinants: empirical evidences from western Nepal. *Heliyon*, **6**(11), e05492, https://doi.org/10.1016/j.heliyon.2020.e05492

Priyank H. and Vikas D. (2019). Water pollution is killing millions of Indians. Here's how technology and reliable data can change that | World Economic Forum, https://www.weforum.org/agenda/2019/10/water-pollution-in-india-data-tech-solution/

Prüss-Ustün A., Wolf J., Bartram J., Clasen T., Cumming O., Freeman M. C., Gordon B., Hunter P. R., Medlicott K. and Johnston R. (2019). Burden of disease from inadequate water, sanitation and hygiene for selected adverse health outcomes: An updated analysis with a focus on low- and middle-income countries. *International Journal of Hygiene and Environmental Health*, **222**(5), 765–777, https://doi.org/10.1016/J.IJHEH.2019.05.004

Shaji E., Santosh M., Sarath K. V., Prakash P., Deepchand V. and Divya B. V. (2021). Arsenic contamination of groundwater: A global synopsis with focus on the Indian Peninsula. *Geoscience Frontiers*, **12**(3), 101079, https://doi.org/10.1016/j.gsf.2020.08.015

Sharma, Baidya M., Poudel P., Panthi S. R., Pote-Shrestha R. R., Ghimire A. and Pradhan S. P. (2021). Drinking water status in Nepal: An overview in the context of climate change. *Journal of Water Sanitation and Hygiene for Development*, **11**(6), 859–866, https://doi.org/10.2166/washdev.2021.045

TWP (2021). Water in crisis – Spotlight Nepal, https://thewaterproject.org/water-crisis/water-in-crisis-nepal

UNICEF (2022). Water, Sanitation and Hygiene: strengthening Sustainable WASH Programming. UNICEF India, https://www.unicef.org/india/what-we-do/water-sanitation-hygiene

Uprety S., Dangol B., Nakarmi P., Dhakal I., Sherchan S. P., Shisler J. L., Jutla A., Amarasiri M., Sano D. and Nguyen T. H. (2020). Assessment of microbial risks by characterization of Escherichia coli presence to analyze the public health risks from poor water quality in Nepal. *International Journal of Hygiene and Environmental Health*, **226**(November 2019), 113484, https://doi.org/10.1016/j.ijheh.2020.113484

US EPA (2023). Strategies for climate change adaptation | US EPA, https://www.epa.gov/arc-x/strategies-climate-change-adaptation

Wolf M. J., Emerson J. W., Esty D. C., Sherbinin A., Wendling Z. A., (2022). 2022 Environmental performance index 2022. Yale Center for Environmental Law & Policy, New Haven, Connecticut, pp. 1–4.

World Bank (2019). Worsening water quality reducing economic growth by a third in some countries: World Bank, https://www.worldbank.org/en/news/press-release/2019/08/20/worsening-water-quality-reducing-economic-growth-by-a-third-in-some-countries

World Bank (2023). How is India addressing its water needs? https://www.worldbank.org/en/country/india/brief/world-water-day-2022-how-india-is-addressing-its-water-needs

doi: 10.2166/9781789063714_0025

Chapter 3

Domestic and industrial wastewater treatment: current status and challenges in India

Arya Vijayanandan[1], *Absar Ahmad Kazmi*[2] *and Ligy Philip*[3]*

[1]*Department of Civil Engineering, Indian Institute of Technology, Delhi, India*
[2]*Department of Civil Engineering, Indian Institute of Technology, Roorkee, India*
[3]*Department of Civil Engineering, Indian Institute of Technology, Madras, India*
**Corresponding author: ligy@iitm.ac.in*

ABSTRACT

Planned water management is inevitable in India's path towards sustainable development. Rapid urbanization and population growth have resulted in significant amounts of wastewater being generated and severe water pollution-related issues in the country. Given the scarcity of freshwater resources and the extent of pollution, it is essential to implement and maximize safe wastewater reuse. The primary impediment to reaching this goal is the enormous disparity between treatment capacity and wastewater generation. Only effective water usage and wastewater reuse practices will help to meet future water demand. Enforcing zero liquid discharge and stringent regulations to ensure treated wastewater quality are necessary to optimally reuse industrial wastewater. The release of untreated industrial wastewater into municipal sewerage system must be prevented. Selection of the appropriate treatment units, operation and maintenance, and real-time monitoring of wastewater quality are the key measures in ensuring the treated wastewater quality. Moreover, the current wastewater treatment methods need to be modified or upgraded to remove the toxic and emerging contaminants too. This chapter discusses the role of domestic and industrial effluents in water pollution, the challenges encountered, and regulations and policies related to water management. Examples of wastewater reuse and zero liquid discharge are also included in the chapter.

Keywords: domestic and industrial wastewater, wastewater management, zero liquid discharge, circular economy

3.1 INTRODUCTION

Water scarcity is an impending problem affecting the lives of many populations globally. The world economic forum has identified water scarcity as one of the major global risks (WEF, 2015). Two thirds of the world population lives in areas facing water scarcity for at least a month in a year and about 50% of this population lives in China and India (WWAP, 2017). Rapid urbanization and industrialization in India have boosted the economic development of the country. However, water management has not improved at par with the water demand and wastewater generation. The existing infrastructure for sewerage network and wastewater treatment plants (WWTPs) is not adequate for the population we have in India and there is a huge disparity in the quantum of wastewater generation and installed treatment capacity. According to the Central Pollution Control Board (CPCB), the total estimated sewage generation from urban centres in India is 72,368 MLD and the available operational treatment capacity is sufficient to treat only 37% of the wastewater generated. Most of the peri-urban and rural areas lack wastewater management.

Diminishing freshwater resources are also another impending problem faced by India. More than half of the Indian cities fall under water stress regions (NITI Aayog, 2019). The agricultural sector accounts for 70% of water demand globally (WWAP, 2017). Industrial and energy sectors also exert increasing water demand. Forty per cent of thermal power plants in India are located in areas with water scarcity (Luo *et al.*, 2018). Global water demand is expected to grow by 50% by 2030 and a 400% increase in global industrial water demand is predicted by 2050 (UNEP, 2016). The scarcity of required quality and quantity of water may hamper industrial growth. Excessive groundwater abstraction has resulted in a lowered groundwater table, inland and coastal salinity, and pollution. The only way to tackle this issue is to encourage the reuse of treated wastewater. Wastewater reuse would cater to the water demand and mitigate the water pollution issues due to untreated or partially treated wastewater discharge. There are several successful examples of wastewater reuse in India. The components essential for large-scale promotion of wastewater reuse are identification and upscaling of suitable technological interventions, policies or guidelines on wastewater reuse, finance and incentives, stakeholder awareness, and capacity building (TERI, 2020).

The world water development report, 2017, considered wastewater as a sustainable resource and highlighted that wastewater management must focus on pollution prevention at source, treatment, and reuse of wastewater and resource recovery (WWAP, 2017). To achieve the sustainability development goal 6 (SDG6), wastewater management is essential and target 6.3 aims at reducing the discharge of untreated wastewater, improving the treatment and disposal of wastewater and encouraging the reuse of reclaimed wastewater. Planned water management with a focus on circular economy is essential for India's mission to achieve SDG6.

3.2 DOMESTIC AND INDUSTRIAL WASTEWATER POLLUTANTS

The composition of wastewater mostly depends on its origin. It is estimated that around 80% of the water consumed is converted into wastewater. The major constituents of wastewater include organic matter, nutrients, dissolved ions, pathogens, heavy metals and emerging and toxic contaminants. The secondary treatment unit of WWTP is designed to remove the organic matter. The average biological oxygen demand (BOD) and chemical oxygen demand (COD) concentrations in domestic wastewater range from 150 to 200 mg/L and 250 to 400 mg/L, and may go as high as 300–400 and 600–800 mg/L, respectively, in water-scarce areas. Industrial wastewater may contain a high concentration of non-biodegradable organic matter, and BOD/COD ratio is used to determine the biodegradability of industrial wastewater. Nutrient-rich wastewater discharge causes eutrophication in water bodies. The primary sources of phosphate in aquatic environments are industrial effluents from the fertilizer, detergent, and soap industries, agricultural run-off, and domestic wastewater containing detergents and cleaning agents. Nitrogen in wastewater is mainly present in the form of organic nitrogen and ammonia nitrogen. Complex organics, colour and high content of dissolved ions are mostly associated with industrial wastewaters.

Conventional WWTPs focus on organic removal and pathogen reduction and little importance is given to the removal of excess nutrients. In recent years, biological nutrient removal (BNR) methods such as simultaneous nitrification–denitrification, Anammox, A2O and so on have been used to produce high-quality effluents for meeting water reuse criteria. Domestic and industrial wastewaters are the major contributors to heavy metal pollution in the environment. Total dissolved solids (TDS) in wastewater are mainly attributed to inorganic salts in wastewater and ionizable organics to a certain extent. These substances escape the conventional filtration process and can cause taste, odour, and colour to the water. The industrial activities which produce high-strength wastewater include steel production, pharmaceutical manufacture, mining operations, oil and gas exploration, agrochemicals, and food processing. High TDS can be reused only by adopting zero liquid discharge schemes employing reverse osmosis. TDS concentrations above 2200 mg/L in the treated wastewaters have shown toxic effects on the aquatic ecosystem. In addition, emerging contaminants (ECs) also pose risk to the environment. ECs are the anthropogenic chemicals released into the environment at trace levels of concentrations and they are found to exert harmful effects to aquatic life. ECs include pharmaceutically active compounds (PhACs), chemicals in personal care products, pesticides, and industrial chemicals. In recent years, the microplastics (MPs) pollution has been of serious concern as these can also act as a carrier for other persistent pollutants such as hexachlorobenzene, triclosan, aldrin, tonalite chlordane, dioxin, among others (Mammo *et al.*, 2020). Hence, the WWTPs need to be upgraded to remove these contaminants.

3.3 CURRENT STATUS OF WASTEWATER TREATMENT IN INDIA

Globally, it is estimated that over 80% of untreated wastewater is discharged into the water bodies (WWAP, 2017). The wastewater generation and the treatment capacity for different regions worldwide are presented in Figures 3.1 and 3.2. There is a significant disparity between the wastewater generated and treated in nations belonging to the Global South. According to the national inventory of sewage treatment plants (STPs) published by CPCB (2021), out of 1093 functional WWTPs in India, 570 WWTPs were found to comply with the prescribed discharge standards, which means that only 17% of the wastewater generated is getting the necessary treatment. Stringent rules and discharge standards are enforced for industrial effluents. According to the report submitted by CPCB to National Green Tribunal (NGT) in 2020, 64,484 industries require effluent treatment plants (ETPs) and 3% of them do not have ETPs. Out of eight CETPs located along the banks of River Ganga, only one was found complying with the discharge standards (CPCB, 2021).

There is a major shift towards the reuse of treated wastewater globally. In India, most states are reusing the wastewater for non-potable purposes such as horticulture, irrigation, washing activities, and non-contact impoundments. The state of Haryana is topmost in terms of the amount of wastewater reused, with approximately 80% of the treated wastewater used for various purposes. The National Capital Territory (NCT) of Delhi has set a target to increase wastewater reuse by 60% and the proposed reuse practices include irrigation, rejuvenation of water bodies, and indirect potable use.

Existing WWTPs have employed either conventional or advanced treatment technologies. The selection of treatment technology suitable for the wastewater characteristics is crucial to ensure

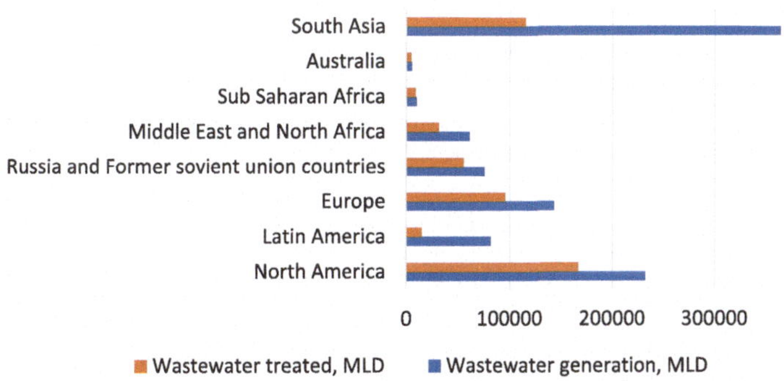

Figure 3.1 Global data on wastewater generation and treatment. (*Source*: Data taken from Sato *et al.*, 2013).

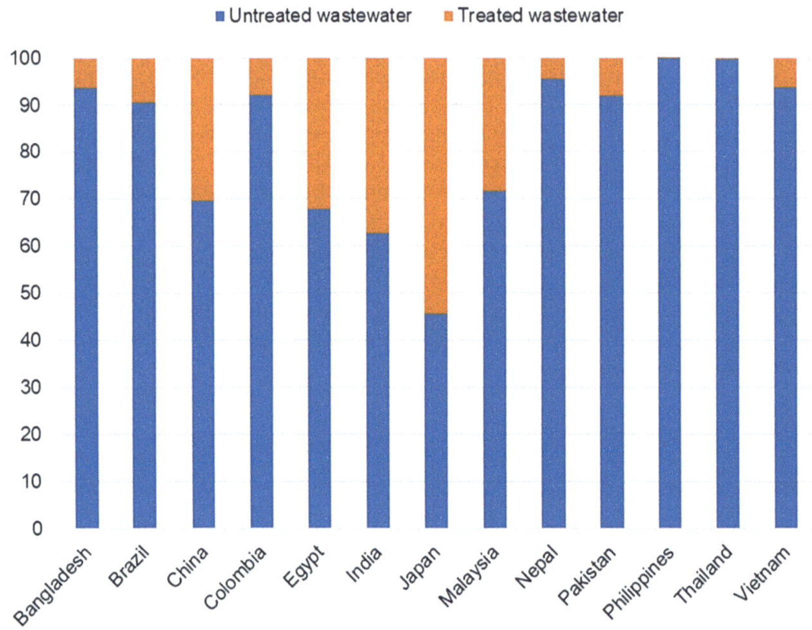

Figure 3.2 Percentage of wastewater treated in the Global South. (*Source*: Data taken from CPCB, 2021; Sato *et al.*, 2013; UN Habitat and WHO, 2021).

performance, along with proper operation and maintenance. Most common aerobic processes such as activated sludge process, sequencing batch reactor, membrane bioreactor, moving-bed biofilm reactor, trickling filter, oxidation ditch, and rotating biological contactors and anaerobic processes such as upflow anaerobic sludge blanket reactor, expanded granular sludge blanket reactor, and anaerobic filters are used as secondary biological treatment units. Sequencing batch reactor (SBR) is the most sought-after technology, followed by the activated sludge process (ASP). Anaerobic treatments are not preferred due to odour, corrosion, and low treatment efficiency issues. Most of the countries in the Global South rely on low-cost systems such as pond-based treatment. The technology-wise distribution of WWTPs in India, Africa, and Latin America are presented in Figure 3.3.

3.4 REGULATIONS AND POLICIES ON WASTEWATER MANAGEMENT

There are laws and policies in place to prevent water pollution. In India, the Water Act, 1974 resulted in the establishment of central and state pollution control boards to curtail water pollution. The pollution control boards have actively enforced the regulations for treating and discharging municipal and industrial effluents. As part of the Environmental Act, 1986, regulations were introduced to specify the discharge standards for particular applications such as irrigation, domestic, industrial, recreation, and so on. There have been measures to curb pollution and rejuvenation of major rivers. The Ganga action plan (GAP) launched in 1985 created infrastructure for the collection and treatment of wastewater. The 'Namami Gange' programme launched in 2014 with an approved budget of 20,000 crores aims at the conservation and rejuvenation of Ganga including infrastructure for wastewater treatment, river front development, cleaning of river surface, and the conservation of biodiversity. In addition, there are other flagship programmes such as the Swachh Bharat Abhiyan (Clean India Mission), the Atal Mission for Rejuvenation and Urban Transformation (AMRUT) scheme, Jawaharlal Nehru National

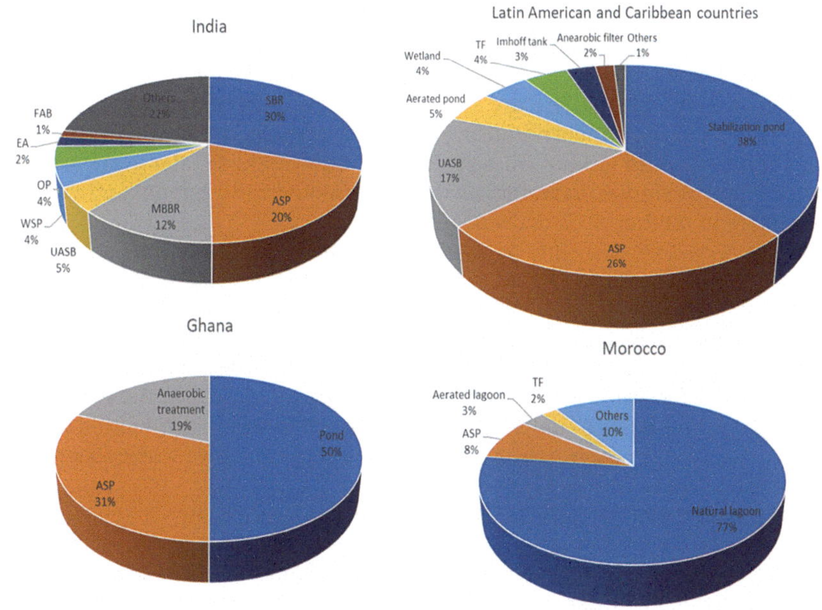

Figure 3.3 Technology-wise distribution of different WWTPs based on installed capacity. (*Source*: Data taken from CPCB report, 2021; Nikiema *et al.*, 2013; Noyola *et al.*, 2012).

Urban Renewal Mission (JNNURM) and the Smart City Initiative focusing on achieving SDG6 goals. The National Water Policy released in 2012 emphasizes the reusing and recycling of water and the provision of sanitation amongst other measures. The recent National Green Tribunal (NGT) order in 2019 highlights the need to reuse and recycle wastewater to meet the water demand. The thermal power plants within the 50 km radius of WWTPs are directed to use the treated wastewater as per the Tariff policy, 2016. There are initiatives to encourage sludge management. Solid waste management rules, 2016 mandate the urban local bodies (ULBs) to treat the sludge and promote reuse in agriculture. Waste to energy initiative is promoted by the Ministry of New and Renewable Energy.

There are no regulations focusing on wastewater management in most of the African countries. Many Arab countries have implemented guidelines for the reuse of treated wastewater. USEPA (National water reuse action plan), EU (Water framework directive), Australia (Guidelines for water recycling) and Singapore (Sewerage and drainage act) have stringent regulations and policies on wastewater management and reuse. These documents would be useful to other countries in policy making.

3.5 SUSTAINABLE WASTEWATER MANAGEMENT

3.5.1 Life cycle analysis

The sustainability of WWTPs is often assessed by life cycle analysis (LCA). LCA helps to understand the environmental impacts of WWTPs during their life cycle. The environmental impacts of WWTPs are mainly attributed to energy consumption, chemical addition, greenhouse gas emissions, treated wastewater, and sludge containing nutrients and heavy metals (Kamble *et al.*, 2019). Fossil fuel-based power consumption is a significant contributor towards environmental impacts such as abiotic depletion potential, global warming potential, acidification potential, and photochemical ozone creation potential. Eutrophication potential (EP) is associated with nitrogen and phosphorus emissions in the treated wastewater. The implementation of nutrient removal technologies can bring down EP impact. Toxicity potential depends on the heavy metal content in the treated wastewater and sludge. The shift to greener electricity production can considerably reduce the global warming potential of WWTP. Decision support tools are primarily employed to select the most suitable wastewater treatment technology. There are several factors to be considered in technology selection. The latest decision support tools incorporate sustainability aspects, scenario-based decision making, land availability, and end-use of treated wastewater. Decision support systems are also being used for wastewater reuse. Several studies have included the technological, regional, economic, social, and environmental factors in multicriteria decision analysis.

The water-intensive sectors can shift to innovative water management practices. Estimating the water footprint in manufacturing a product would reveal the over usage of water, water losses or any other unaccounted ways in which water is consumed. It is gaining popularity in the industrial sector in order to adopt water conservation practices. Two methodologies are used worldwide to estimate the water footprint, one developed by water footprint network and the second recommended by ISO 14046. ISO water footprint tool incorporated the aspects of LCA in the estimation. The water footprint network helps industries to understand their water demand in production stages. Blue water footprint indicates freshwater consumption, grey water footprint shows wastewater generation, and green water footprint refers to the plant uptake of water.

There should be incentives for industries to follow sustainable water management practices. It can also attract investors. For example, the J&K industry is a leading paper manufacturer in India and has established one of the best production practices globally. One metric ton of pulp or paper is produced by consuming 50 m^3 water. The reclaimed wastewater is used as cooling water and is sent to the nearby village for irrigation.

3.5.2 Circular economy

In the circular economy approach, wastewater is considered a valuable reusable water source. Resource recovery is also another vital aspect of a circular economy. It allows extracting the resources from wastewater while minimizing the cost and energy compared to traditional methods. Cradle-to-cradle approach in wastewater focuses on preventing or reducing wastewater generation at source, resource recovery, wastewater recycling or reuse. Wastewater management must follow the six R's concept: reduction, reclamation, reuse, recycling, recovery, and rethink (Smol *et al.*, 2020). An example of a circular economy approach applied in the wastewater sector is shown in Figure 3.4.

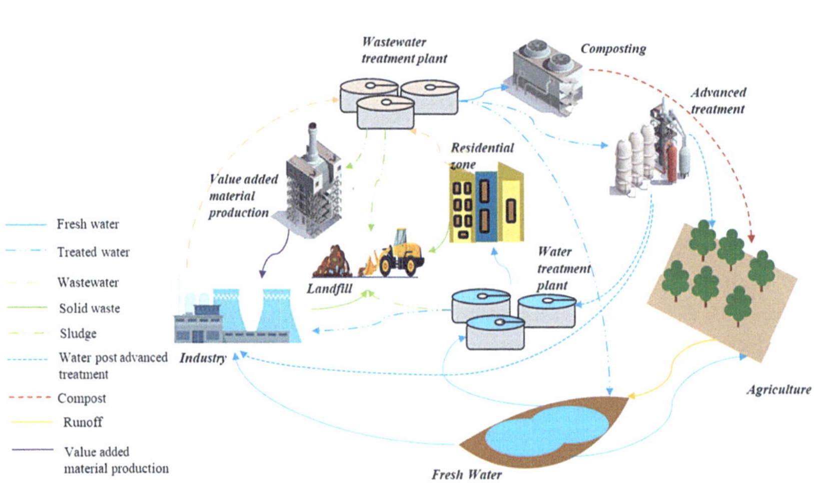

Figure 3.4 Example of circular economy applied in the wastewater sector. (Modified from Mannina *et al.*, 2022).

There is a lot of potential for the circular economy approach for wastewater resource recovery in India. Currently, treated wastewater is being used for non-potable purposes. Non-potable wastewater reuse includes horticulture, irrigation, toilet flushing, industrial use, and groundwater recharge. However, with adequate treatment, the quality can be improved to meet the drinking water quality and with the support of stakeholders involved, it can be used for indirect potable uses. Even though it is a sustainable way to meet the water demand, several challenges such as technical and economic factors, public acceptance, and lack of infrastructure limit the applicability of the treated wastewater. The success story of NEWater, Singapore is an example of how stakeholder interaction, community participation, and technological interventions would help meet the increasing water demand.

Nutrient recovery from wastewater has great significance as the natural sources of phosphorus are getting exhausted due to overexploitation. Struvite precipitation is gaining importance in resource recovery. In this regard, source separation of urine and faecal sludge is a better alternative to conventional treatment in smaller communities. Urine is a rich source of nitrogen and phosphorus. Faecal sludge can be treated to produce organic fertilizer or biogas as well.

3.5.3 Zero liquid discharge

Zero liquid discharge (ZLD) refers to the treatment and reuse of wastewater generated by industry within its premises. It was mandated by the CPCB in 2015 for textile and pulp and paper industries, distilleries, and tanneries. The textile industry in Tirupur is the first one to adopt zero liquid discharge scheme in the country. With the adoption of reverse osmosis (RO), the industry could reuse the treated wastewater and recover the salts. The cost associated with RO and operational difficulties often limit the performance of ZLD scheme. The thermal treatment unit is also expensive, which increases the capital cost of ZLD. Successful case studies on zero liquid discharge are discussed below.

3.5.3.1 Chemplast Sanmar Limited

Chemplast Sanmar Limited is a manufacturing company that produces silicon wafers, PVC resins, caustic soda, chlorinated solvents, and other chlorochemicals. The wastewater produced by this industry contains a high load of oil, grease, volatile organic compounds (VOCs) and COD and TDS in the range of 10,000 mg/L. A high efficiency RO (HERO™) ZLD system is implemented for ZLD by Aquatech at Chemplast (Figure 3.5). In this system, a high pH environment is maintained in the RO section. After the installation of this system, the company reported 88–90% water recovery for RO. The overall water recovery is reported to be above 97% and the treated water is used as cooling water. The sludge produced is sent to landfills and the salt is recovered and reused in the manufacturing processes.

3.5.4 Pharmez Special Economic Zone (SEZ), Ahmedabad

Pharmez Special Economic Zone (SEZ) near Ahmedabad, is operated by Zydus Infrastructure. Here, the common effluent treatment plant (CETP) is designed to treat industrial wastewater from 12 different pharmaceutical sites. The treatment process mainly constitutes a clarification unit and a membrane bioreactor of 750 KLD capacity. The plant was later upgraded to meet ZLD. The treatment scheme followed is presented in Figure 3.6.

3.6 CASE STUDIES ON WASTEWATER REUSE

There are several successful examples of wastewater reuse practices in India and the CPHEEO (Central Public Health & Environmental Engineering Organisation) has released a compendium on the STPs focusing on wastewater reuse. Selected case studies are presented here.

3.6.1 Tertiary treatment plants to meet industrial water demand in India

3.6.1.1 Bangalore Water Supply and Sewerage Board (BWSSB)

For water recycling and reuse, the BWSSB has set up two tertiary treatment plants (TTPs) in Bangalore: a 10 MLD plant in Yelahanka, and a 60 MLD plant in the Vrishabhavathi Valley in May 2003. In the

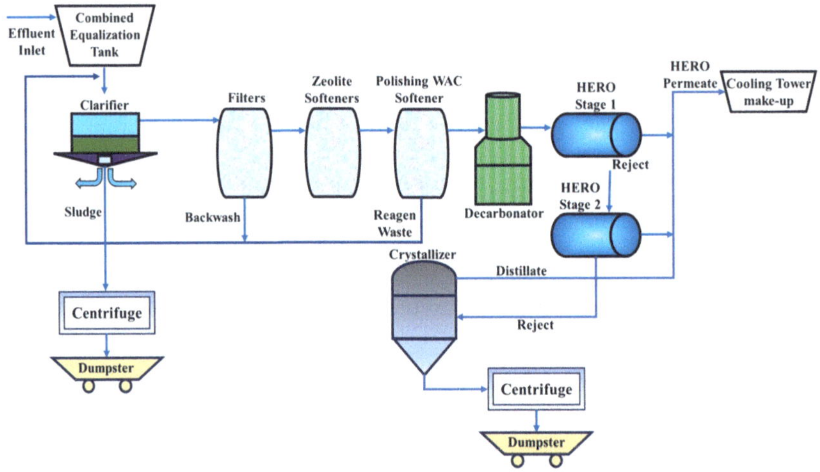

Figure 3.5 Schematic diagram representing unit processes involved in the wastewater treatment adopted by Chemplast group. (Modified from Aquatech, 2023).

10 MLD scheme, primary treatment (screening, grits removal, and grease removal), secondary treatment (primary settling and activated sludge process), and tertiary treatment (filtration using sand and gravel, and coagulation with aluminium sulphate for the removal of suspended particles) are included. The effluent from the secondary stage is used to meet the non-potable water demand by industries and commercial establishments. The TTP supplies the chlorinated recycled water to ITC Ltd., Wheel and Axel Plant, and Bangalore International Devanahalli Airport. The treatment scheme at 60 MLD Vrishabhavathi Valley TTP comprises of trickling filter, DENSADEG high-rate clarifier (combined flash mixer, lamella separators, and counter current flow thickener), FLOPAC aerobic biological filtration unit, and chlorine-based disinfection. The chlorinated recycled water from TTP is sent to the nearby power plants (M/s Karnataka Power Corporation Ltd. in Bidadi and M/s Pulikeshi Power Corporation Ltd. in Kumbalgodu). This scheme generates revenue of about Rs. 18 lakhs per month.

3.6.1.2 Chennai Metropolitan Water Supply and Sewerage Board (CMWSSB)

Two tertiary treatment RO plants having a capacity of 45 MLD each are set up at Kodangaiyor and Koyambedu by CMWSSB. The treatment units include pre-chlorination, rapid sand gravity filters, basket strainers, ultrafiltration, cartridge filters, RO, and ozonation. The treated water from Koyambedu is supplied at the rate of Rs. 65/KL to SIPCOT industries and other small-scale industries in the vicinity. It has led to a revenue generation of Rs. 19.67 crores. Kodangaiyoor plant supplies treated wastewater to industries located in Manali. The water is supplied at a rate of Rs. 80/KL and the revenue generated is Rs. 48.17 crores. The installation of the TTPs has reduced the freshwater demand by 40 MLD.

3.6.1.3 Surat Municipal Corporation (SMC)

The SMC has constructed a 40 MLD TTP at Pandesara industrial estate to treat the secondary treated wastewater from Bamroli WWTP. Treatment units include sand filtration, ultrafiltration, RO and activated carbon filtration and the treated water is supplied to the industries in Pandesara. This has resulted in a decrease in freshwater demand by 40 MLD. The treated wastewater is supplied at a rate of Rs. 23/KL and has generated a revenue of about 48 crores. SMC plans to establish more TTPs to

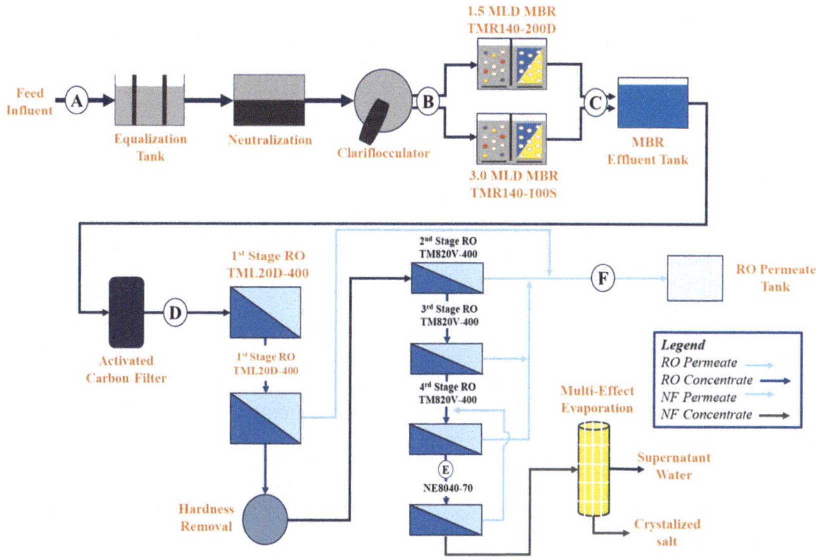

Figure 3.6 Treatment process schematics of ZLD system with MBR and high-recovery RO/NF.

cater to the industrial water demand to save freshwater resources for potable water production and encourage wastewater reuse for more activities.

3.6.1.4 Bhandenwadi STP, Nagpur

Treated wastewater (110 MLD) from Bhandenwadi STP is reused as cooling water in the thermal power plant in Koradi. The treatment scheme comprising of sequencing batch reactor, chlorination, and multimedia filtration is able to achieve the required quality of BOD and TSS less than 5 mg/L for cooling water. Maharashtra has drafted the wastewater reuse policy in 2017. The power plant industry was able to cut down the costs for procuring freshwater while the ULB recovered the O&M costs from the industry.

3.6.1.5 Kodangaiyur STP, Tamil Nadu

Kodangaiyur STP is a successful example of resource recovery from wastewater and sludge. About 36 MLD of the treated wastewater is reused for industrial use by Chennai Petroleum Corporation Limited (CPCL). STP produces about 13 MW of electricity per day by the anaerobic digestion of the sludge and has attained 98% of self-sufficiency in meeting its energy demand.

3.6.2 Examples of Wastewater Reuse from the Global South

3.6.2.1 South Africa

The wastewater recycling plant at Durban having a capacity of 47.5 MLD supplies water to two major industries (paper and oil refinery industries). Advanced tertiary treatments are included to produce high-quality water. The revenue generation from the industry has helped to provide potable water to 0.4 million people. ESKOM, one of the largest electric public utilities has installed RO units for cooling water recycling. The RO plant at the Lethabo power station treats 12 MLD of blow-down water released from cooling equipment containing high TDS. The treated water is further recycled as cooling water.

3.6.2.2 Egypt

New Cairo WWTP of 250 MLD capacity is set up to prevent water pollution. The treated wastewater is sent for irrigation and the compost is sold to cement industry to be used as a fuel.

3.6.2.3 Mexico

Atotolnico WWTP caters to the agricultural demand of 80,000 ha of land. Separate treatment schemes are provided for domestic wastewater and storm water. The biogas generated from anaerobic digesters is used for heat and power production. Tenorio WWTP treats 45% of wastewater generated in San Luis Potosí City. A portion of the treated wastewater is sold to power plants. The treated wastewater is also reused for irrigation and to set up a natural wetland. The revenue generation from the power plants ensures the O&M of the plant.

3.6.2.4 Peru

Enlozada WWTP was constructed to prevent the direct discharge of wastewater into the Chili River. The treated wastewater is used by the copper mining industry for mining operations. The project has created public awareness on the importance of wastewater treatment.

3.7 THE WAY FORWARD

Based on the analysis of the current status and challenges faced in the Indian wastewater sector, the following strategies are proposed for sustainable wastewater management (Eawag, 2023).

(i) The selection of treatment technology is critical in achieving the required quality of treated wastewater. Technology upgrades are required considering the increase in wastewater generation and the occurrence of toxic and contaminants in wastewater.

(ii) Most of the time, the treatment scheme is adopted without considering future expansion and resource recovery. Many a time, untreated or partially treated wastewater and sludge are discharged from poorly maintained WWTPs. Hence, regular and stringent monitoring is required to ensure the proper functioning of existing plants. For some treatment technologies, the O&M cost is very high and the lack of funds results in the poor functioning of the treatment plants. Sustainable and cost-effective treatment schemes must be preferred to ensure long-term performance.

(iii) Untreated industrial wastewater ingress to the municipal sewer should be prevented. Many wastewater treatment plants in industrial towns are not functioning properly due to the mixing of untreated industrial wastewater with sewage.

(iv) There is a need for fit-for-purpose rather than end-of-pipe treatment schemes. There are no clear guidelines or specific standards related to wastewater reuse. Currently, there are norms recommended by CPHEEO, 2013 for different reuse applications and wastewater reuse policies framed by a few states. A national level wastewater reuse standard for specific applications would ensure uniformity and compliance with the expected effluent quality.

(v) Fit-for-purpose treatment schemes, depending on the end use of reclaimed water must be given importance. The use of untreated and diluted wastewater for irrigation must be prevented. Health risk estimation is required to ensure the quality of reclaimed wastewater used for the irrigation of edible crops.

(vi) Decentralized and on-site treatment systems must be promoted in peri-urban and rural areas as short-term pollution control measures. Moreover, most of the popular DEWATS focuses on wastewater reuse and resource recovery.

(vii) In the case of industries, the prime focus must be given to the volume reduction of effluents and the release of toxic and hazardous chemicals into the effluents. The industries must be motivated to look for alternative manufacturing processes involving more eco-friendly or biodegradable raw materials.

(viii) There should be proper planning to deal with the wastewater generated by small- and medium-sized enterprises (SMEs). Wastewater generated from SMEs must be directed to CETPs and the industries must be encouraged to reuse the recovered chemicals in addition to wastewater recycling.

(ix) Sludge treatment is also not given much importance. It is a valuable resource of organics and nutrients. Waste to wealth must be prioritized, and the production of fertilizer and biogas from sewage sludge should be encouraged by ULBs.

(x) There is a need for stricter implementation of policies and monitoring mechanisms. Lack of information is another issue. Record maintenance, regular monitoring and stringent discharge guidelines are inevitable in ensuring the smooth functioning of WWTPs and CETPs.

(xi) Uninterrupted power supply is essential for the smooth functioning of the WWTP and power generation from sludge digestion could be practiced ensuring this. Even among the installed WWTPs, some are non-functional and underperforming, which aggravates the pollution from wastewater discharges.

(xii) Capacity building of various stakeholders is needed in the water management sector. Public awareness is essential for practicing wastewater recycling and reuse on a larger scale. Social stigma and the perception of health risks can only be addressed by conducting training sessions or workshops. The plant operators must have sufficient technical skills to take care of the routine maintenance and regular functioning of the plant. They should be trained to handle emergencies and be aware of the risks.

(xiii) Public–private partnerships could go a long way to create infrastructure for the treatment of wastewater. Various financial models such as design, build, and operate (DBO) and design, build, operate, and transfer (DBOT), viability gap funds are followed to implement the new projects. Proper financial mechanisms should be framed to increase the treatment capacity, sustain the functioning of installed WWTPs and for the upgradation or rehabilitation of the underperforming WWTPs. Incentives for circular economy initiatives would encourage more ULBs and industries to implement such techniques in WWTPs and CETPs.

To conclude, India's move to a circular water economy can be achieved by framing action plans. As we are approaching 2030, many efforts still need to be made to attain SDG6.

REFERENCES

Aquatech. (2023). http://www.aquatech.com/wp-content/uploads/61.-Chemplast-Mettur-HERO-ZLD.pdf (downloaded on 8th May 2023).

CPCB. (2021). National Inventory of STPs in India. Central Pollution Control Board, India.

Eawag. (2023). https://www.eawag.ch/en/department/sandec/projects/sesp/4s-small-scale-sanitation-scaling-up/ (downloaded on 8th May 2023).

Kamble S., Singh A., Kazmi A. and Starkl M. (2019). Environmental and economic performance evaluation of municipal wastewater treatment plants in India: a life cycle approach. *Water Science and Technology*, **79**(6), 1102–1112, https://doi.org/10.2166/wst.2019.110

Luo T., Krishnan D. and Sen S. (2018). Parched Power: Water Demands, Risks, and Opportunities for India's Power Sector, Working Paper. World Resources Institute, Washington, DC.

Mammo F. K., Amoah I. D., Gani K. M., Pillay L., Ratha S. K., Bux F. and Kumari S. (2020). Microplastics in the environment: Interactions with microbes and chemical contaminants. *Science of The Total Environment*, **743**, 140518, https://doi.org/10.1016/J.SCITOTENV.2020.140518

Mannina G., Gulhan H. and Ni B. J. (2022). Water reuse from wastewater treatment: The transition towards circular economy in the water sector. *Bioresource Technology*, **363**, 127951, https://doi.org/10.1016/J.BIORTECH.2022.127951

Nikiema J., Figoli A., Weissenbacher N., Langergraber G., Marrot B. and Moulin P. (2013). Wastewater treatment practices in Africa – experiences from seven countries. *Sustainable Sanitation Practice*, **14**, 26–34.

NITI Aayog. (2019). Composite Water Management Index. NITI Aayog, New Delhi.

Noyola A., Padilla-Rivera A., Morgan-Sagastume J. M., Güereca L. P. and Hernández-Padilla F. (2012). Typology of municipal wastewater treatment technologies in Latin America. *Clean – Soil, Air, Water*, **40**(9), 926–932, https://doi.org/10.1002/clen.201100707

Sato T., Qadir M., Yamamoto S., Endo T. and Zahoor A. (2013). Global, regional, and country level need for data on wastewater generation, treatment, and use. *Agricultural Water Management*, **130**, 1–13, https://doi.org/10.1016/j.agwat.2013.08.007

Smol M., Adam C. and Preisner M. (2020). Circular economy model framework in the European water and wastewater sector. *Journal of Material Cycles and Waste Management*, **22**, 682–697, https://doi.org/10.1007/s10163-019-00960-z

TERI. (2020). Mainstreaming Governance on Wastewater Treatment, Water Re-Use and Resource Recovery: Learnings From India and the European Union, TERI, New Delhi.

Toray. (2023). https://www.water.toray/knowledge/casestudy/pdf/CaseStudies_WWT_RONFMBRCHEM_AhmedabadIND.pdf

WEF. (2015). World Economic Forum Global Risks Report. World Economic Forum, Geneva.

WWAP. (2017). Wastewater: The Untapped Resource. The United Nations World Water Development Report 2017, UNESCO (United Nations World Water Assessment Programme), Paris.

UNEP. (2016). Policy Options for Decoupling Economic Growth From Water Use and Water Pollution. United Nations Environment Programme, Paris, France.

UN Habitat and WHO. (2021). Progress on Wastewater Treatment – Global Status and Acceleration Needs for SDG Indicator 6.3.1. United Nations Human Settlements Programme (UN-Habitat) and World Health Organization (WHO), Geneva.

doi: 10.2166/9781789063714_0037

Chapter 4

Urban water infrastructure: current status and challenges in India

Sridhar Kumara Narasimhan[1], Shankar Narasimhan[1], S. Murty Bhallamudi[1], Ashutosh Das[2] and M. S. Mohankumar[3]*

[1]Indian Institute of Technology Madras, Chennai, India
[2]Prist University, Thanjavur, India
[3]Formerly with Indian Institute of Science, Bengaluru, India
**Corresponding author: bsm@iitm.ac.in*

ABSTRACT

It is projected that by 2025, more than 50% of the population in the Global South will be living in urban areas accounting for a population of 3.75 billion. As the resources are limited, expanding cities in the Global South are facing severe crisis with regard to resource depletion, lack of space, equitable access to critical infrastructure, among other problems. Moreover, the existing infrastructure is aged and is in a dilapidated condition, requiring immediate attention. Adequate supply of water is not only important for living, but also from the perspective of hygiene, public health and control of diseases. However, despite the provision of adequate water, the incommensurate mismanagement of wastewater leads to non-hygienic conditions. Thus, provision of water supply infrastructure should necessarily go hand in hand with appropriate collection and management of wastewater. Further, management of rainwater and prevention of urban flooding through appropriate implementation of storm water drainage is crucial. In this chapter, we review the current status of urban water infrastructure in India. The review includes domestic water supply systems, sewerage systems and storm water drainage systems. The challenges that are being faced towards the evolution of water-sensitive urban areas are identified, and the way forward for surmounting those challenges is proposed.

Keywords: water supply systems, sewerage systems, storm water drainage, water infrastructure in India

4.1 INTRODUCTION

The percentage of urban population in India has increased from 17.9% in 1960 to approximately 35% in 2021 (World Bank, 2023). Not only has the urban population increased but also the number of urban agglomerations, leading to more centres of demand aggregation from the point of view of providing infrastructure. Currently, the number of urban agglomerations with a population of more than a million in India is more than 50. There are more than 300 cities in India, which have a population more than 100,000. According to projections by the UN, percentage of urban population will be more than that of rural population by 2050. Second, as India is making rapid strides in its economy and is poised to become a $5 trillion economy by 2026–2027 (Mint, 2023), the infrastructural needs will also be much greater than what are being provided now. Also, the aspirations of its people for better infrastructure will grow rapidly. In the 2023 Union Budget, capital investment outlay is increased steeply to $122 billion, accounting for approximately 3.3% of GDP (Ministry of Finance, 2023).

The rapid urbanization imposes enormous stress on all the urban infrastructures. Two of the basic components of any urban infrastructure are water supply and sanitation systems. The existing water supply systems in many of the cities are many decades old and are becoming either dysfunctional or highly inefficient from the perspective of service delivery. There is a significant loss of water in the distribution systems. Energy-intensive pumping operations are not optimized. Many of the cities do not have any sewerage system for collecting and treating the domestic wastewater. They also do not have proper storm water management systems. The untreated or partially treated wastewater is causing severe environmental damage. Lack of proper sewerage system is leading to the pollution of both surface and groundwater bodies. Although major floods do not occur frequently and do not last for large number of days, they cause enormous economic damage, besides loss of precious life. It is estimated that the flooding of just the outer ring road in Bengaluru on 30 August 2022 caused an economic loss of around Rs. 2250 million (Indian Express, 2023). The Mumbai floods in 2005 are estimated to have caused an economic damage to the tune of Rs. 200 billion. Thus, significant attention has to be paid to the water infrastructure development in India in order for the country to sustain the rapid economic development. The following sections provide a review of the present status of urban water infrastructure in India and outline the challenges that are being faced towards the evolution of water-sensitive urban areas.

4.2 HISTORY OF WATER INFRASTRUCTURE IN INDIA

4.2.1 Water supply systems

The existence of piped water supply systems in the modern era dates back to the 13th century, when a 5.5 km long pipeline made of lead was laid for supplying water to the city of London (Walski, 2006). In India, the first organized piped water supply system was commissioned in the city of Mumbai (then known as Bombay) in 1858. Originally Vihar Water Works supplied 3.2 million litres of water. In 1872, a parallel pipeline was laid to carry additional 3.7 MLD. Tulsi Water Works project was commissioned in 1879 (Narasimhan *et al.*, 2019). The second city to commission piped water supply in India was Kolkata. A 27.3 MLD capacity system was commissioned in 1870. It comprised of Palta filter plants; 1.067 m diameter cast iron pipe to carry water from Palta to Tallah and a pumping system to deliver water from Tallah to consumers. In 1872, the city of Chennai (then known as Madras) started sourcing water from Kosasthalaiyar River at Tamaraipakkam and transported it to the city through a 28 km long open channel, and then distributed it in nearby areas through cast iron pipes. The Delhi Water Works at Chandrawal, with a capacity of 4.5 MLD, was commissioned in 1890 to cater to approximately 1.9 lakh people. From 1912, the water was drawn directly from the river when the capacity had to be enhanced (Narasimhan *et al.*, 2019). Sizes and capacities of water supply systems have grown significantly since then. For example, Chennai city at present has a capacity to treat 1494 MLD. The total length of water pipes is 6697 km and number of consumers is 876,891. The Municipal Corporation of Greater Mumbai (MCGM) has a capacity of 3800 MLD to treat water, sourced from seven lakes. The distribution system comprises of two master balancing reservoirs, 27 service reservoirs and more than 6000 km long pipe network.

4.2.2 Sewerage systems

Historically, there has been evidence of early water supply and sanitation systems in The Indus Valley civilization, dating back to period between 2350 and 1810 BCE. These systems comprised of private bathing places and latrines, which are connected to a network of underground drains made of bricks that carried the wastewater to outside the city walls to soak pits. The network of sewer systems that existed in ancient cities such as Rome and Constantinople became dysfunctional in the middle ages. Cities started building sewerage systems once again from the mid-nineteenth century onwards in response to public health issues arising out of rapid industrialization. The great stink of London in 1858 led to the commissioning of a vast underground drainage system (UGD).

The sewerage system in Mumbai was first commissioned in 1867 in Colaba and Worli areas. The city prepared a master plan for the first time in 1979, which got completed in 2003. For a long time, the city of Mumbai did not have sewage treatment plants (STPs). In Chennai, the sewerage system was commissioned in 1890. A master plan was made in 1907 to cater to an expected population of 6.5 lakhs in 1961 (CMWSSB, 2023). The system comprising of mixed gravity and forced mains transported to the sea on the north-eastern boundary of the city. The first STP with a capacity of 23 MLD was commissioned in 1974. A combined UDG was introduced in the main city of Kolkata during 1859–1875, which collected sewage and storm water and discharged into marshy lands, ponds and open spaces in East Kolkata area. Even now the inner city does not have a STP and the sewage is treated in the vast East Kolkata Wetland (EKW) network system. The naturally treated wastewater, rich in nutrients, is taken to bheries (ponds) where algae and fish are grown. Since 2002, EKW has been designated as a Ramsar Site. Unlike the other three metropolitan cities, Delhi had an elaborate sub-soil drainage system predating British times since the reign of Shahjahan. Shahjahani drains were filled up by the British and a system of surface drains was constructed, which emptied into two main subsoil drains (Prashad, 2001), which in turn carried the wastewater to Yamuna. They accelerated the drainage and water works project in Delhi from 1881 onwards.

4.2.3 Stormwater drainage systems

In India, in historical times, the stormwater management focused more on rainwater harvesting than on flood protection. Farmers in the Kutch area in present-day Gujarat used rainwater harvesting for agriculture. There is evidence of the existence of tanks since third century BC. The practice of tank irrigation probably started in the early centuries of the first millennium (Shah, 2008). The Chola kings in Southern India also built many rainwater harvesting tanks for supplying water for agriculture in the 11th century. Although most of the tanks were built for irrigation, Coimbatore District Epigraphs mention that the 30 wetlands in Noyyal river basin in Coimbatore were built in the 8th and 9th centuries for both irrigation and flood protection (Pragatheesh & Jain, 2013).

Many of the present-day urban storm water infrastructures are based on the modern urban sewer systems developed after the 1850s in Europe, and for a long time they have been managed by public health officials. Thus, they were based on the concept of fast conveyance of stormwater outside the cities (Bertrand-Krajewski, 2021). In India too, although metropolitan cities have had a long history of municipal drainage systems dating back to the British period, the storm water collection networks were originally designed as a combined system for sewage as well as storm water runoff. The first combined sewer was commissioned in Kolkata in 1876 for a design rainfall of 6.35 mm/h (Gupta, 2005). The provision of a separate storm water drainage system in urban areas has started only in the last couple of decades. For example, the stormwater drainage master plan for the core city was prepared by the Corporation of Chennai only in 2009 (Greater Chennai Corporation, 2023). In fact, a dedicated manual for storm water management was published by the Ministry of Housing and Urban Affairs, Government of India in May 2019.

4.3 CURRENT STATUS AND CHALLENGES WITH WATER INFRASTRUCTURE

4.3.1 Water supply systems

The goal of Atal Mission for Rejuvenation and Urban Transformation (AMRUT 2.0) is to provide tap water connections to households in all the statutory towns by 2026. This means approximately 26.8

million new tap connections have to be provided in the coming four years (Bassi, 2022). This calls for a significant amount of investment into water supply infrastructure. There are other challenges as well, when one considers the availability of water and status of service delivery by the water utilities. Rapid urbanization, population growth, dwindling water sources, ageing water supply infrastructure and inappropriate governance is pushing many towns and cities into a severe water stress situation. For example, the per capita consumption in Chennai and Bengaluru is only 90 and 100 LPCD, respectively. Although it is reported that the per capita consumption is more than 150 LPCD in Hyderabad, only 70% of the supply is through pipes. In these cities, water is either being sourced from faraway places or expensive desalination technology is being sought. Delhi sources water from as far away as 220 km, while Chennai gets its water from far away Krishna and Cauvery rivers. Such systems are not resilient. As the utilities are unable to bridge the gap between demand and supply, households are pumping water from aquifers, many times violating the pumping norms. People are also pushed to procure water from private vendors who supply water through tankers at exorbitant prices. This is resulting in a complex urban water supply ecosystem, which is partly controlled by public utilities, and households, and the rest by private vendors (Narasimhan *et al.*, 2019).

Water supply is highly unreliable and intermittent in majority of the towns and cities. The water flows in the distribution system for only two to three hours a day. Typically, intermittent water supply results in customers paying more to access water services through alternative means (Charalambous & Laspidou 2017). The intermittent operation of piped water supply systems also results in: (1) increased possibility of water contamination; (2) increased energy consumption, (3) low pressures and (4) increased cost to customers due to the need for investing in local storage facilities. Although the Government of India aims to transition from an intermittent system to a 24 × 7 rapidly, only a few cities have made this transition successfully. The lack of funds with utilities and the perception that private companies offer better services than the public organizations, among other reasons, are driving the utilities to go for public–private-partnership (PPP) for making this transition to continuous system, which in turn requires tariff restructuring. This is creating socio-economic problems.

There is a significant loss of water in the distribution system. The unaccounted water is as high as 52% in the city of Delhi and the average value is around 40% across many cities (Satpathy & Jha, 2022). Only some cities such as Mumbai and Bengaluru have a high percentage of metered connections. The percentage of metered connections is less than 50% in Kolkata and Chennai. Although Mumbai boasts of 81% of metering and unaccounted water of only 20%, piped water coverage is only 76%. There is an inequitable supply of water, with slum areas enduring the most of water shortages. Further, water tariffs for domestic consumption in most of the cities are highly subsidized leading to significant water wastage. There are several other issues related to access and water quality, as discussed by Water Aid India (Water Aid, 2018).

4.3.2 Sewerage systems

Central Pollution Control Board (CPCB) has recently carried out surveys for the national inventory of STPs in India (CPCB, 2021). It is estimated that a total of 72,368 MLD of sewage is generated across all the urban centres in India. Compared to this, the installed capacity of STPs is 31,841 MLD and the capacity of proposed STPs is 4827 MLD. Thus, there is a huge 49% gap between the amount of sewage generated and the installed capacity for treating it. Only 1093 STPs are operational out of a total of 1631 STPs. In all, 102 STPs are dysfunctional, 274 are under construction and 162 are proposed to be constructed. Only 578 STPs out of 1093 operational STPs are able to comply with the norms specified by the CPCB/State Pollution control boards. These numbers indicate that only 12,200 MLD of sewage or only 16% of total sewage generated in the country is actually being treated completely. There is a significant variation in the geographical/state-wise distribution in the installed capacity for sewage treatment. While the state of Maharashtra accounts for 26.7% of installed capacity, states such as Bihar and Jharkhand account for less than 2% installed capacity. In fact, there is only one STP of 10 MLD capacity functioning in Bihar when the study was undertaken. A study found that the

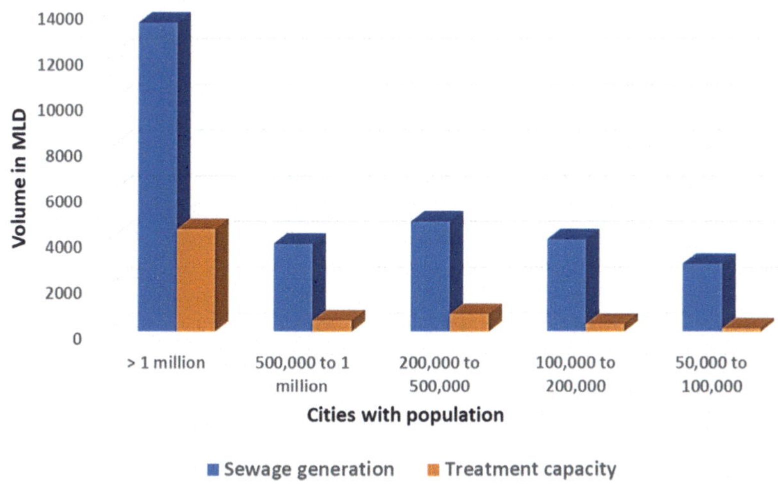

Figure 4.1 Wastewater generation and treatment capacity at city level in India.

most adopted treatment technology is sequential batch reactor (490 STPs accounting for 10,638 MLD capacity), followed by the activated sludge process (321 STPs accounting for 9486 MLD capacity). Wastewater generation and treatment capacity gap is shown in Figure 4.1.

The above-mentioned statistics indicate that the main problem afflicting the wastewater management in India is not only the lack of capacity for treating the sewage, but also the inability to operate and comply with the effluent standards. This mainly stems from the lack of funds for operation and maintenance, prioritization of funds for other purposes and the lack of skilled personnel for efficient operation. One of the major factors behind the underutilization of the capacity of STPs, especially in the case of newly installed plants is the inefficiency in the collection of generated sewage. There have been instances where the treatment plants have been installed but the coverage of the town or city with the UGD to collect and transport it to STP is not 100%. In some cases, although the UGD exists, households refuse to connect to the system because the tariff is high. Also, there have been instances where households practice on-site grey water recycling in their back yards and send only the toilet water to the UGD. These factors are affecting not only the functioning of STPs but also are resulting in frequent blockage of the UGD due to deposition of solids. These issues should be linked to the process of consent to be given by the municipality during house construction.

As per the directive of the National Green Tribunal (NGT), all the states and union territories have drawn plans for reuse of treated wastewater for different purposes. As of now, among others, Delhi (405 MLD; 12.5%), Haryana (192 MLD; 16%) and Tamil Nadu (211 MLD; 6.6%) have drawn up plans for reusing treated sewage for industrial and construction activities, horticulture, irrigation, washing (roads, vehicles, trains) and so on. There is a significant potential for increasing the percentage of treated wastewater reuse in all the states. However, challenges exist for retrofitting the existing sewerage systems for reuse because it may be required to separate grey and black waters. Also, on-site grey water treatment and reuse may reduce the flow rate in the sewerage system leading to solids deposition and frequent maintenance problems.

The rapid expansion of cities in India and other nations in the Global South poses challenges to the provision of conventional sewerage systems in per-urban areas during the transition period, that is, before the peri-urban areas are brought under the city corporation. Therefore, on-site sanitation systems such as septic tanks are commonly adopted in peri-urban areas. However, there may not be proper disposal of septic tank effluent due to lack of space. Also, there may not be a proper disposal of grey water. In such cases, it may be advisable to go for collection and transport of grey water and septic tank effluent using small bore system, and subsequent treatment by nature-based systems.

4.3.3 Stormwater drainage systems

Historically, among the three water infrastructure systems, the provision of stormwater drainage (SWD) facilities has been getting the least attention and priority from the urban local bodies. In parallel, rapid and unplanned urbanization has led to disappearance of natural drainage channels and many water bodies, which used to act as detention ponds to reduce peak flows. These factors, along with the impact due to climate change, have increased the frequency of both fluvial and pluvial flooding in almost all the urban agglomerations in India.

Mega cities as well as small towns in India severely lack effective SWD facilities. The major factors contributing to the failure of SWD systems in Indian cities, among others, are: (a) inadequate coverage, (b) inadequate capacity, (c) bad maintenance and choking due to indiscriminate dumping of solid waste and (d) conventional design. For example, the total stormwater drains length in Chennai is approximately 1894 km, as compared to the total road length of about 2847 km. Besides, although the core city has integrated storm water drain facilities, added areas included within the Greater Chennai Corporation do not have these facilities. Similarly, the total SWD length in Hyderabad is approximately 1302 km, while the total road length is around 9013 km. On a national scale, at present only 20% of the road network is covered with storm water drainage network and the associated catchment (MoHUA, 2019). Many of the existing SWD systems are very old, have not considered appropriate design rainfall intensity, and are not designed for handling high-intensity storms. For example, core of the SWD in Mumbai city is 70 years old and has been designed for a storm intensity of just 25 mm/h at low tide. Therefore, the city gets flooded even when 25 mm/h rainfall occurs, especially during high tide period. Similarly, Chennai city had adopted a rainfall intensity of 31.39 mm for SWDs for a length of 345 km as opposed to a rainfall intensity of 68 mm/h, as suggested in 2014 by Tamil Nadu Urban Infrastructure Financial Services Limited (TNUIFSL). The rapid assessment report prepared by Narasimhan *et al.* (2016) in the aftermath of 2015 Chennai floods has noted that drainage system was inadequate because of: (a) construction of many bridges across major drainages and (b) blockage of system due to indiscriminate dumping of solid waste into open surface drains, a very common phenomenon in SWD systems in India. As in the case of Chennai city, the micro drainage system is disjointed in many instances and it is not connected properly with macro-drainage system.

Another important factor contributing to the failure of SWD systems is that they are based on conventional design procedures. For example, a good urban SWD should be designed based on hourly or sub-hourly precipitation data. It is also required to have high spatial resolution data. However, such high-resolution spatial data for sub-hourly rainfalls is not available most of the times and designs are typically based on interpolations, leading to uncertainties. Rupa and Mujumdar (2018) state that the SWD design should be based on in-depth understanding of spatial distribution of short-duration rainfalls and uncertainty quantification. One cannot over emphasize the impact of climate change on the intensity–duration–frequency (IDF) curve. Therefore, the design should be based on design rainfall intensity picked up from climate change adjusted IDF curves (Rupa & Mujumdar, 2018). In general, SWDs in India are designed based on the old paradigm of clearing the stormwater from a location as fast as possible. It has now been well recognized that this concept does not lead to sustainable urban drainage systems.

4.4 THE WAY FORWARD

4.4.1 Water circularity

Currently, the water supply systems are mostly based on linear water balance framework and are finding difficult to bridge the gap between demand and supply without going in for expensive solutions. One way of achieving source sustainability in water supply systems is by transitioning into a circular

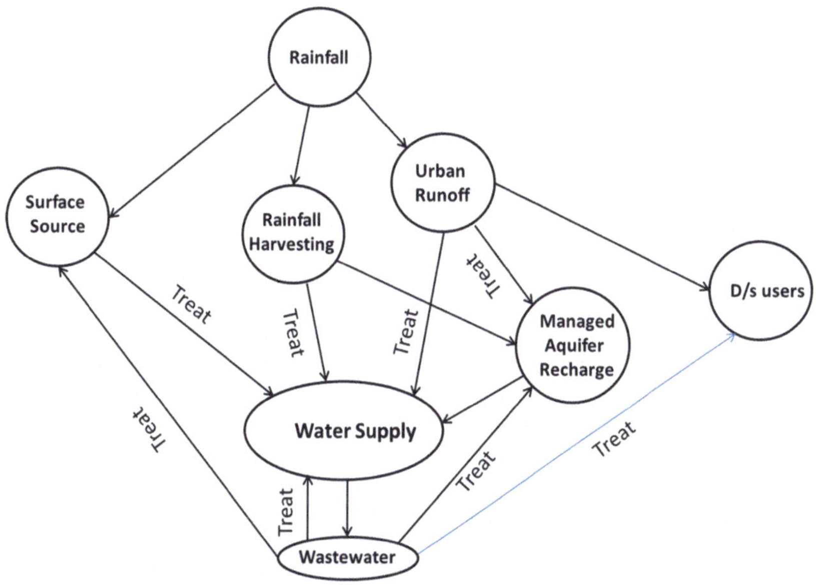

Figure 4.2 Schematic for circular water balance.

water balance framework. This framework will need adaptation of several novel technologies and practices such as: (a) managed aquifer recharge for increasing either groundwater potential for future exploitation or preservation of groundwater quality from getting contaminated due to salt water intrusion in case of coastal cities; (b) rainwater harvesting for both direct use in households and for groundwater recharge; (c) reducing the water demand by adopting novel water saving processes (e.g. using waterless urinals; storing the once used water from sinks for flushing etc.) and (iv) reusing and recycling the treated wastewater for both potable and non-potable purposes (Figure 4.2).

It is obvious from Figure 4.2 that the adaptation of circular water balance framework will require development of novel water and wastewater treatment technologies, tailor-made based on the influent water quality and the water quality required based on the fit for use. For example, treatment technology for treating urban runoff for household use will be different than that is needed for other situations. It is also obvious that adaptation of differential treatment of grey water and black water may result in better returns from the system, depending on situation. It may be also necessary to adopt not just dual pipeline systems but multiple pipeline systems to carry different quality water for different types of use. Water supply systems will be a hybrid of centralized and decentralized systems. Data analysis by Chatterjee and Roy (2021) suggests that a circular balance framework increases gross water availability from 350 to 950 Mm^3 in Chennai city. It has been shown that similar enhancements can be achieved in Bengaluru, Delhi and Coimbatore.

4.4.2 Leakage reduction

It has been mentioned earlier that there is a significant loss of water from water distribution networks, sometimes as high as 50% on input water does not reach the customers. It does not make any economic sense to not address this issue and go for expensive alternatives to procure extra water required for meeting the demand. The detection of leakage and taking actions to reduce the leakage in large water distribution networks is not a simple task. It will require significant amount of time and it will require money. The concept of District Metered Areas (DMA) should be considered in the planning and design, which will make the water auditing in a given network easy, reduce the time for localizing the leak and ease in operation for reducing the leaks (Narasimhan *et al.*, 2019). The conquer and divide methodology proposed by Rajeswaran *et al.* (2018) can be adopted for already existing water distribution networks for leak detection.

4.4.3 Sustainable urban drainage systems

Over the years, the experience with SWD systems designed and implemented using the conventional paradigm has indicated that these systems are not sustainable. New paradigms such as low impact development, source control, sustainable urban drainage systems, among others have been introduced during 1960s–1990s in order to realize the other goals of an urban drainage system such as increased aquifer recharge, ecological protection and water quality improvement. Sponge city paradigm has been introduced in China since 2014. These sustainable urban drainage systems (SuDS) reduce the peak flow to be carried through channels and pipes and hence the size of the system required to prevent frequent flooding of the area. Several cities across the world have already adopted these concepts in one way or the other, and have addressed the problem of urban flooding. There are many SuDS measures such as swales, bio-retention cells, porous parking lots, roof gardens, rain gardens, detention tanks and so on for reducing the peak flow and for obtaining a desired runoff water quality. This topic is addressed in much more detail in Chapter 23 of this book.

4.4.4 Integrated planning

It can be observed from Figure 4.2 that the adaptation of circular economy will have significant impact on the planning and design of all the three water infrastructures, namely, water supply, wastewater management and storm drainage systems. Rainwater harvesting to increase the water availability to households locally and diverting overland flow for using it for managed aquifer recharge through injection wells will obviously reduce the amount of water that flows into storm drainage system as well as the flood peak. This will influence the sizing of drains. However, catching every drop of water where it falls for some local use will not leave water for downstream purposes. This may affect the ecological services of downstream channels and water bodies which have been receiving this storm water. If one knows how much water can be sourced from rainwater harvesting and treated stormwater facilities *a priori*, then one can decide on the amount of water that needs to be sourced either from surface or groundwater sources and accordingly plan, design and implement the water supply infrastructure. Similarly, if on-site grey water treatment for local recycling is used, then the size of the sewage system will reduce as the amount of water to be carried reduces. However, if all the grey water is captured and recycled, it may have long-term repercussions on the maintenance of the sewerage system due to reduced flows and settlement of solids. Also, the necessity of dual piping system and associated cost in case of centralized wastewater treatment system and recycling may offset the economic benefits accrued from reducing the need for sourcing fresh water from other sources. It is obvious that the recycling and reuse of treated wastewater may work in cities such as Chennai which are water starved as compared to a city in the Gangetic plains where surface water is available in abundance. In those cases, the recycling of treated wastewater should be adopted based on other environmental benefits. Thus, it can be seen that the adaptation of water circularity will necessarily require integrated planning, implementation and operation of the entire urban water infrastructure.

4.4.5 Others

Appropriate government policies are required to make the projects under PPP structure successful. One of the most contentious issues is tariff. Proper pricing for water supplied to households should be arrived at, keeping in mind the difficulties the disadvantaged sections face. There should be a differential price structure for the fresh water and recycled water to promote the use of recycled water on a large scale. There should be a policy for equitable distribution of potable water, especially during periods of scarcity, which many cities in India face due to vagaries of monsoon rainfall. Chapter 25 of this book discusses governance and policy issues in more detail. Government of India intends to move towards 24 × 7 water supplies. However, 24 × 7 systems exist only in a few cities, and it will take some time before we will have 24 × 7 systems in all the municipalities. During the transition period, there should be a proper operation of current intermittent systems to deliver the services optimally. It is essential to automate all the components of water infrastructure (water supply, sewerage system and storm water drainage) using the Internet of Things (IoT) platform, to make these systems smart and wise.

4.5 SUMMARY

In this chapter, we have discussed the present status of urban water infrastructure in India. We have considered water supply systems, sewerage systems and urban drainage systems. The deficiencies in existing infrastructure are brought out. The way forward for addressing these issues through adaptation of water circularity, sustainable urban drainage systems and integrated implementation are discussed. More details on sustainable urban drainage systems, water infrastructure and challenges for implementation of new projects and technologies are provided later in Chapters 23, 24 and 25, respectively.

REFERENCES

Bassi N (2022). Pathways for Achieving Drinking Water Security in Urban India, https://egov.eletsonline.com/2022/05/pathways-for-achieving-drinking-water-security-in-urban-india/

Bertrand-Krajewski J. L. (2021). Integrated urban stormwater management: evolution and multidisciplinary perspective. *Journal of Hydro-environment Research*, **38**, 72–83, https://doi.org/10.1016/j.jher.2020.11.003

Central Pollution Control Board (CPCB) (2021). National Inventory of Sewage Treatment Plants.

Charalambous B. and Laspidou C. (2017). Dealing with the Complex Interrelation of Intermittent Supply and Water Losses. IWA Publishing, London, UK.

Chatterjee B. and Roy A. (2021). Creating Urban Water Resilience in India: A Water Balance Study of Chennai, Bengaluru, Coimbatore, and Delhi, https://www.orfonline.org/research/creating-urban-water-resilience-in-india/

CMWSSB (2023). https://chennaimetrowater.tn.gov.in/seweragesystem.html (downloaded on 3 February 2023).

Greater Chennai Corporation (2023). https://chennaicorporation.gov.in/gcc/department/storm-water/ (downloaded on 3 February 2023).

Gupta K. (2005). The drainage systems of India's cities. *Waterlines*, **23**(4), 22–24, https://doi.org/10.3362/0262-8104.2005.022

Indian Express (2023). https://indianexpress.com/article/cities/bangalore/floods-caused-rs-225-crore-loss-outer-ring-road-companies-associations-8126614/ (Downloaded on 10 May 2023).

Ministry of Finance (2023). https://pib.gov.in/PressReleasePage.aspx?PRID=1895289 (Downloaded on 10 May 2023).

Mint (2023). https://www.livemint.com/news/india/india-would-become-5-trillion-economy-by-2026-27-cea-anantha-nageswaran-11655202333724.html (Downloaded on 10 May 2023).

MoHUA (2019). Manual on Stormwater drainage systems, CPHEEO, Ministry of Housing and Urban Affairs.

Narasimhan B., Bhallamudi S. M., Mondal A., Ghosh S. and Mujumdar P. (2016). Chennai floods 2015: A rapid assessment. Report prepared for Interdisciplinary Centre for Water Research, Indian Institute of Science, Bengaluru.

Narasimhan S., Narasimhan S. K. and Bhallamudi S. M. (2019). Urban water distribution. In: Water futures of India: Status of Science and Technology, P. P. Mujumdar and V. M. Tiwari (eds), Indian National Science Academy, New Delhi, pp. 321–360.

Pragatheesh A. and Jain P. (2013). Environmental Degradation of the Coimbatore Wetlands in the Noyyal River Basin. EIA Resource and Response Centre (ERC), Nilgiri, Tamil Nadu, India, Legal Initiative for Forest and Environment (LIFE), New Delhi.

Prashad V. (2001). The technology of sanitation in colonial Delhi. *Modern Asian Studies*, **35**(1), 113–155, https://doi.org/10.1017/S0026749X01003626

Rajeswaran A., Narasimhan S. and Narasimhan S. (2018). A graph partitioning algorithm for leak detection in water distribution networks. *Computers & Chemical Engineering*, **108**, 11–23, https://doi.org/10.1016/j.compchemeng.2017.08.007

Rupa C. and Mujumdar P. P. (2018). Quantification of uncertainty in spatial return levels of urban precipitation extremes. *Journal of Hydrologic Engineering*, **23**(1), 04017053, https://doi.org/10.1061/(ASCE)HE.1943-5584.0001583

Satpathy S. and Jha R. (2022). Intermittent water supply in Indian cities: considering the intermittency beyond demand and supply. *AQUA – Water Infrastructure, Ecosystems and Society*, ***71***(12), 1395–1407.

Shah E. (2008). Telling otherwise: a historical anthropology of tank irrigation technology in South India, technology and culture. *Water*, **49**(3), 652–674.

Walski T. M. (2006). A history of water distribution. *Journal – American Water Works Association*, **98**(3), 110–121, https://doi.org/10.1002/j.1551-8833.2006.tb07611.x

Water Aid (2018). State of Urban Water Supply in India. Wateraid India, New Delhi.

World Bank (2023). https://data.worldbank.org/indicator/SP.URB.TOTL.IN.ZS?locations=IN (Down loaded on 10 May 2023).

doi: 10.2166/9781789063714_0047

Chapter 5

Designing water policy in India as adaptive governance for sustainability

Sudhir Chella Rajan[1,2] *and Christoph Woiwode*[1,2*]

[1]*Department of Humanities and Social Sciences, Indian Institute of Technology Madras, Chennai 600036, India*
[2]*Indo-German Center for Sustainability, Indian Institute of Technology Madras, Chennai 600036, India*
**Corresponding author: woiwode@igcs-chennai.org*

ABSTRACT

India's governance regimes for water have always been important, given its vast size, development challenges, and administrative and ecological complexity. Their effective and legitimate implementation has a new urgency in the context of climate change, leading to widespread interest in 'adaptive governance'. The term 'adaptive water governance' clearly takes on different meanings across contexts, ranging from transboundary institutions and riverine institutions, through national-level wetlands coastal, groundwater and wastewater policy to state-level, metropolitan and village-level institutions. These diverse governance constellations are crucial for managing natural hazards such as floods and storm surges, irrigation, industrial uses, drinking water supply, wastewater and local water bodies and their hydrology. Furthermore, in addition to formal (governmental) forms of regulation and management, there are many other non-formal or semi-formal collective forms of water management involving non-governmental organizations, farmer groups, water user associations, corporations, and self-help groups.

Keywords: water policy, water governance, climate change, adaptation, peri/urbanization, Chennai, India

5.1 INTRODUCTION

Governance is understood as a set of processes set in motion among vast groups of the population, if not the entire population, by allowing certain rules and their enforcement to hold sway for a specified time. Formal governmental processes, as well as large and small organizations and their network are lead actors in setting the rules and governing populations. Water governance relates to the set of formal (statutes, rules) and informal (customs, traditions) institutions that manage the regulation: organization, supply, delivery, and the pricing of freshwater as well as irrigation (Bevir, 2012).

In the context of climate change, adaptive water governance has taken on worldwide urgency (Schultz *et al.*, 2015). In this chapter, we review the development and multi-scalar conditions of water governance in India and propose patterns of adaptive governance that could help build resilience in climate-constrained conditions. India's existing drinking water and sanitation challenges and opportunities provide a special context for this study. On the one hand, it highlights the need for expanding investment in WASH (Water, Sanitation and Hygiene) programmes that result in infrastructure that is accessible, which means that it has good connectivity and services that are adequate, affordable, equitably distributed, and effectively maintained (Woodhouse & Muller, 2017). On the other, at the systems level and at different scales of demand and supply, it provides evidence of fiscal instability, corruption and compromised institutions and otherwise poor management of scarcity and ecosystem integrity (Shadabi & Ward, 2022). We follow existing literature by proposing an adaptive pathway to water governance in India that bears in mind the multi-scalar and existing fragmented institutions and their challenge (Ravetz & Connelly, 2018). Adaptive governance has networked and overlapping systems of governance that work in cooperation with each other through deliberative measures.

We draw on the vast literature on governance deficits as a resource for developing adaptive pathways. Scholars have documented emerging formal–informal linkages and agents making worlds on their own that largely escape state scrutiny or are moulded into each other's plots in complex ways. Each of these syndromes or patterns has the potential to intensify beyond administrative boundaries and formal spatial units (Roy, 2005). Urban geography itself was for long barred from paying attention to the urbanized hinterland, with a few notable exceptions.

The chapter is divided into the following sections. Section 5.2 provides the special context of India, especially the remarkable archaeological evidence of a well-developed and possibly adaptive water governance to changing social and climatic conditions in ancient Harappa (2400–1900 BCE). In Section 5.3, we describe the current status of governance and existing challenges at national and subnational levels. Section 5.4 uses planetary evidence to show that fragmentation in governance need not mean breakdown but could help cement exactly the type of adaptive or 'flexible' governance that is needed for finding sustainable solutions. In Section 5.5, we propose, as an example, pathways to sustainability that could constitute adaptive governance in an existing peri-urban landscape.

5.2 BACKGROUND

India has had a mixed form of management and patterns of water supply and sanitation since prehistoric times. Indeed, its most impressive water governance probably took place during the late phase of the Harappan civilization (2600–1900 BCE). Water drainage systems were remarkably and similarly engineered in each of the hundreds of towns and cities across over a million square kilometres in a manner that suggests coordination of careful planning and attention to the grade and sizing of drains, including successful separation of wastewater from freshwater. To the extent that the mature period of the Harappan civilization lasted for well over 700 years, it must be deemed an administrative success and possibly sustainable arrangement, even if there was rapid decline and depopulation of its cities after 1900 BCE (McIntosh, 2007). The region's complex socio-economic landscape in which similar types of water and infrastructure management took place across diverse scales can be illustrated in typical observations from the archaeologists of Harappa:

> The number of wells and their association with neighbourhoods could indicate a culturally based need for discrete and relatively private water sources. The cities and smaller settlements also had carefully designed and well-maintained drainage systems. Wells and bathing platforms were lined with bricks, and small drains carried water away from the wells or living areas to larger street drains. The street drains were equipped with sump pits, and bins were available

for the disposal of nonliquid waste, which was presumably collected and dumped outside the settlement (Shinde, 2016: 131).

> Water played a vital part in the life of the Harappan people, and they were skilled hydraulic engineers... One of the most impressive rooms of the Harappan house was the bathroom.... In a few households, there may have been the refinement of a kind of shower: a small stair along one side of the bathroom allowing another person to ascend and pour water over the bather. The watertight bathroom floor was constructed sometimes of stone or of pottery sherds but usually of baked bricks, sawn and ground to achieve a perfect fit; the floor sloped slightly to allow the water to flow into one corner, where it ran away into the efficient drainage system that served the city, via terra-cotta drainpipes or drainage chutes. Drains from upper-story rooms were often built into the walls so that they discharged near street level. (McIntosh, 2007: 235–236)

There are several other rich accounts of waterscapes across the ancient Indus Valley. Almost every household had a bath and occasionally a toilet, sometimes with clay bricks and a seat, whose outflows and those from bathing platforms were transported through terracotta drains, with brick-lined clay pits at junctions to prevent clogging. There were also wider receiving drains and partial treatment conducted through sedimentation pits. Nearly every house in Lothal, Mohenjo-Daro and Harappa had a connection to the municipal wastewater and sewer system in a checkerboard configuration (Khan *et al.*, 2021). Other sites had decentralized wastewater collection systems, including U-shaped channels whose effluents were poured into a jar for further processing, including sludge separation (Bisht *et al.*, 1984). Similar processes of freshwater collection took place through rainwater harvesting, wells, and tanks (Figure 5.1).

There is no clear evidence of the exact form of governance that was specific to Harappan society. We do understand, however, that many possible forms fit the archaeological evidence, including a federation of decentralized councils, a strictly ascetic lifestyle led by the elite priestly class, and rule by non-farming traders managing long-distance transport in luxury items like beads and timber (McIntosh, 2007).

The subsequent major innovations in water management involved oxbow lakes in Mahanadi and other river basins, temple wells and their drainage systems, step wells in the arid region of the Thar and its surrounding areas, and the elaborate hydraulic economies of the Chola period and beyond. Each of these involved significant patterns of coordination spanning multiple actors and gave rise to productive agrarian and trade economies over the centuries (Singh *et al.*, 2020).

Modern water engineering developed during the colonial period out of a plethora of pre-existing forms at various scales. For instance, in peninsular South India, the cascading network of 'erys', surface well-temple tank irrigation and drinking water supply complex, was built on relatively stable hydro-ecological principles and has survived in many parts to this day. Around large cities like Chennai, however, large-scale municipal water and sanitation systems replaced existing forms fairly rapidly in the 20th century (Arabindoo, 2017).

Post-independence, many water supply programmes in India were a combination of traditional, colonial-era, and privatized systems, sometimes with separate usage rules. Altogether, there is a mixture of evidence about how water and commons were and continue to be governed, including hierarchical and exploitative forms of governance as well as more democratic models (Amrith, 2018).

In this chapter, we give special attention to the Harappan case because of the extraordinary scale and superior technology that it represented and also because it offers tantalising potential for speculation about what counts for sustainable governance across large urban and peri-urban landscapes. As we note in Sections 5.3 and 5.4, a plausible 'backcast' of the Harappan case gives us some useful ideas for proposing intelligent design for peri-urban futures in India.

Figure 5.1 Corbelled drain and gateway, Mohenjo Daro and bathing platform (https://www.harappa.com/slide/drain-harappa; https://www.harappa.com/indus6/bathingfloor82.html).

5.3 CURRENT STATUS OF WATER GOVERNANCE IN INDIA

According to the Constitution of India, water is a state responsibility, which renders the National Water Policy (NWP) not legally binding but left to be implemented through state water policies and other related regulations. Woodhouse and Muller (2017: 228) acknowledge that federal arrangements 'add further complexity to the political processes inherent in water governance' by highlighting the issues of trans-boundary governance with its neighbours and shared rivers (e.g. Ganga, Brahmaputra, and Kauvery) among various Indian states.

In such a framework, the extensive powers of the national government limit its ability to design inter-state water resource management or negotiate agreements with neighbouring countries. Despite these issues, a recent study by Ahmed and Araral (2019) of eight Indian states that stresses the importance of water governance for sustainable development, suggests an overall improvement from before the introduction of the sustainable development goals (SDG 6 water and sanitation) surveyed in 2014/2015 and afterwards (surveyed in 2017/2018).

In India, the Department of Water Resources, River Development and Ganga Rejuvenation under the Ministry of Jal Shakti (Water Resources) are the apex bodies at the national level. The first NWP of India came into force in 1987. In its latest revision of 2012, the policy reflects various paradigm shifts, including watershed restoration, attempts at legal implementation at the state level and more multidisciplinary approaches compared to its 2002 predecessor that was more technocratic (Shah, 2013). The central government nevertheless initiates and leads various grand programmes and schemes of national importance in the water sector, such as Inter Linking of Rivers, Dam Rehabilitation and Improvement Programme, National Groundwater Management and Improvement Scheme, or National Hydrology Project. Similarly, the NWP is framed by a number of complementary statutes.

In the rest of this section, we illustrate water governance in the state of Tamil Nadu to exemplify the framework at the state level. Tamil Nadu's legislative framework has evolved in an incremental fashion over the last 130 years. Consequently, post-independence, this has led to disparate, often outdated water management statutes until recent times. Since the 2000s though, some momentum generated by national legislation and local initiatives such as water harvesting produced additional legislation and governing agencies.

The various elements of the water administration in Tamil Nadu are outlined in Figure 5.2.

Legislative reform also introduced new instruments of water management with the objective for more rigid planning and regulatory control. One such innovative tool, introduced in the 2000s, is the river basin planning approach, often considered the key characteristic of planning in Integrated Water Resources Management (IWRM) (Biswas, 2008). An example is the Palar River Basin Management Board, consisting of 23 members including all major stakeholders. This basin authority was the first to be formed in the Asian region (PWD, 2003).

For example, Sriperumbudur taluk (Figure 5.3) is an administrative unit part of two major river basins (Chennai river, specifically the Adyar sub-basin and the Palar river), which makes the management of its water resources complicated (Arabindoo, 2017). Notably, the administrative boundaries are overlapping, for the taluk belongs partly to the Chennai Metropolitan Area (CMA) with the remaining area under jurisdiction of the Department of Town and Country Planning (DTCP). This configuration makes the region subject to multiple planning statutes and authorities. Venot *et al.* (2011), in a study on the Krishna River basin management, similarly highlight challenges of multiple scales in decision-making and non-water issues such as socio-economic realities, arguing that these must be factored into river basin governance frameworks as well as to be effective.

Another important regulatory and monitoring instrument was the introduction of groundwater block categorization of the Panchayat Union Blocks in Tamil Nadu as 'over-exploited' 'critical', 'semi critical', and 'safe' blocks for ground water development in 2004. It enables the Government of Tamil Nadu to direct that no schemes should be formulated in blocks that are identified as overexploited or critical, whereas in semi-critical and safe blocks all the schemes should be formulated in consultation with the State Ground and Surface Water Resource Data Centre (SG&SWRDC). Moreover, the

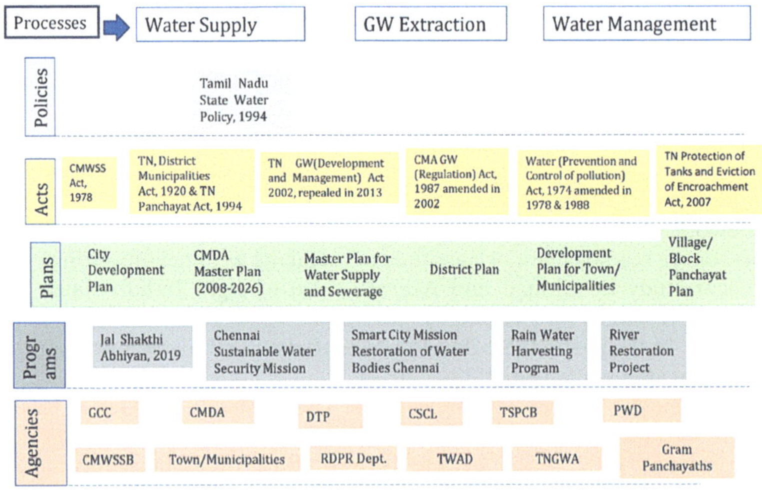

Figure 5.2 Elements of water governance in Tamil Nadu. (*Source*: authors' representation).

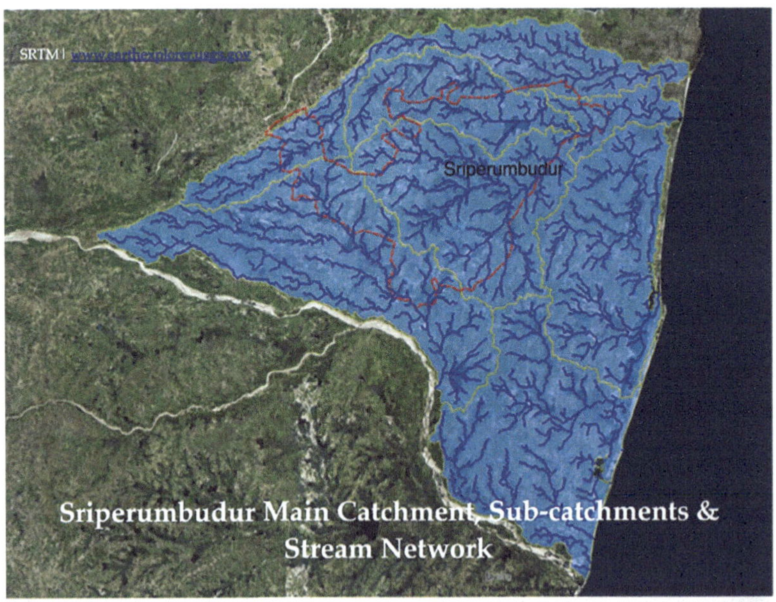

Figure 5.3 Sriperumbudur taluk intersects with sub-catchments within the main Chennai basin Indo-German Centre for Sustainability IGCS.

Government of India stipulated the implementation of appropriate rainwater harvesting and artificial recharge schemes in all the categories of blocks and during implementation of these schemes prioritize overexploited and critical blocks to avoid further depletion Chinnaswamy and Agoramoorthy (2015).

Furthermore, an obligatory no-objection-certificate (NOC) for industrial and large-scale domestic water extraction was introduced as well. Depending on the quantity, it is mandatory to be acquired from either the SG&SWRDC or the Water Utilization Committee. However, given the continued unsustainable use by agriculture and sale out to urban agglomerations for domestic use, the realities of groundwater management seem to be rather bleak, bordering on 'anarchy' (Shah, 2010), for water governance is deeply entrenched in administrative hierarchies within the government (Molle *et al.*, 2009) and Indian society (Mollinga, 2010).

The specific challenges and constraints faced by the sector are even more apparent at the level of local urban bodies, particularly metropolitan water governance of a city like Chennai. Water is managed by multiple government agencies with different jurisdictional territorial boundaries that are partly overlapping. In conjunction with the differentiated and often vague mandates of these various stakeholders that have different roles, water governance is fragmented and characterized by incoherent coordination between agencies at different levels (Rockefeller Foundation and Greater Chennai Corporation, 2019).

This volatile situation and the hitherto predominant approach to urban development that has for a long time ignored the ecosystemic functions of water in a rapidly urbanizing context are significant contributors to haphazard land conversions that are accelerating risks of flooding and water scarcity (Coelho, 2016). This phenomenon has been prevalent for long across many cities in India, including Bengaluru, Mumbai, Hyderabad, and Ahmedabad.

Adding to this peri-urbanized growth is the increasing privatization of infrastructure provisions, including water supply in industrial and especially residential estates granting exclusive access thus creating 'splintering urbanism' as suggested by Kennedy and Sood (2016). These and other scholars (Bhattacharya & Sanyal, 2011) point out the inherent 'bypass mechanisms' of such governance structures, which are territorialized in spatial inequities in the provision of basic infrastructure, but also access to housing and employment opportunities.

In much of peninsular India, the 1000-year-old 'ery' network is designed to capture the slow and gradual movement of water across the landscape through strings of cascading retention and irrigation tanks. These seasonal water bodies cum water-retention tanks are currently centrally managed by the Public Works Department. They were originally built for irrigation purposes, but have increasingly been brought under urban forms of management and until very recently were allowed to be overrun by urban expansion. Upland erys offer special opportunities to buffer surface water flows and transform sheet flows into contained flows. In Section 5.4, we provide an assessment of how adaptive governance strategies might be used to integrate decentralized water management with state-level governance resources (Jameson & Baud, 2016).

5.4 ADAPTIVE GOVERNANCE: FRAGMENTATION DOES NOT IMPLY BREAKDOWN

Adaptive governance is understood as a mode of organizing through social networks collective strategies of adapting to the climate crisis equitably and responsively. These have been proposed in ways that involve elements of collaboration, co-design, co-production, participation, and deliberation (Anguelovski & Carmin, 2011).

Adaptive governance often seeks synergies across various spatial and institutional combinations, systems, and levels. It is closely related to three modes of urban and regional governance (Ravetz, 2020).

- Mode 1 is a *functional* approach that involves command and control government. This is a traditional approach to governing that involves a strong central government with policies, regulations, and bureaucratic procedures.
- Mode 2 is an *evolutionary* approach that is informed by market-based competition and an embrace of innovation. There is a conscious effort by stakeholders to go beyond traditional modes of governance (mode 1) and provide fresh ways of organizing society. However, mode 2 continues to rely on existing societal structures related to economics, politics, and culture.

- Mode 3 can be characterized as '*co-evolutionary*' and emphasizes the importance of collaborative partnerships to co-produce decisions and actions. This mode of governance varies significantly from the previous two because of its emphasis on democratic inclusion and distributed agency of involved stakeholders.

In addition to the mode of governance, peri-urban climate adaptation involves various spatial and government scales of analysis. Of particular relevance to one of our recent research projects, the Peri-cene project, are four different scales (Ravetz, 2022). An **area-based** approach is geographical that focuses on various sectors in a particular location. An **agenda** or **sector** approach focuses on the activities of a particular group of stakeholders such as farmers and how they are implicated in peri-urban governance issues. An **organizational** approach zeroes in on groups of stakeholders and how they influence governing processes. Finally, **case studies** provide a tangible, context-rich approach to interpreting and assessing how climate governance unfolds in a particular place (Winter & Karvonen, 2022).

For each type of governance outlined in Table 5.1, there are important questions to guide further analysis. For deliberative governance, important questions are what types of expertise/knowledge are used? Is there an integrative (cross-sectoral) multi-hazard approach? For multi-level governance, important questions are whether there is top-down or bottom-up conflict or synergy? Does it respond to local needs and opportunities as well as anticipate global forces? Central to associative governance is a concern of how stakeholder conflicts are managed and how stakeholder synergies are formed and maintained. There are several concerns for responsive governance: is it responsive, innovative, and knowledge-based? How is risk managed? Is there a sharing of costs and benefits? Finally, key questions for collaborative governance include: how are informal claims on land and resources managed? Is there negative informality or corruption? What positive kinds of informality can be seen?

To illustrate adaptive governance with a fictional case study, we reasonably speculate that the success of water governance in Harappa through evident climatic and societal stresses (Sarkar *et al.*, 2016) might be mapped in the way represented in Figure 5.4. In Harappa, as reported above, it seems likely that multiple scales of overlapping and synergistic practice were replicated from household to village/town to region, with design-centred and engineered customs of water extraction, distribution,

Table 5.1 Organizational analysis by adaptive/'collective governance intelligence'.

Organizational Qualities	Causal Model/Negative Issues (Linear/Evolutionary)	Synergistic/Adaptive Model (Co-evolutionary/ Collective Intelligence)
Deliberative governance: 'Deeper' integration of policy agenda formation and competing values	Linear problem-fixing, materialist, myopic	Deliberative/responsive
Multi-level governance: 'Vertical' multi integration of spatial/ systems levels	Command and control/power and conflict	Multi-level
Associative governance: 'Wider' integration of stakeholder interests and potentials	Command and control/power and conflict	Associative/inclusive
Responsive governance: 'Further' integration of policy and service value chains	Fragmented and privatized services/infrastructure	Co-production, social learning
Collaborative governance: 'Deeper' dynamics of informal/self-organized/co-creative actions	Inequality, exploitation, corruption	Collaborative/creative

Source: Ravetz (2022).

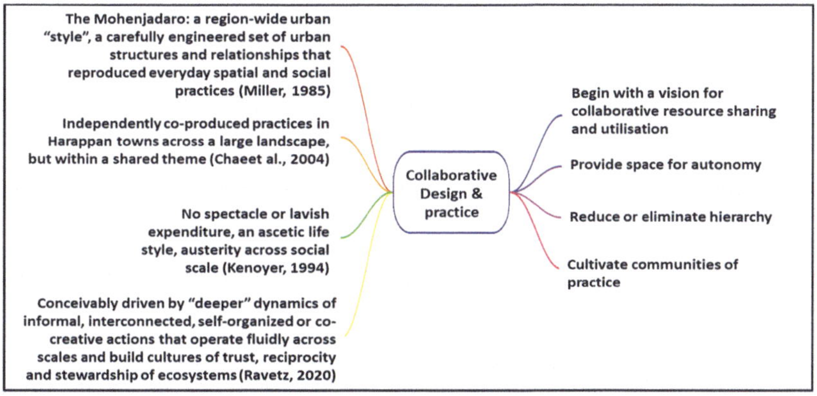

Figure 5.4 A speculative model of Harappa's collaborative design and practice. (*Source*: authors).

use, and disposal. Flood control may have been similarly managed through a combination of careful eco-design and neighbourhood-level management of drains and soil.

5.5 PATHWAYS TO SUSTAINABILITY

A vision for sustainability in water governance needs to pay attention to the full life cycle of water, especially upstream–downstream impacts and sharing of responsibility and rewards. These need to be operationalized with principles of adaptive governance to develop a network of water and sanitation services that include decentralized systems with mutually supportive, networked connections across social and ecological systems (Rjike *et al.*, 2012). Here, we provide a simple template for developing adaptive governance strategies at three scales for the context of greater Chennai (Figure 5.5).

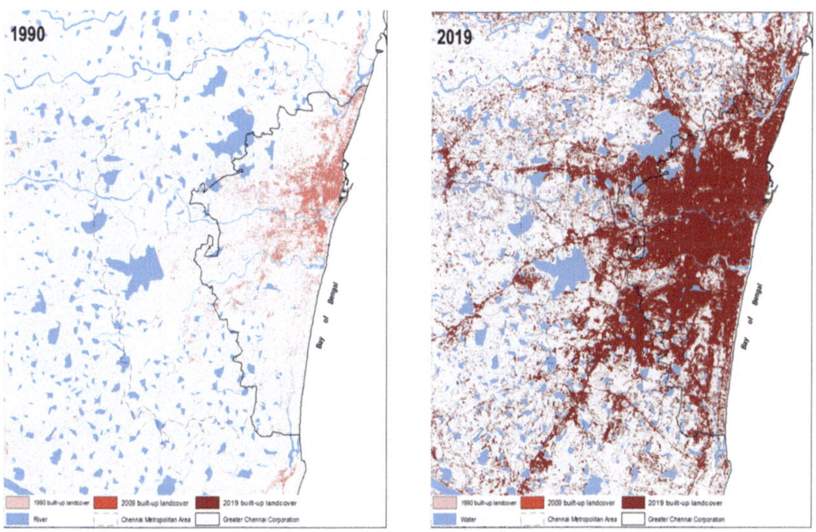

Figure 5.5 Urbanization in the Chennai region has encroached on water bodies and disrupted the drainage network Indo-German Centre for Sustainability IGCS.

At **landscape and community scales**, collaborative design initiatives need to be developed through the strengthening of existing grassroots networks and the building of collaborative relationships across class and other social barriers. These barriers may be the most challenging in conditions where there is a lack of communication, distrust and disbelief in government agents and agencies, complexity and uncertainty, inadequate signals for the need for collective action, bounded rationality, and narrow self-interest (Moser, 2010).

At the **macro level**, infrastructure support and protective/enabling rules need to be established by central, state, and local government bodies based on extending maximum autonomy to end-users and providing special attention to unserved populations. Government support is essential for providing resources for training, collaborative design, and local innovation by youth and other stakeholders, NGOs, university-led research groups and students. Corporate social responsibility initiatives might also be included in these forms of support, along with private sector investment and financing for long-term expansion and operations.

The focus and criteria for these investments should not be to maximize financial returns but amplify other indicators relating to restoring commons, reviving local water bodies, developing enterprises

Sriperumbudur has Special Economic Zones (SEZs) housing global giants like Hyundai and Saint Gobain which were set up to create thousands of jobs. Areas like Katchipattu, a village on the outskirts of Sriperumbudur, an industrial town that's part of the Chennai economic boom story, have become "bypass" spaces

Figure 5.6 Neighbourhood scale: Katchipattu, outside Sriperumbudur.

20 by 10 km grid made up of multiple wards within Chennai city limits and village panchayats just beyond. High flood risk, fishing communities prone to coastal hazards.

Pathway envisaged: Eco-tourism, blue-green design

Figure 5.7 Landscape scale: Muttukadu and surrounding coastal area. (*Source*: Water as Leverage, 2018).

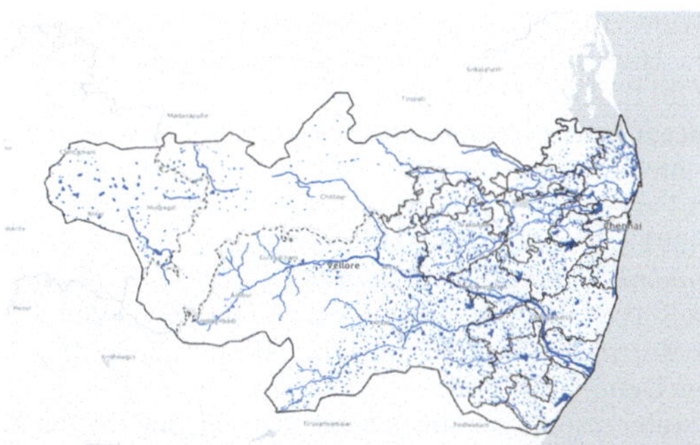

The Chennai basin extends more than 200 kilometres inland and transects several districts in 3 states. It is vulnerable to both floods and drought.

Pathway envisaged: bioregionalism, including food sovereignty using agroecological strategies

Figure 5.8 Bioregional or macro scale: Muttukadu and surrounding coastal area. (*Source*: Water as Leverage, 2018).

and other initiatives, including building capacity through peer-to-peer learning for NGOs, youth groups and community organizations.

(1) ***Neighbourhood scale:*** Pathways envisaged for a neighbourhood scale are illustrated in Figure 5.6 for example the case of Katchipattu in Tamil Nadu.

The use of a community gardening initiative helps to develop social capital amongst disenchanted youth groups and marginalized communities by providing a supplementary source of income and potentially rebuilding and rekindling a lost sense of community and arresting social unrest within the community.

(2) ***Meso or landscape scale:*** Pathways envisaged for a meso scale are illustrated in Figure 5.7 for example the case of Muttukadu in Tamil Nadu.

Future policies could use design tools to incorporate 'blue-green' and 'sponge-city' concepts to avoid further social and ecological fragmentation and raise the potential for resilience at such a scale.

(3) ***Bioregional scale:*** Pathways envisaged for a macro scale are illustrated in Figure 5.8 for the example of Muttukadu and surrounding coastal area in Tamil Nadu.

Bioregional strategies would mean re-imagining current food supply chains, shortening them, involving local, marginalized communities, alternative socio-economic models such as FPOs (farmer producer organizations) that will over time ensure food sovereignty and safeguard ecological assets like erys.

REFERENCES

Ahmed M. and Araral E. (2019). Water Governance in India: evidence on water law, policy, and administration from eight Indian states. *Water*, **11**, 2071 https://doi.org/10.3390/w11102071

Amrith S. (2018). Unruly Waters: How Rains, Rivers, Coasts, and Seas Have Shaped Asia's History. Basic Books.

Anguelovski I., and Carmin J. (2011). Something borrowed, everything new: Innovation and institutionalization in urban climate governance. *Current Opinion in Environmental Sustainability*, **3**, 169–175.

Arabindoo P. (2016). Unprecedented natures? *City*, **20**, 800–821. https://doi.org/10.1080/13604813.2016.1239410

Bevir M. (2012). Governance: A Very Short Introduction. OUP Oxford, Oxford, UK.

Bhattacharya R. and Sanyal K. (2011). Bypassing the squalor: new towns, immaterial labour and exclusion in post-colonial urbanisation. *Economic & Political Weekly*, **46**(31), 18–33.

Bisht R. S., Lal B. B. and Gupta S. P. (1984). Structural remains and town planning of Banawali. In: Frontiers of the Indus Civilization: Sir Mortimer Wheeler Commemoration Volume B.B. Lal, S.P. Gupta and S. Asthana (eds). Books & Books, New Delhi, pp. 89–98.

Biswas A. K. (2008). Integrated water resources management: is it working? *Water Resources Development*, **24**(1), 5–22, https://doi.org/10.1080/07900620701871718

Coelho K. (2016). Tenements, ghettos, or neighbourhoods? Outcomes of slum-clearance interventions in Chennai. *Review of Development & Change*, **XXI**(1), 111–136, https://doi.org/10.1177/0972266120160106

Chinnasamy P. and Agoramoorthy G. (2015). Groundwater storage and depletion trends in Tamil Nadu State, India. *Water Resources Management*, **29**(7), 2139–52, https://doi.org/10.1007/s11269-015-0932-z.

Jameson S. and Baud I. (2016). Varieties of knowledge for assembling an urban flood management governance configuration in Chennai, India. *Habitat International*, **54**(2), 112–123, https://doi.org/10.1016/j.habitatint.2015.12.015

Kennedy L. and Sood A. (2016). Greenfield development as tabula rasa: rescaling, speculation and governance on India's urban frontier. *Review of Urban Affairs, EPW*, **2**(17), 41–49.

Khan S., Yilmaz N., Valipour M. and Angelakis A. N. (2021). Hydro-technologies of Mehrgarh, Baluchistan and Indus Valley civilisations, Punjab, Pakistan (ca. 7000–1500 BC). *Water*, **13**(20), 2813, https://doi.org/10.3390/w13202813

McIntosh J. (2007). The Ancient Indus Valley: New Perspectives. Abc-Clio, Santa Barbara, CA.

Molle F., Mollinga P. P. and Wester P. (2009). Hydraulic bureaucracies: flows of water, flows of power. *Water Alternatives*, **2**(3), 328.

Mollinga P. (2010). The material conditions of a polarised discourse. Clamours and silences in critical analysis of agricultural water use in India. *Journal of Agrarian Change*, **10**(3), 414–436, https://doi.org/10.1111/j.1471-0366.2010.00277.x

Moser S. C. (2010). Communicating climate change: History, challenges, process and future directions. *WIREs Climate Change*, **1**(1), 31–53, https://doi.org/10.1002/wcc.11.

Public Works Department. (2003). Workshop on Action Planning for Sustainable Management of Palar River Basin: A Report. Government of Tamil Nadu, Chennai.

Ravetz J. (2020). Deeper City. Collective Intelligence and the Pathways From Smart to Wise. Routledge, London.

Ravetz J. (2022). Peri-Cene Synthesis Report: Overview. A First Exploration of the 'Peri-Eco-Urban Anthropocene. Deliverable D3-3, March 2023, University of Manchester, Manchester, UK.. Available from: https://peri-cene.net/

Ravetz J. and Connelly A. (eds) (2018). Water Governance in Greater Manchester. Manchester Urban Institute, University of Manchester, Manchester, UK. https://naturalcourse.co.uk/uploads/2018/10/Water-governance-in-GM-final-report1.pdf

Rijke J., Brown R., Zevenbergen C., Ashley R., Farrelly M., Morison P. and van Herk S. (2012). Fit-for-purpose governance: A framework to make adaptive governance operational. *Environmental Science & Policy*, **22**, 73–84, https://doi.org/10.1016/j.envsci.2012.06.010

Rockefeller Foundation and Greater Chennai Corporation. (2019). Discovery Area Report: Metro Governance. 100 Resilient Cities Program, Chennai.

Roy A. (2005). Urban informality: toward an epistemology of planning. *Journal of the American Planning Association*, **71**(2), 147–158, https://doi.org/10.1080/01944360508976689

Sarkar A., Mukherjee A. D., Bera M. K., Das, B., Juyal N., Morthekai P., Deshpande R. D., Shinde V. S. and Rao L. S. (2016). Oxygen isotope in archaeological bioapatites from India: Implications to climate change and decline of bronze age harappan civilization. *Scientific Reports*, **6**(1), 26555.

Schultz L., Folke C., Österblom H. and Olsson P. (2015). Adaptive governance, ecosystem management, and natural capital. *Proceedings of the National Academy of Sciences*, **112**(24), 7369–7374, https://doi.org/10.1073/pnas.1406493112

Shadabi L. and Ward F. A. (2022). Predictors of access to safe drinking water: policy implications. *Water Policy*, **24**(6), 1034–1060, https://doi.org/10.2166/wp.2022.037

Shah T. (2010). Taming the Anarchy: Groundwater Governance in South Asia. Routledge, Abingdon.

Shah M. (2013). Water: towards a paradigm shift in the twelfth plan. *Economic and Political Weekly*, **48**(3), 40–52.

Shinde V. (2016). Current perspectives on the Harappan civilisation. In: A Companion to South Asia in the Past Schug G. R. and Walimbe S. R. (eds.) Wiley Blackwell, London, UK. pp. 125–144.

Singh P.K., Dey P., Jain S. K. and Mujumdar P. P. (2020). Hydrology and water resources management in Ancient India. *Hydrology and Earth System Sciences*, **24**, 4691–4707, https://doi.org/10.5194/hess-24-4691-2020.

Venot J.-P., Bharati L., Giordano M. and Molle F. (2011). Beyond water, beyond boundaries: Spaces of water management in the Krishna river basin, South India. *Geographical Journal*, **177**(2), 160–170, https://doi.org/10.1111/j.1475-4959.2010.00384.x

Water as Leverage (2018) – RISE Chennai Raising Waters, Rising Fortunes, Deltares-IGCS IIT Madras, Chennai.

Winter A. K. and Karvonen A. (2022). Climate governance at the fringes: Peri-urban flooding drivers and responses. *Land Use Policy*, **117**, 106124, https://doi.org/10.1016/j.landusepol.2022.106124

Woodhouse P. and Muller M. (2017). Water governance – An historical perspective on current debates. *World Development*, **92**, 225–241, https://doi.org/10.1016/j.worlddev.2016.11.014

Section 2

New-Age Material for Water and Wastewater Treatment

INTRODUCTION

Water sources are getting contaminated with a multitude of pollutants due to natural and anthropogenic activities. Apart from the conventional pollutants, many emerging contaminants are also being detected in various concentrations across different water bodies. Though access to an adequate quantity of good quality water is a fundamental right, it remains inaccessible for a large part of the population due to the rampant pollution of dwindling water resources. Several conventional methods are available for the treatment of water and wastewater. However, most of these methods are unable to remove the pollutants to the acceptable level for producing drinkable water. Hence, there is a need to develop new materials and technologies to address these issues. It is equally important to assess the toxicity and environmental impact of the developed materials once they are let into the environment.

This section has five chapters addressing the aspects mentioned earlier. Porous organic polymers (POPs) have attracted significant attention in the last few years due to their structural and functional tunability, superior hydrothermal stability, lighter weight, tuneable pore size, and high specific surface area. The chapter on **New framework solids for water purification: porous organic materials for water treatment** deals with various types of emerging amorphous and crystalline POPs for wastewater treatment and the challenges related to low-cost, large-scale fabrication, and processibility of the materials and their applications for the removal of emerging persistent contaminants from water. The chapter on **New materials for arsenic and fluoride removal** gives an overview of simple, efficient, and affordable materials that, combined with appropriate technologies, provide promising scalable solutions for mitigating arsenic and fluoride in water. Carbon-based materials (CBM) have unique electrical, mechanical, and physicochemical properties, which make them ideal environmental adsorbents, sensors, membranes, and catalysts. The chapter on **Emerging carbon-based nanocomposites (CBNCs) for the removal of hazardous materials** discusses the effective employment of CBNCs for water purification, especially for the removal of textile dyes, volatile organic substances, toxic metals, oil, and biological contaminants. Biopolymer-reinforced nanocomposites (BPNCs) are new-generation materials that can be tailored into various forms like nanoparticles, granular materials, hydrogels, membranes, and coated substrates. The chapter on **Bio-polymer-reinforced nanocomposites for water and wastewater treatment: applications and future prospects** provides an insight into the application of BPNCs for removing contaminants in water and wastewater, along with associated challenges and prospects of BPNCs in the field application.

The use of new and improved materials spanning from macro to micro and to nano-range for applications in varied fields has increased exponentially. Although the scientific community's interest has been inclined towards nanomaterials due to their unique properties, the use of these innovative materials has given rise to concerns regarding their potential toxicity on living beings and the environment. The chapter on **A holistic approach to assess the toxic behaviour of emerging nanomaterials in aquatic system** elaborates on how emerging nanomaterials behave in relation to dynamic microenvironments at the nano–bio–eco interface level and how this affects their toxicity, fate, and exposure potential. It also provides a brief account of the exposure pathways and different models used for toxicity assessment/evaluation. This chapter also discusses the fate and toxicity of these materials.

doi: 10.2166/9781789063714_0063

Chapter 6

Function-led design of porous organic materials for water treatment

*Arkaprabha Giri, G. Shreeraj, Aniket Sahoo and Abhijit Patra**

Department of Chemistry, Indian Institute of Science Education and Research Bhopal, Bhopal, India
**Corresponding author: abhijit@iiserb.ac.in*

ABSTRACT

Porous organic polymers (POPs) have received substantial research attention in the last two decades owing to their modular functionality, tailorable pore size, excellent hydrothermal stability, and high specific surface area. Thus, POPs have emerged as potential adsorbent materials for efficiently removing diverse classes of contaminants ranging from organic micropollutants to heavy metal ions from water. Incorporating guest-responsive supramolecular cavitands and cages into the porous polymers enhances their activity and facilitates the selective scavenging of organic micropollutants from water. On the contrary, crystalline covalent organic framework-based thin films/membranes with precise one-dimensional nanochannels are highly suitable for nanofiltration-based water treatment for the size-selective removal of organic micropollutants and metal salts. In this chapter, we focus our discussion on the various types of emerging amorphous and crystalline POPs for wastewater treatment. Additionally, we will discuss the challenges related to the processability of the materials as well as shed light on some of the future avenues of POPs for the removal of emerging persistent micropollutants from water.

Keywords: porous organic polymers, covalent organic frameworks, micropollutants, adsorption, nanofiltration

6.1 INTRODUCTION

The ecotoxicity of various contaminants on the environment and human health is now a mounting concern faced by the society. In India, the primary concern is the unrestricted release of anthropogenic waste to the surface as well as groundwater by the domestic, agricultural, and industrial sectors (Sackaria & Elango, 2020; Schwarzenbach *et al.*, 2006). Contaminants such as microplastics, coal, microorganisms, bio-macromolecules, and viruses can be separated via various chemical/photochemical degradation, disinfection, sedimentation, and ultra/microfiltration techniques. However, the separation of highly water-soluble micropollutants having small molecular dimensions is highly challenging. In surface water bodies, like rivers, lakes, or ponds, the major micropollutants range from various pharmaceuticals products for example, endocrine disruptors, antibiotics, steroids, to industrial chemicals like dyes, oxo-anions, food additives, plastic precursors, and agricultural disposals and industrial chemicals like pesticides, herbicides, and fertilizers (Alsbaiee *et al.*, 2016). On the contrary, the groundwater suffers from heavy metal contaminations, such as iron and arsenic. Sorption is one of the most energy-efficient techniques to purify water from such micropollutants. However, several bottlenecks are associated with the commonly used carbonaceous adsorbents, such as slow uptake kinetics, low adsorption capacity towards polar micropollutants, and tedious regeneration process (Alsbaiee *et al.*, 2016). At this juncture, porous organic polymers (POPs), due to their high porosity, tuneable pore size, modular pore functionalities, and excellent hydrothermal and chemical stability, have emerged as advanced adsorbent materials for wastewater

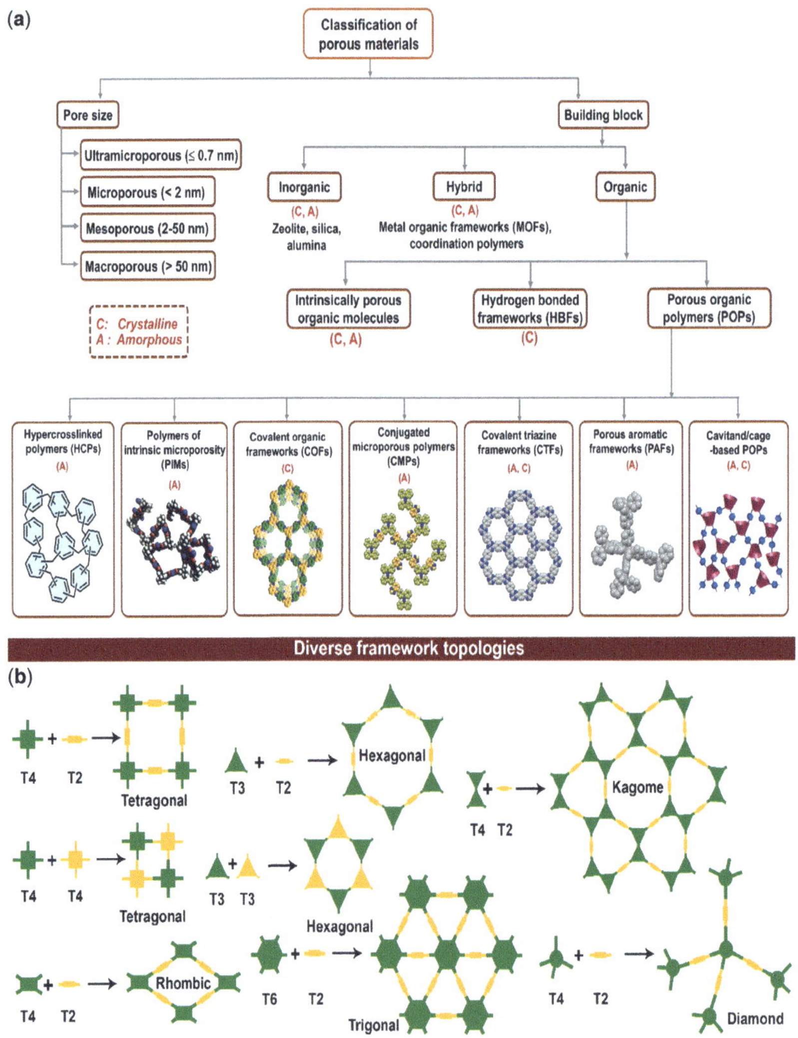

Figure 6.1 (a) A schematic depiction of the general classification of porous materials based on the pore diameter, the types of building units (inorganic, organic, hybrid), and the structural periodicity (amorphous and crystalline). A pictorial illustration of the various subcategories of POPs; (b) pictorial representation of diverse network topologies obtained by employing the building blocks with different propagating sites.

treatment (Giri *et al.*, 2022a, 2022b; Slater & Cooper, 2015; Sun *et al.*, 2020). Besides, membrane-based nanofiltration is another highly flourishing technology for water purification (Werber *et al.*, 2016). Processable POP-based membranes have recently been explored for the effective separation of organic micropollutants along with metal salts from water. In this chapter, a brief appraisal of the various classifications, fabrication strategies of the POP-based adsorbents and membranes, and their applications in wastewater treatment through sorption and filtration is delineated.

6.2 CLASSIFICATION OF POROUS ORGANIC MATERIALS

According to IUPAC (International Union of Pure and Applied Chemistry), porous materials are broadly classified based on the pore diameter into four categories, namely, (1) ultra-microporous ($\leq$0.7 nm), (2) microporous ($<$2 nm), (3) mesoporous (2–50 nm), and (4) macro-porous ($>$50 nm, Figure 6.1a, Thommes *et al.*, 2015). Depending on the types of building units, porous materials are of three types, such as inorganic (e.g. zeolites), hybrid [for example metal organic frameworks (MOFs)], and organic porous materials (Giri *et al.*, 2022b; Slater & Cooper, 2015). Furthermore, organic porous materials are of two types, such as porous organic molecules and POPs (Giri *et al.*, 2022b). Based on the distinct structural and functional features, like long-range order, types of building blocks, bonding motif, and extended π-electron conjugation, POPs can further be subcategorized namely hypercrosslinked polymers (HCPs), polymers of intrinsic microporosity (PIMs), conjugated microporous polymers (CMPs), covalent organic frameworks (COFs), covalent triazine frameworks (CTFs), porous aromatic frameworks (PAFs), and cavitand/cage-based POPs (Figure 6.1a, Diercks and Yaghi, 2017; Giri *et al.*, 2020, 2022b; Lee and Cooper, 2020).

Out of these, COFs and a few CTFs are crystalline in nature, having distinct pore topologies (Nguyen *et al.*, 2021). In the present chapter, we focus our discussion on the task-specific designing of POPs for water purification.

6.3 DESIGN AND FABRICATION OF POPS

The judicious design of monomers is crucial for the template-free bottom-up fabrication of POPs. A wide variety of monomeric building blocks are reported in the literature with different geometries and propagating sites (Xu *et al.*, 2013). Connecting such building units (represented in green and yellow cartoon building block, Figure 6.1b) through kinetically driven irreversible polymerizations often leads to the formation of amorphous POPs. On the other hand, the thermodynamically driven reversible polycondensation reaction leads to the construction of crystalline POPs, such as COFs. A large variety of network topologies, for example, tetragonal, hexagonal, trigonal, rhombic, Kagome, and tetrahedral can be predesigned through judicious assimilation of the building blocks for the task-specific applications of resultant COFs (Figure 6.1b, Xu *et al.*, 2013). The highly explored polymerization reactions to fabricate amorphous POPs like, CMPs and PAFs, HCPs are the transition metal-catalysed polymerizations, like Suzuki–Miyaura, Sonogashira–Hagihara, Buchwald–Hartwig, Yamamoto coupling, Friedel–Crafts alkylation, Glaser–Hey coupling, and Click reaction (Bandyopadhyay *et al.*, 2016; Giri *et al.*, 2022a, 2022b). On the other hand, reversible chemical reactions, for instance, boronic ester and Schiff base polycondensation, provide a facile defect-healing process to obtain crystalline porous frameworks (Giri *et al.*, 2022b). Therefore, altering the building units and polymerization strategies leads to facile modulation of the structural and functional properties of POPs.

6.4 ADSORPTION-BASED WATER PURIFICATION

Adsorption-based sequestration is advantageous because of its cost-effectiveness, easy operation, and also faster removal. Organic micropollutants are a diverse set of analytes, including toxic dyes, herbicides, pesticides, plastic components, drugs, and small amine and phenolic components whose presence in water, even in small amounts, may cause serious health hazards to human and aquatic lives (Alsbaiee *et al.*, 2016). The removal of such organic micropollutants from water through adsorption is highly challenging because of their high-water solubility (log $K_{OW} < 3.0$, K_{OW} = partition coefficient in *n*-octanol/water, Giri *et al.*, 2022a).

Patra and co-workers employed a paddlewheel-like molecular building block, triptycene, for the development of HCPs for rapid scavenging of broad-spectrum polar organic micropollutants (POMs, Giri *et al.*, 2022a). Three distinct polymerization routes were explored, namely, Friedel–Crafts crosslinking, Scholl reaction, and solvent knitting methods to obtain diverse nanoscale morphologies, such as Friedel–Crafts triptycene polymer (FCTP) with irregular aggregates, Scholl triptycene polymer (SCTP) with rigid spheres, and solvent-knitted triptycene polymer (SKTP) with nanosheet-like morphology, respectively (Figure 6.2a, b). Post-synthetic modification with chlorosulphonic acid led to the formation of sulphonic acid-functionalized HCPs, such as FCTPS, SCTPS, and SKTPS, to improve the adsorption performance (Figure 6.2a). The highly porous, dispersible, and sheet-like solvent-knitted triptycene-based HCP (SKTP, 2385 m^2/g) exhibited superior performance compared to the HCPs obtained through Friedel–Crafts reaction [FCTP (1626 m^2/g)] and Scholl reaction (SCTP, 1247 m^2/g) in removing the micropollutants, like cationic methylene blue (MEB), anionic rose Bengal

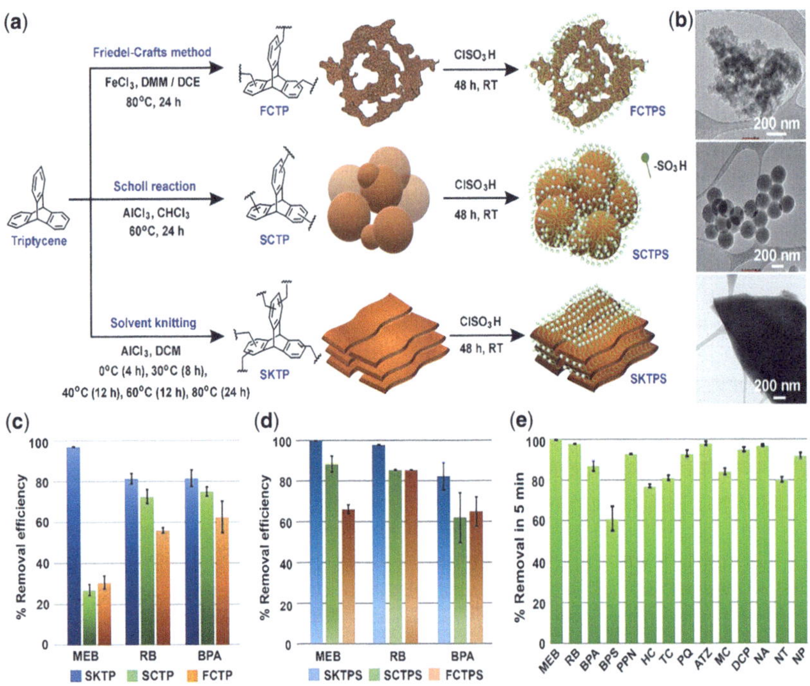

Figure 6.2 (a) Schematic representation of the fabrication of triptycene-based HCPs, such as FCTP, SCTP, and SKTP, through Fiedel–Crafts alkylation, Scholl reaction, and solvent knitting, respectively. The post-synthetic sulphonation of the respective HCPs through chlorosulphonic acid treatment to obtain FCTPS, SCTPS, SKTPS; (b) the FESEM images of FCTP, SCTP, and SKTP showing irregular, spherical, and nanosheet-like morphologies, respectively. Per cent removal efficiency in 2 min of cationic methylene blue (MEB, 0.1 mM), anionic Rose Bengal (RB, 0.1 mM), and neutral bisphenol A (BPA, 0.1 mM); (c) pristine HCPs, SKTP, SCTP, and SKTP (1 mg/mL); (d) sulphonated HCPs, SKTPS, SCTPS, and FCTPS (1 mg/mL); (e) broad-spectrum polar organic micropollutant (0.1 mM) removal by SKTPS (1 mg/mL) in 5 min. (*Source*: adapted with the copyright permission from Giri *et al.*, 2022a).

(RB) and neutral bisphenol A (BPA) (Figure 6.2c). The sulphonated counterparts such as SKTPS (1444 m^2/g), FCTPS (916 m^2/g), and SCTPS (1122 m^2/g) showed superior performance compared to the pristine HCPs because of the electrostatic and H-bonding interactions with the POMs. SKTPS showed fast adsorption kinetics (k_2 for methylene blue: 17.6 g/g/min), high sorption capacity (q_{max} for methylene blue: 374, rose Bengal: 885, bisphenol A: 333, paraquat: 305; tetracycline: 209 mg/g) with a wide working range of pH (3–9), and excellent recyclability. SKTPS could sequester a broad spectrum of micropollutants from water within 5 min (Figure 6.2e). SKTPS was also demonstrated for column-based micropollutant separation. Since SKTP was synthesized using cheap catalysts ($AlCl_3$) and chloroform as the crosslinker, the materials can be made on a large scale. Further, replacing triptycene with cheap monomers may open the scope of fabricating low-cost, highly efficient HCP-based adsorbent materials.

Macrocyclic cavitands, such as cyclodextrins, calix[n]arenes, and pillar[n]arenes, having well-defined cavities, versatile functionality, and excellent guest recognition properties have been explored in supramolecular chemistry for molecular separation (Giri *et al.*, 2020). Dichtel and co-workers first employed β-cyclodextrin (β-CD) as a cavitand building unit for the fabrication of mesoporous polymer (P-CDP; 263 m^2/g, Alsbaiee *et al.*, 2016). P-CDP was fabricated from a mild base-catalysed aromatic nucleophilic substitution reaction (S_NAr) between β-CD and tetrafluoroterephthalonitrile (TFN, Figure 6.3a). P-CDP showed faster sequestration of organic micropollutants, including bisphenol A from water compared to its non-porous counterparts (NP-CDP, EPI-CDP) and also commercial carbonaceous adsorbents, such as NAC, GAC, Brita AC (Figure 6.3b). The same group developed Red-P-CDP by reducing the nitrile groups into amine to remove per- and polyfluorinated alkyl substances (PFASs, Figure 6.3c, Klemes *et al.*, 2019). The amine functionalities in Red-P-CDP were partially protonated at pH 7 (zeta potential: $+1.7 \pm 0.8$ mV). As a result, Red-P-CDP displayed augmented binding affinity towards anionic PFASs. Patra and co-workers used *C*-phenylresorcin[4]arene (RN4) as a model macrocyclic cavitands and TFN as a fluorine-rich linker for the fabrication of RN4-F polymer, which exhibited the specific surface area 1230 m^2/g) nearly eight-fold higher than that of the pristine RN4 cavitand (156 m^2/g; Figure 6.3d, Giri *et al.*, 2019). RN4-F having a high specific surface area, and F-rich network showed excellent selectivity towards cationic micropollutant separation owing to the cation–π interaction originating from electron-rich resorcin[4]arene core and the fluorine–cation interaction provided by TFN linker (Figure 6.3e). Furthermore, the intrinsic pores of the resorcin[4]arene cavitand in RN4-F are responsible for the preferable adsorption of smaller-sized (<1.5 nm) micropollutants leading to the size-selective separation. The efficient separation of cationic methylene blue (MEB) from an equimolar mixture of MEB and anionic methyl orange (MO) was demonstrated using RN4-F.

Not only for organic micropollutants, POPs and POP-based composite materials have been demonstrated for efficient sequestration of heavy metals and oxo-anion from water. Selective removal of heavy metals and oxo-anions is based on two different working principles (Sun *et al.*, 2020). For heavy metal ions like mercury (Hg^{2+}), the soft–soft interaction is the key component for selective chelation, whereas an anion-exchange mechanism is a strategy for fabricating an oxo-anion sequestering material. Ghosh and co-workers developed a new class of hybrid anion-exchange material by embedding a water-stable, amino functionalized, cationic Zr-based metal-organic polyhedral (MOP) ($\{[Cp_3\ Zr_3O(OH)_3]_4(NH_2{-}BDC)_6\}{\cdot}Cl_4$) inside a porous COF (Figure 6.3f, Mollick *et al.*, 2020). The grafted MOP inside the COF matrix is the main active site for anion exchange, having Cl^- counter anions, free NH_2 groups, and Zr(IV) in the secondary building unit (SBU). The exchangeable Cl^- ions in the hybrid material take part in the anion-exchange process, whereas the additional free NH_2 groups stabilize the exchanged toxic oxo-anions through strong hydrogen bonding. Along with this, the Cp_3Zr_3O-$(OH)_3$ SBU of the MOP has labile hydroxyl groups, which can be replaced by oxo anions. The hybrid material was tested for its sequestration efficiency of $HAsO_4^{2-}$ and ReO_4^- in the presence of excess interfering ions (Cl^-, NO_3^-, SO_4^{2-}, ClO_4^-, etc., generally present in brackish water) and was found to be highly effective in comparison with the pristine COF and MOP (Figure 6.3g, h). The COF

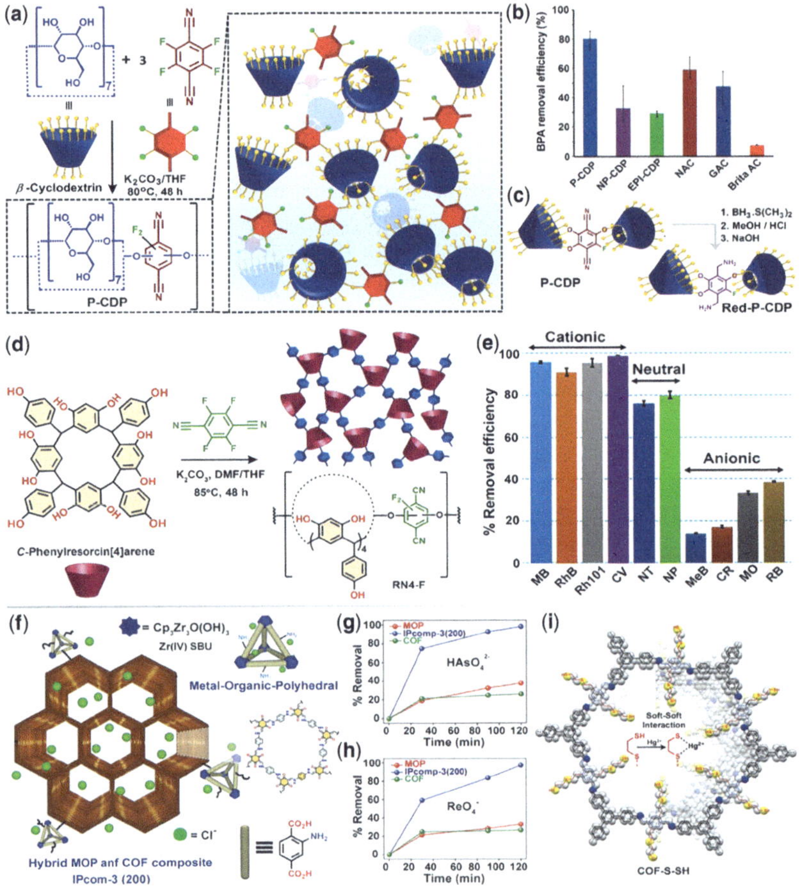

Figure 6.3 (a) Schematic representation of the fabrication of β-cyclodextrin-based POP (P-CDP) through aromatic nucleophilic substitution reaction (SNAr); (b) bisphenol A (BPA) removal efficiency of P-CDP compared to non-porous β-cyclodextrin-based polymers (NP-CDP, EPI-CDP) and the commercial adsorbent materials (NAC, GAC, Brita AC); (c) post-synthetic modification of P-CDP polymer into Red-P-CDP through the reduction of nitrile groups to effectively remove polyfluorinated compounds; (d) schematic representation of the fabrication of C-phenylresorcin[4]arene-based POP; (e) per cent removal efficiency of RN4-F for the organic micropollutants (MB: methylene blue, RhB: rhodamine B, Rh101: rhodamine 101, CV: cresyl violet, NT: β-napthol, NP: p-nitrophenol, MeB: methyl blue, CR: Congo red, MO: methyl orange, RB: Rose Bengal) with different molecular dimensions and charges; (f) schematic illustration of the anion-exchangeable MOP–COF hybrid material; (g) per cent removal of arsenate; (h) rhenate in the presence of excess (1000-fold) interfering anions, namely, Cl^-, NO^{3-}, Br^-, SO_4^{2-} and ClO^{4-}; (i) space-filled model of thioether–thiol functionalized imine-linked COF (COF–S–SH) for Hg(II) removal. (*Source*: Fig. (a, c), (b), (e), (f–h) are adapted from Alsbaiee *et al.* (2016); Giri *et al.* (2020), Giri *et al.* (2019), and Mollick *et al.* (2020), respectively, with copyright permission from the American Chemical Society and Springer Nature).

backbone restricts them from aggregation, exposes the exchangeable functional groups compared to the pristine MOP, and drastically increases the sorption efficiency. The MOP–COF hybrid [IPcom-3(200)] exhibited fast and selective sequestration of high as well as trace levels of oxo-anions, especially As(V) from highly contaminated groundwater below the WHO recommended level (<10 ppb) in the presence of excessive (~1000 times) interfering anions.

Ma and co-workers developed vinyl-functionalized mesoporous COF (COF-V) via a thiol-ene 'click' reaction followed by treating with 1,2-ethanedithiol to form COF-S-SH for the removal of mercury (Figure 6.3i, Sun *et al.*, 2017). The efficiency of COF-S-SH was further compared with the available adsorbing materials like mesoporous silica (2.5 times), thiol-functionalized mesoporous carbon (5 times), thiol-laced MOF (7 times), and various other commercially available adsorbents. Such a strong affinity of the material towards Hg was attributed to the flexible thiol arms, which could chelate with Hg^{2+} through soft–soft interactions (Sun *et al.*, 2017). Ma and co-workers also designed a highly robust amorphous thiol functionalized PAF as a nano-trap for Hg^{2+} (Li *et al.*, 2014). The densely functionalized covalently linked thiol chelating group, uniformly distributed throughout the single-walled pore surface, increases the Hg^{2+} affinity towards the solid adsorbent (Li *et al.*, 2014). Therefore, the tailorable functionalities, high porosity, and excellent stability make the POPs emerging solid-state adsorbents for efficient and selective scavenging of organic micropollutants as well as toxic metal ions from water.

6.5 NANOFILTRATION-BASED WATER PURIFICATION

Nanofiltration membranes made from amorphous POPs have been explored for molecular sieving applications. Livingston and co-workers fabricated ultrathin and microporous polymer nanofilms through an interfacial polymerization route (Figure 6.4a, Jimenez-Solomon *et al.*, 2016). The polyarylate thin films of thickness as low as 20 nm could be obtained by fine-tuning the monomer concentrations. The polyarylate films supported on P84 polyimide (PI) ultrafiltration support (PAR-PI) showed high rejection values for different dyes in methanol (CV: ~97%, BB: ~98%). The excellent organic solvent nanofiltration performance of the membrane composites could be attributed to the enhanced micro-porosity and interconnectivity of voids due to the presence of contorted monomers. Guest-responsive cavitands/macrocycles are often introduced in amorphous polymer membranes to improve selectivity. Liu and co-workers fabricated a mesoporous membrane (pTC-P5A) with a pore size of 3.4 nm using benzoyl chloride and pillar[5]arene (P5A) in an interfacial polymerization method (Figure 6.4, Zhao and Liu, 2018). Owing to the electron-rich cavity of P5A, the membrane afforded an excellent charge-selectivity wherein the cationic malachite green oxalate got effectively separated from anionic sulphorhodamine B in aqueous phase nanofiltration (Figure 6.4c).

Banerjee and co-workers adopted the interfacial crystallization strategy to fabricate highly crystalline and stable COF thin films (Dey *et al.*, 2017). A series of COF films of varying pore sizes were synthesized at a liquid–liquid interface using triphenylphloroglucinol (in dichloromethane) and different amines (in aqueous phase along with *p*-toluenesulphonic acid, Figure 6.4d). The representative COF film, TpBpy showed high fluxes towards both protic (water: 211 L/m²/h/bar) and aprotic (acetonitrile: 339 L/m²/h/bar) solvents owing to its ordered and continuous mesoporous nanochannels. The acetonitrile and the water permeance of the COF film are found to be higher than many polyamide-based membranes reported in the literature. It also displayed a high solute rejection performance for dyes (rhodamine B: 98%, acid fuchsin: 97%, and brilliant blue G: 94%, Figure 6.4e), along with the size-selective molecular separation (Figure 6.4f). Recently, Shaffer and co-workers demonstrated the fabrication of large area (64 cm²) ultrathin (24 nm) COF films for the molecular sieving application with high solvent permeance (Shevate & Shaffer, 2022).

COFs with microporous one-dimensional (1D) channels are also explored as desalination membranes. Huang and co-workers employed the post-synthetic modification methodology to tune the pore aperture size of the COF membranes for efficient metal ion rejection (Figure 6.4g, Liu *et al.*, 2017). They have functionalized the 1D channels of the pristine IISERP-COF-1 membrane with carboxylic acid groups to obtain the membrane, IISERP-COOH-COF-1 (Figure 6.4h). The pore size of the IISERP-COOH-COF-1 membrane was reduced to 0.65 nm from 1.3 nm and is suitable for ion rejection. The IISERP-COOH-COF-1 membrane afforded a high mono- and divalent ion rejection (Na_2SO_4: 96.3%, $MgSO_4$: 97.2%, $FeCl_3$: 99.6%, $MgCl_2$: 90.6% and NaCl: 82.9%).

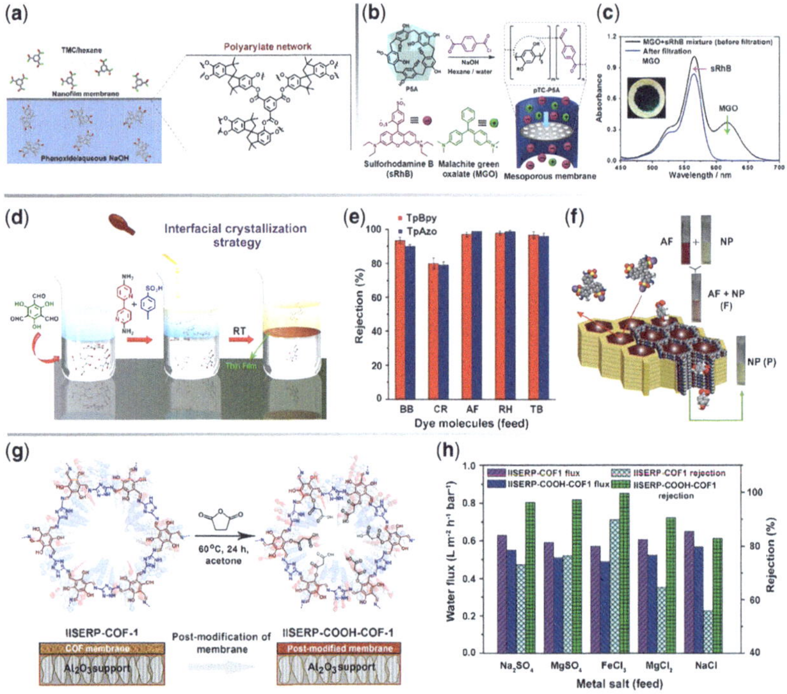

Figure 6.4 (a) Schematic representation of fabrication of polyarylate nanofilms through interfacial polymerization between trimesoyl chloride (TMC) and aromatic phenols; (b) synthetic scheme of the interfacially fabricated pTC-PSA membrane for charge-selective molecular separation; (c) UV–Vis spectra of MGO and sRhB mixture before (black) and after (blue) filtration with pTC-P5A membrane; (d) schematic representation of the fabrication of COF thin films using interfacial crystallization strategy at the liquid–liquid interface; (e) molecular sieving performance of TpBpy, TpAzo membranes towards brilliant blue-G (BB), Congo red (CR), acid fuchsin (AF), rhodamine B (RH), and thymolphthalein blue (TB); (f) size-selective molecular separation from the mixture of AF and nitrophenol (NP) using TpBpy membrane; (g) post-synthetic modification of IISERP-COF-1 via an in situ ring-opening reaction of phenolic-OH groups in the IISERP-COF-1 membrane with maleic anhydride; (h) ion rejection and water flux performance of the pristine and post-modified membranes for different aqueous salt solutions. (*Source*: Fig. (b), (c), (d–f), (g, h) are adapted from Giri *et al.* (2020); Zhao and Liu (2018), Dey *et al.* (2017), and Shevate and Shaffer (2022), respectively, with copyright permission from the Royal Society of Chemistry and the American Chemical Society).

6.6 CONCLUSION

Over the years, nanoporous polymers have become efficient materials for sorption- and nanofiltration-based water purification applications. Despite the significant advances made in the current field, some seminal challenges require the critical attention of the scientific community. Guest selectivity is the key concern associated with the amorphous polymeric adsorbents, which can be augmented with the use of guest-responsive building blocks towards removing specific contaminants. Fabrication of ultra-microporous (<0.7 nm) COFs for highly efficient molecular sieving has been challenging due to the size constraint of the molecular building units. Approaches, such as strategic pore engineering of COFs through post-modification of inner walls and design of AB stacked or interpenetrated 2D/3D COFs, could circumvent the existing limitation. Overall, the cheap and scalable design routes without sophisticated monomers for fabricating processable porous organic materials can significantly improve their prospects in real-time wastewater treatment and large-scale seawater desalination.

REFERENCES

Alsbaiee A., Smith B. J., Xiao L. L., Ling Y. H., Helbling D. E. and Dichtel W. R. (2016). Rapid removal of organic micropollutants from water by a porous β-cyclodextrin polymer. *Nature*, **529**(7585), 190–194, https://doi.org/10.1038/nature16185

Bandyopadhyay S., Anil A. G., James A. and Patra A. (2016). Multifunctional porous organic polymers: tuning of porosity, CO_2, and h_2 storage and visible-light-driven photocatalysis. *ACS Applied Materials & Interfaces*, **8**(41), 27669–27678, https://doi.org/10.1021/acsami.6b08331

Dey K., Pal M., Rout K. C., Kunjattu H. S., Das A., Mukherjee R., Kharul U. K. and Banerjee R. (2017). Selective molecular separation by interfacially crystallized covalent organic framework thin films. *Journal of the American Chemical Society*, **139**(37), 13083–13091, https://doi.org/10.1021/jacs.7b06640

Diercks C. S. and Yaghi O. M. (2017). The atom, the molecule, and the covalent organic framework. *Science*, **355**, aal1585, https://doi.org/10.1126/science.aal1585

Giri A., Hussain W., Sk B. and Patra A. (2019). Connecting the dots: knitting *c*-phenylresorcin[4]arenes with aromatic linkers for task-specific porous organic polymers. *Chemistry of Materials*, **31**(20), 8440–8450, https://doi.org/10.1021/acs.chemmater.9b02563

Giri A., Sahoo A., Dutta T. K. and Patra A. (2020). Cavitand and molecular cage-based porous organic polymers. *ACS Omega*, **5**(44), 28413–28424, https://doi.org/10.1021/acsomega.0c04248

Giri A., Biswas S., Hussain M. W., Dutta T. K. and Patra A. (2022a). Nanostructured hypercrosslinked porous organic polymers: morphological evolution and rapid separation of polar organic micropollutants. *ACS Applied Materials & Interfaces*, **14**(5), 7369–7381, https://doi.org/10.1021/acsami.1c24393

Giri A., Khakre Y., Shreeraj G., Dutta T. K., Kundu S. and Patra A. (2022b). The order-disorder conundrum: a trade-off between crystalline and amorphous porous organic polymers for task-specific applications. *Journal of Materials Chemistry A*, **10**(33), 17077–17121, https://doi.org/10.1039/D2TA01546C

Jimenez-Solomon M. F., Song Q. L., Jelfs K. E., Munoz-Ibanez M. and Livingston A. G. (2016). Polymer nanofilms with enhanced microporosity by interfacial polymerization. *Nature Materials*, **15**(7), 760–767, https://doi.org/10.1038/nmat4638

Klemes M. J., Ling Y. H., Ching C., Wu C. Y., Xiao L. L., Helbling D. E. and Dichtel W. R. (2019). Reduction of a tetrafluoroterephthalonitrile-beta-cyclodextrin polymer to remove anionic micropollutants and perfluorinated alkyl substances from water. *Angewandte Chemie International Edition*, **58**(35), 12049–12053, https://doi.org/10.1002/anie.201905142

Lee J. S. M. and Cooper A. I. (2020). Advances in conjugated microporous polymers. *Chemical Reviews*, **120**(4), 2171–2214, https://doi.org/10.1021/acs.chemrev.9b00399

Li B. Y., Zhang Y. M., Ma D. X., Shi Z. and Ma S. Q. (2014). Mercury nano-trap for effective and efficient removal of mercury(ii) from aqueous solution. *Nature Communications*, **5**, 5537, https://doi.org/10.1038/ncomms6537

Liu C. Y., Jiang Y. Z., Nalaparaju A., Jiang J. W. and Huang A. S. (2017). Post-synthesis of a covalent organic framework nanofiltration membrane for highly efficient water treatment. *Journal of Materials Chemistry A*, **7**(42), 24205–24210, https://doi.org/10.1039/C9TA06325K

Mollick S., Fajal S., Saurabh S., Mahato D. and Ghosh S. K. (2020). Nanotrap grafted anion exchangeable hybrid materials for efficient removal of toxic oxoanions from water. *ACS Central Science*, **6**(9), 1534–1541, https://doi.org/10.1021/acscentsci.0c00533

Nguyen H. L. (2021). Reticular design and crystal structure determination of covalent organic frameworks. *Chemical Science*, **12**(25), 8632–8647, https://doi.org/10.1039/D1SC00738F

Sackaria M. and Elango L. (2020). Organic micropollutants in groundwater of India – a review. *Water Environment Research*, **92**(4), 504–523, https://doi.org/10.1002/wer.1243

Schwarzenbach R. P., Escher B. I., Fenner K., Hofstetter T. B., Johnson C. A., von Gunten U. and Wehrli B. (2006). The challenge of micropollutants in aquatic systems. *Science*, **313**(5790), 1072–1077, https://doi.org/10.1126/science.1127291

Shevate R. and Shaffer D. L. (2022). Large-area 2D covalent organic framework membranes with tunable single-digit nanopores for predictable mass transport. *ACS Nano*, **16**(2), 2407–2418, https://doi.org/10.1021/acsnano.1c08804

Slater A. G. and Cooper A. I. (2015). Function-led design of new porous materials. *Science*, **348**, aaa8075, https://doi.org/10.1126/science.aaa8075

Sun Q., Aguila B., Perman J., Earl L. D., Abney C. W., Cheng Y. C., Wei H., Nguyen N., Wojtas L. and Ma S. Q. (2017). Postsynthetically modified covalent organic frameworks for efficient and effective mercury removal. *Journal of the American Chemical Society*, **139**(7), 2786–2793, https://doi.org/10.1021/jacs.6b12885

Sun Q., Aguila B., Song Y. P. and Ma S. Q. (2020). Tailored porous organic polymers for task-specific water purification. *Accounts of Chemical Research*, **53**(4), 812–821, https://doi.org/10.1021/acs.accounts.0c00007

Thommes M., Kaneko K., Neimark A. V., Olivier J. P., Rodriguez-Reinoso F., Rouquerol J. and Sing K. S. W. (2015). Physisorption of gases, with special reference to the evaluation of surface area and pore size distribution (IUPAC technical report). *Pure and Applied Chemistry*, **87**(9–10), 1051–1069, https://doi.org/10.1515/pac-2014-1117

Werber J. R., Osuji C. O. and Elimelech M. (2016). Materials for next-generation desalination and water purification membranes. *Nature Reviews Materials*, **1**, 16018, https://doi.org/10.1038/natrevmats.2016.18

Xu Y. H., Jin S. B., Xu H., Nagai A. and Jiang D. L. (2013). Conjugated microporous polymers: design, synthesis and application. *Chemical Society Reviews*, **42**(20), 8012–8031, https://doi.org/10.1039/c3cs60160a

Zhao Q. and Liu Y. (2018). Macrocycle crosslinked mesoporous polymers for ultrafast separation of organic dyes. *Chemical Communications*, **54**(53), 7362–7365, https://doi.org/10.1039/C8CC04080J

doi: 10.2166/9781789063714_0073

Chapter 7

New materials for arsenic and fluoride removal

Tanvi Gupte[1,2] *and Thalappil Pradeep*[1,3]*

[1]*DST Unit of Nanoscience (DST UNS) and Thematic Unit of Excellence, Department of Chemistry, Indian Institute of Technology Madras, Chennai, India*
[2]*Department of Metallurgical and Materials Engineering, Indian Institute of Technology Madras, Chennai, India*
[3]*International Centre for Clean Water, IIT Madras Research Park, Chennai, India*
**Corresponding author: pradeep@iitm.ac.in*

ABSTRACT

Arsenic and fluoride are two major global water pollutants. They are introduced into the environment via various natural and anthropogenic activities. Several countries in the world are facing the detrimental health and environmental effects of these contaminants. Many technologies have been proposed and implemented for the removal of arsenic and fluoride from water. However, there is an ever-increasing demand for affordable and reliable water purification approaches. All the technologies are primarily based on any one of the following purification techniques: membrane filtration, reverse osmosis, electrodialysis, adsorption, precipitation, coagulation, oxidation, and ion exchange. The contaminant removal efficiency depends on the properties of the material used. Various materials have been tested and are used to deliver arsenic and fluoride-free water. In developing countries, affordability plays an important role in the implementation of water treatment strategies. This chapter gives an overview of simple, efficient, and affordable materials that combined with appropriate technologies, provide promising scalable solutions for the mitigation of arsenic and fluoride in water.

Keywords: arsenic, fluoride, removal technologies, adsorbents, sustainability indices

7.1 INTRODUCTION

Water is vital for the sustenance of life on earth. Though groundwater comprises only 0.6% of the total water resources on earth, it is the major and favoured source of potable water in developing countries' rural and urban areas. In rural India, groundwater caters to 80% of drinking water and 50% of agricultural demand (Meenakshi & Maheshwari, 2006). Several geogenic and anthropogenic activities have contaminated aquatic ecosystems with metals, non-metals, and other toxic compounds. Arsenic and fluoride are both naturally occurring elements that can be found in the Earth's crust. However, when present in high concentrations in water, they can cause harm to human health. The World Health Organisation (WHO) has categorized both as critical chemicals that can cause large-scale health effects as a result of drinking contaminated water (World Health Organization, 2017).

7.1.1 Arsenic and fluoride contamination

Groundwater contamination with arsenic and fluoride at hazardous levels has been reported in several countries including Bangladesh, China, India, Mexico, and Argentina (Mukherjee *et al.*, 2019b). Arsenic and fluoride contamination sources are natural and anthropogenic. Arsenic and fluoride come to be present in ground and surface waters primarily through the dissolution of natural arsenic and fluoride-containing minerals and ores present in rocks and soil (Maity *et al.*, 2021). Anthropogenic sources of fluorides are attributed to the production of phosphate fertilizers, bricks, tiles, and ceramics. Fluoride is essential in minute quantities for the normal mineralization of bones and the formation of dental enamel.

Arsenic is a Group VA metalloid. Various species of As are arsenic trioxide (AsO_3), arsenite As(III)), arsenate (As(V)), methylated species, arsenobetaine (AB), and arsenocholine (AC). It exists of four oxidation states: arsenite (As^{3+}), arsenate (As^{5+}), arsenic (As^0), and arsenide (As^{3-}). The mobility of arsenic depends on redox potential (Eh), pH, biological activity, and adsorption/desorption from aquifer rocks and sediments. In general, the mobility of As(III) is higher than As(V) in aquifers due to the lower probability of adsorption of neutral As(III) on mineral surfaces (Hao *et al.*, 2018). The anthropogenic sources of arsenic are mining, ore production and processing, electronic device fabrication, combustion of fossil fuels, synthesis and use of dyes and pigments, pharmaceuticals, and agricultural insecticides and pesticides (Mukherjee *et al.*, 2019a).

In India, the acceptable limit of arsenic in drinking water is 10 ppb, and in the absence of any alternate source, the limit is 50 ppb (IS 10500, n.d.). For fluoride, the permissible limit in India is 1 ppm, and in the absence of an alternate source, the limit is 1.5 ppm (IS 10500, n.d.). Chronic exposure to arsenic can cause various health problems, such as arsenicosis, reproductive disorders, peripheral neuropathy, and it also has a carcinogenic potential (Mukherjee *et al.*, 2019a). Fluoride can cause discolouration of the teeth (dental fluorosis), weakening of the bones (skeletal fluorosis), defects in knees, hips, bones, and in extreme cases, it may also cause paralysis (Patel *et al.*, 2019). Hence, restricting the presence of arsenic and fluoride within stipulated limits in water supply is essential.

7.1.2 The current scenario in India

Arsenic and fluoride contamination is a significant problem in certain regions of India. They are the two major contaminants of Indian groundwater in terms of affected areas and the number of people. The worst affected areas are the middle and lower Gangetic planes and some parts of Central and South India. Fluoride contamination, on the other hand, is more widespread, affecting several states. Table 7.1 shows the state-wise list of the number of districts affected by fluoride and arsenic, as reported by the Central Ground Water Board (CGWB) of India (CGWB, n.d.).

The Indian government and various organizations have been implementing various measures to address the problem. The Jal Jeevan Mission (JJM) is one of the initiatives taken by the government under which community water purification plants (CWPPs) have been installed in arsenic and fluoride-affected regions. However, implementing these measures on a large scale remains a challenge, and more needs to be done to tackle the problem.

Table 7.1 Occurrence of high fluoride and arsenic in groundwater in some states of India.

S. No.	State	Districts Affected by Fluoride	Districts Affected by Arsenic
1	Andhra Pradesh	16	3
2	Assam	2	8
3	Bihar	5	21
4	Chattisgarh	2	1
5	Delhi	7	2
6	Daman & Diu	–	1
7	Gujarat	18	12
8	Haryana	11	8
9	Himachal Pradesh	–	1
10	Jammu & Kashmir	1	3
11	Jharkhand	4	1
12	Karnataka	14	2
13	Kerala	2	–
14	Maharashtra	8	–
15	Madhya Pradesh	13	8
16	Manipur	–	2
17	Odisha	18	1
18	Punjab	9	7
19	Rajasthan	32	1
20	Tamil Nadu	8	–
21	Telangana	–	1
22	Uttar Pradesh	7	13
23	West Bengal	7	8

7.2 MATERIALS FOR ARSENIC AND FLUORIDE REMOVAL

The selection of material(s) for removing fluoride and arsenic using any of the aforementioned purification techniques depends on the removal efficiency, affordability, scalability, and regeneration capacity.

7.2.1 Metal oxides and hydroxides

Metal oxides and hydroxides in the form of nanoparticles or composites are very effective for removing arsenic and fluoride from water due to their affinity. The majority of them are used as adsorbents. However, few of them can be used with other treatment methods.

Iron oxide/hydroxide/oxyhydroxide, such as ferric oxide (Fe_2O_3), ferric hydroxide ($Fe(OH)_3$), akaganèite (β-FeOOH), and ferrihydrites ($Fe_{10}O_{14}(OH)_2$) have been used for removing arsenic from water due to their high affinity for arsenic under neutral conditions. These compounds can effectively remove both inorganic and organic forms of the contaminant. Moreover, they are non-toxic, low-cost, and available in significant quantities. The adsorption of arsenic and the effectiveness of the adsorbents depend on arsenic species, pH, and potential. Aluminium oxide and hydroxide, such as

Table 7.2 Metal-oxides and hydroxides-based arsenic and fluoride removal.

Metal-oxide/Hydroxide/Oxyhydroxide	Uptake Capacity (mg/g)		Reference
Arsenic Removal	**As(III)**	**As(V)**	
Confined metastable 2-line ferrihydrite	100	100	Anil Kumar *et al.* (2017)
α-FeOOH nanoparticles	–	76	Ghosh *et al.* (2012)
Zr-doped β-FeOOH nanoparticles	120	60	Sun *et al.* (2013)
Magnetic γ-Fe_2O_3 nanoparticles	74.83	105.25	Lin *et al.* (2012)
Fe/AlO(OH)	–	102–129	Muedi *et al.* (2021)
Mesoporous iron oxide	136.89	31.82	Bae *et al.* (2020)
Hydrous cerium oxide (HCO) nanoparticles	170	107	Li *et al.* (2012)
Iron–zirconium (Fe–Zr) binary oxide	120	46.1	Ren *et al.* (2011)
Iron hydroxide nanopetalines	91.74	217.76	Wang *et al.* (2022)
Fluoride Removal			
Fe–Al–Ce trimetal oxide	178		Wu *et al.* (2007)
Cellulose–Al–Fe nanocomposite (AlOOH and FeOOH in carboxymethyl cellulose matrix)	75.2		Egor *et al.* (2021)
Boron-doped biochar/Al_2O_3	196.1		Jiang *et al.* (2019)
Inorganic polymeric coagulant made by etching of aluminium trihydrate ($Al_2O_3{\cdot}3H_2O$) with acid	87.68		Solanki *et al.* (2020)
Nanoporous biochar-supported magnesium oxide (MgO-BC)	83.05		Wan *et al.* (2019)

alumina (Al_2O_3) and aluminium hydroxide ($Al(OH)_3$), have been widely used for removing fluoride from water. These compounds can adsorb fluoride ions from water by forming surface complexes. Table 7.2 lists some metal-oxides/hydroxides/oxyhydroxides used for arsenic and fluoride removal.

7.2.2 Biopolymers and biominerals

Various biopolymers and biominerals have been explored to remove contaminants from water. Reports based on hydroxyapatite, cellulose, alginate, chitosan, and gelatin are described here. Hydroxyapatite (HA; $Ca_5(PO_4)_3OH$), a member of the calcium phosphate family, is a widely researched adsorbent for treating contaminated water and soil. The surface of HA has tunnels and active sites that aid in improving the adsorption. The adsorption on HA can occur through either ion pair formation ($F^- \cdots OH_2^+$), ion exchange (F^- with OH^-), or H-bonding ($F^- \cdots H \cdots O$) in aqueous conditions. These mechanisms result in the formation of mixed fluoridated HA ($Ca_5(PO_4)_3(OH,F)$) or stable fluorapatite ($Ca_5(PO_4)_3F$) (Nagaraj *et al.*, 2018). Nayak *et al.* synthesized nanocrystalline HA from egg shells and its fluoride adsorption capacity was 22.3 mg/g (Nayak *et al.*, 2017). He *et al.* developed an HA nanowires-based membrane with a fluoride adsorption capacity of 40.65 mg/g at neutral pH (He *et al.*, 2016b). Nagaraj *et al.* developed a mineral (Al^{3+}, La^{3+}, and Ce^{3+})-substituted HA nanocomposite material by hydrothermal method with a fluoride removal capacity of 8.36 mg/g (Nagaraj *et al.*, 2018).

Sharma *et al.* used ZnO nanocrystal decorated regenerated micro fibrillated cellulose for As(V) removal with the capacity of 4421 mg/g at neutral pH (Sharma *et al.*, 2019). Sinha *et al.* developed a filter device made of three-dimensional (3D) macroporous alginate/akaganeite nanorod composite (MAAC) scaffolds for arsenic removal with the capacity of 70.5 mg/g for As(III) and 24.4 mg/g for As(V) (Sinha *et al.*, 2018). Guo and Chen fabricated bead cellulose loaded with iron oxyhydroxide for adsorption and removal of arsenic with an adsorption capacity for As (III) and As (V) of 99.6 and

33.2 mg/g at neutral pH. The composite can be regenerated using 2M NaOH (Guo & Chen, 2005). Mukherjee *et al.* reported functionalized microcellulose-reinforced 2-line ferrihydrite composite that showed exceptional As(III) and As(V) adsorption capacities of 143 and 83 mg/g, respectively (Mukherjee *et al.*, 2019c).

Researchers are also investigating biopolymeric carrier mediums along with sorbent material to generate practical filter materials. Scientists have developed biopolymeric beads made of hydroxyapatite-implanted cerium-based metal organic frameworks incorporating alginate and chitosan for fluoride removal. These beads displayed an adsorption capacity of 4.8 mg/g in 20 min (Jeyaseelan & Viswanathan, 2022). Fernando *et al.* investigated a biopolymer-based nanohydroxyapatite (n-HAp) material for the removal of fluoride, arsenic, cadmium, and lead from water. They explored the composites of HA with chitosan, carboxymethyl cellulose, sodium alginate, and gelatin. Their study revealed the HA–chitosan composite as the most versatile sorbent (Fernando *et al.*, 2021).

7.2.3 Biological origin

Recently, researchers have been exploring the application of natural materials to remove various pollutants due to their efficiency, recyclability, and environmental friendliness. Calcium-containing adsorbents are a good choice for fluoride adsorption due to their favourable characteristics, including high ionic exchangeability, availability, adsorption affinity, and capacity to form bonds with various organics of varied sizes (Hashemkhani *et al.*, 2022). Scientists have used oyster shells (Hashemkhani *et al.*, 2022; Kim *et al.*, 2020b) and egg shells (Lee *et al.*, 2021) as a sorbent for fluoride removal from water. Hashemkhani *et al.* (2022) obtained the maximum adsorption capacity of 7.53 mg/g using oyster shell at pH 5.5 and contact time of 85 min (Hashemkhani *et al.*, 2022). Lee *et al.* reported maximum adsorption capacity of 258.28 mg/g using thermally treated egg shells at pH 7 and 25°C (Lee *et al.*, 2021). Fluoride adsorption reduced in the presence of anions in the following order: $HPO_4^{3-} > HCO_3^{-} >> SO_4^{2-} > Cl^{-}$. According to them, fluoride removal is attributed to the formation of calcium fluorite (CaF_2). Sinha *et al.* (2003) used *Eichhornia crassipes* (water hyacinth) and its activated carbon for defluoridation. H type activated carbon (activated at 600°C) showed better performance than the non-carbonized plant (Sinha *et al.*, 2003). Fox *et al.* studied the removal of arsenic by flocculation–coagulation system using cactus mucilage and ferric (Fe(III)) salt. The system was capable of removing 75–96% arsenic in 30 min, with the majority of the removal occurring in 10–15 min (Fox *et al.*, 2016).

7.2.4 Carbon based materials

Carbon-based materials have been widely used to efficiently remove environmental contaminants owing to their high surface area, surface functional groups, and large number of active sites. Moreover, they can be easily coupled with metal oxides, nanoparticles, and so on, to enhance their adsorption capacities. Some carbon-based composites, such as activated carbon, carbon nanotubes (CNTs), carbon nanowires, graphite, graphene oxide, and carbon nanofibres, along with their uptake capacity for fluoride and/or arsenic, are listed in Table 7.3.

7.2.5 Biochar

One of the relatively new approaches for the decontamination of water is the use of biochar. Biochars are sustainable and renewable carbonaceous compounds synthesized by pre-/post-pyrolysis of biomass derived from feedstock under limited oxygen supply. The micropores of biochar offer high surface area along with several functional groups that are beneficial for the remediation of contaminants. Kumar *et al.* (2019) reported sustainable fluoride remediation using biochars produced by slow pyrolysis of okra stem and black gram straw with sorption capacities of 20 and 16 mg/g. Biochar-based adsorbents for arsenic removal from water have also been researched extensively. Strategies for enhancing the adsorption of arsenic include modifying biochar with metal salts, like $FeCl_3$, $MnCl_4$, $FeSO_4$, and $AlCl_3$, and acids (H_2SO_4, HCl, and HNO_3) (Sharma *et al.*, 2022).

Table 7.3 Carbon-based materials and their uptake capacities.

S. No.	Carbon Material	Contaminant	Uptake Capacity (mg/g)	Reference
1	Aluminium-impregnated hierarchal web of micro-nano-activated carbon fibres	F^-	17	Gupta *et al.* (2009)
2	La/Mg/Si-activated carbon	F^-	220	Kim *et al.* (2020a)
3	Activated carbon–aluminium oxide composite	F^-	14	Iwar *et al.* (2022)
4	ZrO_2 embedded in carbon nanowires	As(III) As(V)	28.61 106.57	Luo *et al.* (2016)
5	Mesoporous graphene oxide-lanthanum fluoride nanocomposite	As(V)	18.52	Lingamdinne *et al.* (2019)
6	Nano-alumina wrapped carbon microspheres	As(V)	68	Raj *et al.* (2023)

7.2.6 Metal organic frameworks

The development of metal organic frameworks (MOF), a new class of three-dimensional crystalline inorganic–organic porous hybrid materials with tuneable porosity and exceptional surface area, has made them promising candidates for a variety of applications, including gas storage, catalysis, sensor, drug delivery, ion exchange, and adsorption and removal of hazardous materials. Fe_2Co_1MOF-74 exhibited exceptionally high As(III) and As(V) removal capacities of 266.5 and 292.2 mg/g,

Table 7.4 Non-conventional technologies for arsenic and fluoride removal from water.

S. No.	Material	Removal Technology	Contaminant	Removal Efficiency	Reference
1	Aluminium form of phosphomethylated resin	Electrodeionization	F^-	73.2%	Gahlot *et al.* (2015)
2	Imidazolium-based ionic liquids grafted on thin-film composite forward osmosis membranes	Forward osmosis	As(V)	99.5%	Yang *et al.* (2019)
3	Hypochlorite oxidation, adsorption on Fe(II) coagulates, and low pressure membrane	Oxidation, adsorption, low-pressure membrane	As(III) and As(V)	>95%	Elcik *et al.* (2013)
4	Fe anode and air cathode with in situ generation of H_2O_2	Air cathode-assisted iron electrocoagulation (ACAIE)	As(III) and As(V)	~100%	Bandaru *et al.* (2020)
5	Polyhedral oligomeric silsesquioxane-functionalized graphene oxide/ polyvinylidene difluoride (POSS-rGO/ PVDF) electrospun membranes	Membrane distillation	As(III) and As(V)	99.9%	Leaper *et al.* (2021)

respectively (Sun *et al.*, 2019). UiO-66 has also shown good arsenic adsorption capacities (Singh *et al.*, 2022). Aluminium fumarate (AlFu) MOF has been reported as a superabsorbent for fluoride with an adsorption capacity of 600 mg/g at 293 K (Karmakar *et al.*, 2016). However, there are not many industrial applications of MOFs for the removal of heavy metals. To address this, the membrane integration of MOFs is being explored by researchers. An adsorption membrane made of zirconium metal-organic frameworks (Zr-MOFs) was prepared for rapid fluoride removal from drinking water. The maximum adsorption capacity was 102.40 mg/g at pH 7.0 (He *et al.*, 2016a). Wang *et al.* reported an iron mesh-based MOF (MIL-100(Fe)) filter with capacity for As(III) and As(V) of 35.2 and 19.2 mg/g, respectively (Wang *et al.*, 2018).

7.2.7 Other technologies

Apart from the conventional methods discussed in this chapter, researchers are exploring several other technologies for the removal of arsenic and fluoride. A few of these methods are listed in Table 7.4.

7.3 EVALUATING SUSTAINABILITY INDICES OF TECHNIQUES

Several factors need to be considered to evaluate the sustainability of any fluoride and arsenic removal technique:

- Efficiency: The technique should be strongly able to effectively remove fluoride and arsenic from water.
- Energy: The amount of energy needed to operate, including the cost and availability of energy sources, should be economical. The carbon footprint of the process can be reduced by using renewable energy sources, such as solar, wind, and hydropower. Moreover, the amount of contaminant removed per unit of energy consumed must be as high as possible.
- Cost: The total expenses of the technique, including initial investment, operation, maintenance, and waste disposal cost, must be considered.
- Material: The materials used in water treatment must be affordable, sustainable, and effective in contaminant removal. Its environmental and health impacts must be thoroughly evaluated. The local and global availability of the material also plays an important role in determining the overall cost of the removal method. Moreover, the ability of the material to be regenerated or recycled and its potential for environmental degradation must also be considered.
- Scalability: The ability of the technique to operate at the household as well as community level.
- Environmental impact: This considers the generation of waste and its method of disposal, emissions during the operation, and the use of natural resources. The life cycle approach can be implemented to evaluate the energy and environmental sustainability of the water treatment method. Figure 7.1 shows the variables for environmental sustainability assessment (ESA) of adsorption and ion-exchange treatment methods. The life cycle inventory (LCI) can be estimated on the basis of the amount of final resources and pollutants for calculating the natural resources sustainability (NRS) and environmental burdens sustainability (EBS), respectively.
- Social impact: The technique's effects on the cultural and socio-economic aspects should be considered.
- Reject/residue management: Fluoride and arsenic removal plants can generate waste in the form of backwash wastewater, sludge, used adsorbents, or concentrated reject water. These need to be treated and/or stabilized and safely disposed of to prevent contamination of land or water. Various disposal options include landfills, stabilization to reduce toxicity and mobility, mixing with cow dung to promote microbial methylation of arsenic, passive aeration systems, and soil.

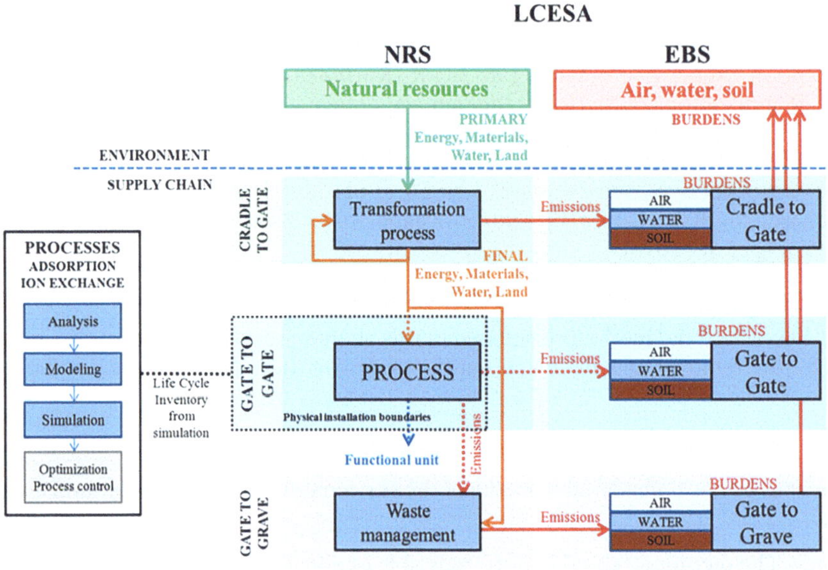

Figure 7.1 Block diagram of the life cycle environmental sustainability assessment (LCESA) for typical adsorption and ion-exchange methods. NRS: natural resources sustainability, EBS: environmental burdens sustainability (Reprinted with permission from Dominguez-Ramos et al. 2014. Copyright 2014 American Chemical Society). (Reprinted with permission from Dominguez-Ramos *et al.*, 2014. Copyright 2014 American Chemical Society).

7.4 CONCLUSION

Arsenic and fluoride-contaminated water have become a global health issue. According to statistics by CGWB, the groundwater of 20 and 19 Indian states is affected by arsenic and fluoride, respectively. It is essential to determine the best management practices to control these contaminants to adhere to water quality standards. This chapter discusses various water treatment techniques and materials for addressing the problem of high concentrations of these toxic contaminants in water. Among the various techniques discussed, adsorption and ion-exchange-based technologies are promising and economical and have found abundant field applications. Extensive research over several decades has resulted in the development of adsorbents with improved removal efficiencies and adsorption capacities. However, all treatment methods produce certain waste materials that must be managed carefully and disposed of properly.

REFERENCES

Anil Kumar A., Som A., Longo P., Sudhakar C., Bhuin R. G., Sen Gupta S., Anshup, Udhaya Sankar M., Chaudhary A., Kumar R. and Pradeep T. (2017). Confined metastable 2-line ferrihydrite for affordable point-of-use arsenic-free drinking water. *Advanced Materials*, **29**, 1604260, https://doi.org/10.1002/adma.201604260

Bae J., Kim S., Kim K. S., Hwang H. K. and Choi H. (2020). Adsorptive removal of arsenic by mesoporous iron oxide in aquatic systems. *Water 2020*, **12**(11), 3147, https://doi.org/10.3390/W12113147

Bandaru S. R. S., Van Genuchten C. M., Kumar A., Glade S., Hernandez D., Nahata M. and Gadgil A. (2020). Rapid and efficient arsenic removal by iron electrocoagulation enabled with in situ generation of hydrogen peroxide. *Environmental Science and Technology*, **54**(10), 6094–6103, https://doi.org/10.1021/ACS.EST.0C00012/ASSET/IMAGES/LARGE/ES0C00012_0004.JPEG

CGWB (n.d). 'Central Ground Water Board, Ministry of Water Resources, River Development & Ganga Rejuvenation, Government of India.' Accessed 16 January 2023. http://cgwb.gov.in/wqreports.html

Dominguez-Ramos A., Chavan K., García V., Jimeno G., Albo J., Marathe K. V., Yadav G. D. and Irabien A. (2014). Arsenic removal from natural waters by adsorption or ion exchange: an environmental sustainability assessment. *Industrial and Engineering Chemistry Research*, **53**(49), 18920–18927, https://doi.org/10.1021/IE4044345/SUPPL_FILE/IE4044345_SI_001.PDF

Egor M., Anil Kumar A., Ahuja T., Mukherjee S., Chakraborty A., Sudhakar C., Srikrishnarka P., Bose S., Ravindran S. J. and Pradeep T. (2021). Cellulosic ternary nanocomposite for affordable and sustainable fluoride removal. *ACS Sustainable Chemistry and Engineering*, **9**(38), 12788–12799, https://doi.org/10.1021/ACSSUSCHEMENG.1C03272/ASSET/IMAGES/MEDIUM/SC1C03272_M002.GIF

Elcik H., Cakmakci M., Sahinkaya E. and Ozkaya B. (2013). Arsenic removal from drinking water using low pressure membranes. *Industrial and Engineering Chemistry Research*, **52**(29), 9958–9964, https://doi.org/10.1021/IE401393P/ASSET/IMAGES/LARGE/IE-2013-01393P_0005.JPEG

Fernando M. S., Wimalasiri A. K. D. V. K., Dziemidowicz K., Williams G. R., Koswattage K. R., Dissanayake D. P., De Silva K. M. N. and De Silva R. M. (2021). Biopolymer-based nanohydroxyapatite composites for the removal of fluoride, lead, cadmium, and arsenic from water. *ACS Omega*, **6**(12), 8517–8530, https://doi.org/10.1021/ACSOMEGA.1C00316/ASSET/IMAGES/LARGE/AO1C00316_0010.JPEG

Fox D. I., Stebbins D. M. and Alcantar N. A. (2016). Combining ferric salt and cactus mucilage for arsenic removal from water. *Environmental Science and Technology*, **50**(5), 2507–2513, https://doi.org/10.1021/ACS.EST.5B04145/SUPPL_FILE/ES5B04145_SI_001.PDF

Gahlot S., Sharma S. and Kulshrestha V. (2015). Electrodeionization: an efficient way for removal of fluoride from tap water using an aluminium form of phosphomethylated resin. *Industrial and Engineering Chemistry Research*, **54**(16), 4664–4671, https://doi.org/10.1021/ACS.IECR.5B00369/SUPPL_FILE/IE5B00369_SI_001.PDF

Ghosh M. K., Poinern G. E. J., Issa T. B. and Singh P. (2012). Arsenic adsorption on goethite nanoparticles produced through hydrazine sulfate assisted synthesis method. *Korean Journal of Chemical Engineering*, **29**(1), 95–102, https://doi.org/10.1007/S11814-011-0137-Y/METRICS

Guo X. and Chen F. (2005). Removal of arsenic by bead cellulose loaded with iron oxyhydroxide from groundwater. *Environmental Science and Technology*, **39**(17), 6808–6818, https://doi.org/10.1021/ES048080K/ASSET/IMAGES/LARGE/ES048080KF00010.JPEG

Gupta A. K., Deva D., Sharma A. and Verma N. (2009). Adsorptive removal of fluoride by micro-nanohierarchal web of activated carbon fibers. *Industrial and Engineering Chemistry Research*, **48**(21), 9697–9707, https://doi.org/10.1021/IE801688K/ASSET/IMAGES/IE-2008-01688K_M004.GIF

Hao L., Liu M., Wang N. and Li G. (2018). A critical review on arsenic removal from water using iron-based adsorbents. *RSC Advances*, **8**(69), 39545–39560, https://doi.org/10.1039/C8RA08512A

Hashemkhani M., Ghalhari M. R., Bashardoust P., Hosseini S. S., Mesdaghinia A. and Mahvi A. H. (2022). Fluoride removal from aqueous solution via environmentally friendly adsorbent derived from seashell. *Scientific Reports*, **12**(1), 1–13, https://doi.org/10.1038/s41598-022-13756-3

He J., Cai X., Chen K., Li Y., Zhang K., Jin Z., Meng F., Liu N., Wang X., Kong L., Huang X. and Liu J. (2016a). Performance of a novelly-defined zirconium metal-organic frameworks adsorption membrane in fluoride removal. *Journal of Colloid and Interface Science*, **484**(December), 162–172, https://doi.org/10.1016/J.JCIS.2016.08.074

He J., Zhang K., Wu S., Cai X., Chen K., Li Y., Sun B., Jia Y., Meng F., Jin Z., Kong L. and Liu J. (2016b). Performance of novel hydroxyapatite nanowires in treatment of fluoride contaminated water. *Journal of Hazardous Materials*, **303**(February), 119–130, https://doi.org/10.1016/J.JHAZMAT.2015.10.028

IS 10500 (n.d). 'IS 10500: Drinking Water: Bureau of Indian Standards.' Accessed 17 January 2023. https://archive.org/details/gov.in.is.10500.2012/page/n3/mode/2up

Iwar R. T., Ogedengbe K. and Ugwudike B. O. (2022). Groundwater fluoride removal by novel activated carbon/aluminium oxide composite derived from raffia palm shells: optimization of batch operations and field-scale point of use system evaluation. *Results in Engineering*, **14**(June), 100407, https://doi.org/10.1016/J.RINENG.2022.100407

Jeyaseelan A. and Viswanathan N. (2022). Investigation of hydroxyapatite-entrenched cerium organic frameworks incorporating biopolymeric beads for efficient fluoride removal. *Industrial and Engineering Chemistry Research*, **61**(23), 7911–7925, https://doi.org/10.1021/ACS.IECR.2C00487/ASSET/IMAGES/LARGE/IE2C00487_0010.JPEG

Jiang X., Xiang X., Hu H., Meng X. and Hou L. (2019). Facile fabrication of biochar/Al_2O_3 adsorbent and its application for fluoride removal from aqueous solution. *Journal of Chemical and Engineering Data*, **64**(1), 83–89.

Karmakar S., Dechnik J., Janiak C. and De S. (2016). Aluminium fumarate metal-organic framework: a super adsorbent for fluoride from water. *Journal of Hazardous Materials*, **303**(February), 10–20, https://doi.org/10.1016/J.JHAZMAT.2015.10.030

Kim M., Choong C. E., Hyun S., Park C. M. and Lee G. (2020a). Mechanism of simultaneous removal of aluminum and fluoride from aqueous solution by La/Mg/Si-activated carbon. *Chemosphere*, **253**(August), 126580, https://doi.org/10.1016/J.CHEMOSPHERE.2020.126580

Kim W., Singh R. and Smith J. A. (2020b). Modified crushed oyster shells for fluoride removal from water. *Scientific Reports*, **10**(1), 1–13, https://doi.org/10.1038/s41598-020-60743-7

Kumar H., Patel M. and Mohan D. (2019). Simplified batch and fixed-bed design system for efficient and sustainable fluoride removal from water using slow pyrolyzed okra stem and black gram straw biochars. *ACS Omega*, **4**(22), 19513–19525.

Leaper S., Cáceres E. O. A., Luque-Alled J. M., Cartmell S. H. and Gorgojo P. (2021). POSS-functionalized graphene oxide/PVDF electrospun membranes for complete arsenic removal using membrane distillation. *ACS Applied Polymer Materials*, **3**(4), 1854–1865, https://doi.org/10.1021/ACSAPM.0C01402/ASSET/IMAGES/MEDIUM/AP0C01402_M002.GIF

Lee J. I., Hong S. H., Lee C. G. and Park S. J. (2021). Fluoride removal by thermally treated egg shells with high adsorption capacity, low cost, and easy acquisition. *Environmental Science and Pollution Research*, **28**(27), 35887–35901, https://doi.org/10.1007/S11356-021-13284-Z/FIGURES/5

Li R., Li Q., Gao S. and Shang J. K. (2012). Exceptional arsenic adsorption performance of hydrous cerium oxide nanoparticles: part A. Adsorption capacity and mechanism. *Chemical Engineering Journal*, **185–186**(March), 127–135, https://doi.org/10.1016/J.CEJ.2012.01.061

Lin S., Lu D. and Liu Z. (2012). Removal of arsenic contaminants with magnetic γ-Fe_2O_3 nanoparticles. *Chemical Engineering Journal*, **211–212**(November), 46–52, https://doi.org/10.1016/J.CEJ.2012.09.018

Lingamdinne L. P., Koduru J. R., Chang Y. Y., Kang S. H. and Yang J. K. (2019). Facile synthesis of flowered mesoporous graphene oxide-lanthanum fluoride nanocomposite for adsorptive removal of arsenic. *Journal of Molecular Liquids*, **279**(April), 32–42, https://doi.org/10.1016/J.MOLLIQ.2019.01.103

Luo J., Luo X., Hu C., Crittenden J. C. and Qu J. (2016). Zirconia (ZrO_2) embedded in carbon nanowires via electrospinning for efficient arsenic removal from water combined with DFT studies. *ACS Applied Materials and Interfaces*, **8**(29), 18912–18921, https://doi.org/10.1021/ACSAMI.6B06046/ASSET/IMAGES/LARGE/AM-2016-060463_0010.JPEG

Maity J. P., Vithanage M., Kumar M., Ghosh A., Mohan D., Ahmad A. and Bhattacharya P. (2021). Seven 21st century challenges of arsenic-fluoride contamination and remediation. *Groundwater for Sustainable Development*, **12**(February), 100538, https://doi.org/10.1016/J.GSD.2020.100538

Meenakshi and Maheshwari R. C. (2006). Fluoride in drinking water and its removal. *Journal of Hazardous Materials*, **137**(1), 456–463, https://doi.org/10.1016/J.JHAZMAT.2006.02.024

Muedi K. L., Brink H. G., Masindi V. and Maree J. P. (2021). Effective removal of arsenate from wastewater using aluminium enriched ferric oxide-hydroxide recovered from authentic acid mine drainage. *Journal of Hazardous Materials*, **414**(July), 125491, https://doi.org/10.1016/J.JHAZMAT.2021.125491

Mukherjee S., Gupte T., Jenifer S. K., Thomas T. and Pradeep T. (2019a). Arsenic in water: speciation, sources, distribution, and toxicology. In: Encyclopedia of Water: Science, Technology, and Society, John Wiley & Sons, Inc, pp. 1–17, https://doi.org/10.1002/9781119300762.wsts0053

Mukherjee S., Gupte T., Jenifer S. K., Thomas T. and Pradeep T. (2019b). Arsenic in water: fundamentals of measurement and remediation. In: Encyclopedia of Water, John Wiley & Sons, Ltd, pp. 1–11, https://doi.org/10.1002/9781119300762.WSTS0054

Mukherjee S., Kumar A. A., Sudhakar C., Kumar R., Ahuja T., Mondal B., Srikrishnarka P., Philip L. and Pradeep T. (2019c). Sustainable and affordable composites built using microstructures performing better than nanostructures for arsenic removal. *ACS Sustainable Chemistry and Engineering*, **7**(3), 3222–3233.

Nagaraj A., Munusamy M. A., Ahmed M., Kumar S. S. and Rajan M. (2018). Hydrothermal synthesis of a mineral-substituted hydroxyapatite nanocomposite material for fluoride removal from drinking water. *New Journal of Chemistry*, **42**(15), 12711–12721, https://doi.org/10.1039/C8NJ02401D

Nayak B., Samant A., Patel R. and Misra P. K. (2017). Comprehensive understanding of the kinetics and mechanism of fluoride removal over a potent nanocrystalline hydroxyapatite surface. *ACS Omega*, **2**(11), 8118–8128.

Patel R. K., Kumar S., Chawla A. K., Mondal P. and Neelam Teychene B. and Pandey J. K. (2019). Elimination of fluoride, arsenic, and nitrate from water through adsorption onto nano-adsorbent: a review. *Current Nanoscience*, **15**(6), 557–575.

Raj S. K., Sharma V., Yadav A., Indurkar P. D. and Kulshrestha V. (2023). Nano-alumina wrapped carbon microspheres for ultrahigh elimination of pentavalent arsenic and fluoride from potable water. *Journal of Industrial and Engineering Chemistry*, **117**(January), 402–413, https://doi.org/10.1016/J.JIEC.2022.10.028

Ren Z., Zhang G. and Chen J. P. (2011). Adsorptive removal of arsenic from water by an iron–zirconium binary oxide adsorbent. *Journal of Colloid and Interface Science*, **358**(1), 230–237, https://doi.org/10.1016/J.JCIS.2011.01.013

Sharma P. R., Sharma S. K., Antoine R. and Hsiao B. S. (2019). Efficient removal of arsenic using zinc oxide nanocrystal-decorated regenerated microfibrillated cellulose scaffolds. *ACS Sustainable Chemistry and Engineering*, **7**(6), 6140–6151.

Sharma P. K., Kumar R., Singh R. K., Sharma P. and Ghosh A. (2022). Review on arsenic removal using biochar-based materials. *Groundwater for Sustainable Development*, **17**(May), 100740, https://doi.org/10.1016/J.GSD.2022.100740

Singh S., Naik T. S. S. K., Basavaraju U., Khan N. A., Wani A. B., Behera S. K., Nath B., Bhati S., Singh J. and Ramamurthy P. C. (2022). A systematic study of arsenic adsorption and removal from aqueous environments using novel graphene oxide functionalized UiO-66-NDC nanocomposites. *Scientific Reports*, **12**(1), 1–15, https://doi.org/10.1038/s41598-022-18959-2

Sinha S., Pandey K., Mohan D. and Singh K. P. (2003). Removal of fluoride from aqueous solutions by *Eichhornia crassipes* biomass and its carbonized form. *Industrial and Engineering Chemistry Research*, **42**(26), 6911–6918.

Sinha A., Cha B. G. and Kim J. (2018). Three-dimensional macroporous alginate scaffolds embedded with akaganeite nanorods for the filter-based high-speed preparation of arsenic-free drinking water. *ACS Applied Nano Materials*, **1**(4), 1940–1948.

Solanki Y. S., Agarwal M., Maheshwari K., Gupta S., Shukla P. and Gupta A. B. (2020). Investigation of plausible mechanism of the synthesized inorganic polymeric coagulant and its application toward fluoride removal from drinking water. *Industrial and Engineering Chemistry Research*, **59**(20), 9679–9687.

Sun X., Hu C., Hu X., Qu J. and Yang M. (2013). Characterization and adsorption performance of Zr-doped akaganéite for efficient arsenic removal. *Journal of Chemical Technology & Biotechnology*, **88**(4), 629–635, https://doi.org/10.1002/JCTB.3878

Sun J., Zhang X., Zhang A. and Liao C. (2019). Preparation of Fe–Co based MOF-74 and its effective adsorption of arsenic from aqueous solution. *Journal of Environmental Sciences*, **80**(June), 197–207, https://doi.org/10.1016/J.JES.2018.12.013

Wan S., Lin J., Tao W., Yang Y., Li Y. and He F. (2019). Enhanced fluoride removal from water by nanoporous biochar-supported magnesium oxide. *Industrial and Engineering Chemistry Research*, **58**(23), 9988–9996.

Wang D., Gilliland S. E., Yi X., Logan K., Heitger D. R., Lucas H. R. and Wang W. N. (2018). Iron mesh-based metal organic framework filter for efficient arsenic removal. *Environmental Science and Technology*, **52**(7), 4275–4284.

Wang Y., Zhang L., Guo C., Gao Y., Pan S., Liu Y., Li X. and Wang Y. (2022). Arsenic removal performance and mechanism from water on iron hydroxide nanopetalines. *Scientific Reports*, **12**(1), 1–15, https://doi.org/10.1038/s41598-022-21707-1

World Health Organization (2017). 'Guidelines for Drinking-water Quality: FOURTH EDITION INCORPORATING THE FIRST ADDENDUM.'

Wu X., Zhang Y., Dou X. and Yang M. (2007). Fluoride removal performance of a novel Fe–Al–Ce trimetal oxide adsorbent. *Chemosphere*, **69**(11), 1758–1764.

Yang Q., Lau C. H. and Ge Q. (2019). Novel ionic grafts that enhance arsenic removal via forward osmosis. *ACS Applied Materials and Interfaces*, **11**(19), 17828–17835.

doi: 10.2166/9781789063714_0085

Chapter 8

Emerging carbon-based nanocomposites for the removal of hazardous materials

Chetna Tewari[1,2], *Sumit Kumar*[3], *Biswajit Saha*[4], *Yong Chae Jung*[2] *and Nanda Gopal Sahoo*[1*]

[1]*Professor Rajendra Singh Nanoscience and Nanotechnology Centre, Department of Chemistry D.S.B. Campus, Kumaun University Nainital, Uttarakhand, India*
[2]*Institute of Advanced Composite Materials, Korea Institute of Science and Technology (KIST), Jeonllabuk-do, Republic of Korea*
[3]*Department of Civil Engineering, Indian Institute of Technology Madras, Chennai, India*
[4]*Indian Institute of Technology Gandhinagar, Palaj, India*
**Corresponding author: ngsahoo@yahoo.co.in*

ABSTRACT

Even though water covers 75% of the earth's surface, only 0.0067% of it is available for human use. These figures continue to deteriorate as the world's population grows, resulting in an increase in the amount of waste created every year. Worldwide, the desire for clean and safe water has always been a major concern. The pursuit of efficient materials for environmental remediation is a critical scientific and technological concern. Carbon-based materials (CBM) have unique electrical, mechanical, and physicochemical properties, making them ideal for use as environmental adsorbents, sensors, membranes, and catalysts. Basically, CBM includes activated carbon, carbon nanofibre, biochar, carbon aerogel, graphene, carbon nanotubes, and so on. These materials can easily be engineered and functionalized, depending on the chemical nature of the target contaminants, and shows great efficiency towards the removal of hazardous materials. Carbon-based nanocomposites (CBNC) play a vital role in water treatment and environmental remediation due to their higher adsorption capacity, improved permeation, porous nature, and selectivity towards pollutants. This book chapter discusses the effective employment of CBNCs, including their future prospects in the field of water purification for the removal of textile dyes, volatile organic substances, toxic metals, oil, and biological contaminants.

Keywords: carbon-based nanomaterials, nanocomposites, water purification.

8.1 INTRODUCTION

In this modern era, every person's craving for a lavish life with all kinds of amenities has led to huge growth in industrialization and subsequently, great demands for water. Every industry needs water for several processes while generating plenty of wastewater with high amounts of toxic substances. This drained water pollutes the surface water and water bodies and creates a serious problem for safe drinking water (Kumar *et al.*, 2022). There are many anthropogenic activities responsible for water pollution, but industrial activities are major pollutants. Water is essential for life and safe drinking water is a basic need of humans. Therefore, it is essential to find a way for water treatment/purification. Basically, polluted water contains heavy metals, organic dyes, biological impurities, and so on (Akhtar *et al.*, 2021). Consumption of this pollutant can cause several diseases such as diarrhoea, cholera, dysentery, typhoid, and polio; in some cases, it may cause death (Moosa & Abed, 2021). According to the report of the World Health Organization, about 2 billion people live in water-stressed countries, which is expected to be exacerbated in some regions as a result of climate change and population growth. This can create a huge scarcity of safe water and demands cost-effective materials for water remediation. This issue can be solved by taking advantage of the easy availability and tuneable surface functionality of carbon-based materials. Carbon-based materials and nanocomposites have great adsorption capacity, porosity, and easy surface modification ability making them capable of removing a wide range of water contaminants like heavy metals, dyes, pharmaceuticals, drugs, and so on (Damiri *et al.*, 2022).

8.2 SYNTHESIS OF CARBON-BASED NANOMATERIALS

8.2.1 Carbon nanotubes

The synthesis of carbon nanotubes (CNTs) has been carried out mainly by arc discharge, laser ablation, microwave, and chemical vapour deposition (CVD) techniques. All these methods have their own advantages and disadvantages. The arc discharge is the oldest method, which requires a vacuum chamber with a continuous supply of inert gas in which two electrodes produce an arc connected to a direct current source. Both single-walled nanotubes (SWCNTs) and multi-walled nanotubes MWCNTs are produced using this method. Utilizing the right catalyst, SWCNTs are produced using this method, but MWCNTs are produced by just maintaining the pressure of the inert gas, with no need for a catalyst. This method allows us to generate nanotubes with a 2–20 nm diameter and several microns in length (Shoukat & Imran, 2022).

Laser ablation is another technique for the synthesis of SWCNTs and MWCNTs and producing CNTs in large amounts. For the synthesis, a pulsed laser is used to ablate the graphite target in the presence of a catalyst and inert gas. The synthesis of CNTs is also done using a domestic microwave and CVD technique. CNTs can be formed by the CVD process using ethanol, xylene, and toluene mixed with hydrogen (Shoukat & Imran, 2022). There is another technique that is inductively coupled with chemical vapour deposition (ICP-CVD), where the nanotubes are synthesized in the presence of plasma created by the ICP system. In this method, the CNTs are crystalline at the temperature of 500°C. When the carbon nanotubes are synthesized by CVD, it has two disadvantages: (1) too many precursors are consumed, (2) high-temperature requirement. Compared to this, in plasma-enhanced CVD, low temperature is required for the nanotube synthesis. Moreover, the nanotubes synthesized by the PECVD are less contaminated as compared to the CVD synthesis.

8.2.2 Graphene

Numerous methods have recently been developed for the synthesis of graphene. The most widely utilized techniques for graphene synthesis are mechanical cleaving (exfoliation), chemical exfoliation, pyrolysis, and thermal CVD (Moosa & Abed, 2021). Microwave synthesis and unzipping nanotubes are two other approaches that have been documented. However, it was discovered that mechanical exfoliation using an atomic force microscopy cantilever could produce graphene with a thickness of 10 nm, which is comparable to graphene with approximately 30 layers. By introducing large alkali ions between the graphite layers, graphite dispersion is exfoliated using the chemical exfoliation process. Similar procedures include creating graphite oxide, dispersing it in a solution, and then reducing it using hydrazine. Also, the cost-effective and bulk synthesis of graphene-based materials has been reported by a two-stage pyrolysis approach using solid waste like plastic (Pandey *et al.*, 2019) and agricultural waste (Tewari *et al.*, 2022) as carbon source with the use of different metal oxides (like Al_2O_3, ZnO) and nano clay as a catalyst and exfoliating agent, as shown in Figure 8.1a and b.

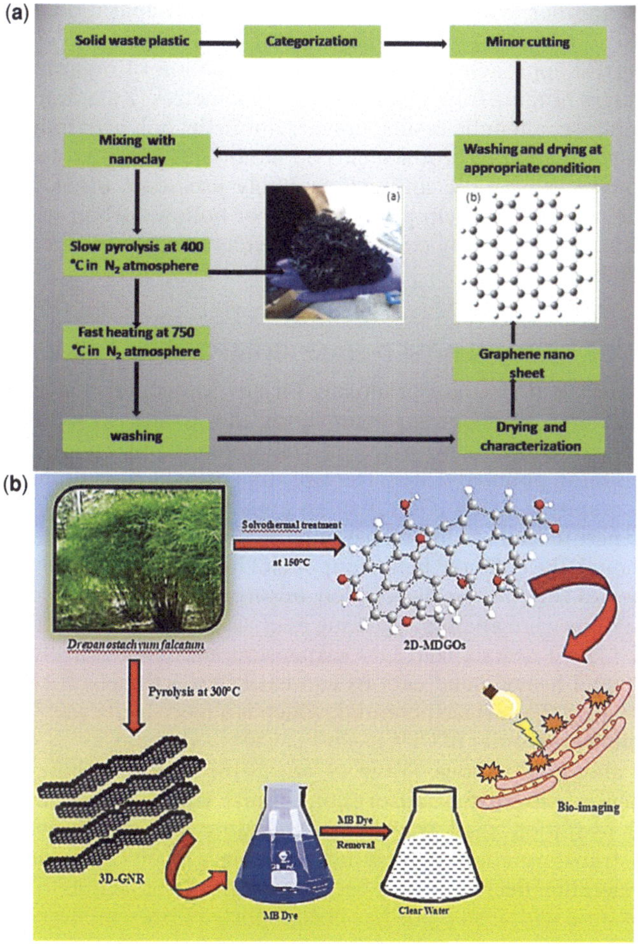

Figure 8.1 (a) synthesis of graphene nanosheets from waste plastic; (b) synthesis of 2D/3D graphene-based nanomaterials from *Drepanostachyum falcatum* plant.

8.2.3 Carbon nanofibres

CVD, electrospinning, templating, drawing, and phase separation are the main methods used to create carbon nanofibres (CNFs). C_2H_2 can be used as the carbon source to create CNFs. To dissolve carbon and create metal carbide, a variety of metals and alloys have been used as catalysts, including iron, cobalt, nickel, chromium, and vanadium. Additionally, in the temperature range of 700–1200 K, molybdenum, methane, carbon monoxide, synthesis gas (H_2/CO), or ethane are employed to act as carbon sources (Jong & Geus, 2007). Only relatively short fibres that are challenging to align, combine, and process into applications are suitable for production using the CVD approach (Zhou *et al.*, 2009a, 2009b). The geometric surfaces of a metallic catalyst particle (10–100 nm) and the gaseous carbon feedstock provided during CNF processing determine how the CNF grows. The outer diameter of the developed CNFs generally depends on the size of the catalyst particles. Carbon nanofibres (CNFs) made by vapour growth have an extremely unique structure resembling annular rings (Poveda & Gupta, 2016).

Electrospinning is another widely used method for the preparation of CNFs (Avila *et al.*, 2014). Owing to its ease of control and environmental compatibility, electrospinning is regarded as a flexible and powerful approach for the massive production of organic polymer or composite nanofibres with diameters ranging from submicron to nanometres (Zussman *et al.*, 2005). Polymeric precursors such as cellulose, phenolic resins, polyacrylonitrile, polybenzimidazole, and pitch-based materials have been electro-spun to produce carbon nanofibres (Yoshimoto *et al.*, 2003). Drawing, template synthesis, phase separation, and self-assembly are some other methods for producing nanofibres. A template process can help produce solid or hollow carbon nanofibres from a variety of raw materials, such as electronically conducting polymers, metals, semiconductors, and carbons (Zhou *et al.*, 2009a, 2009b).

8.3 DEVELOPMENT OF CARBON-BASED NANOCOMPOSITES FOR WATER TREATMENT

Carbon-based nanomaterials (CBN) are promising for environmental remediation applications due to the ease of synthesis and the scope for engineering the surface structure. The very small pore size of these materials can allow them to adsorb contaminants, like dyes, toxic metals, and other organic pollutants present in polluted water. The two-dimensional class of the CBN with oxygenated functionality, that is, graphene oxide (GO), has an amphiphilic nature. When GO is dispersed in water, its amphiphilic nature leads to its self-assembling into micelles or liposomes, which decreases its available sites. Although the colloidal behaviour of GO has been extensively studied in the presence of common environmental cations, the aggregation, adsorption, and morphological transformation of GO under heavy metal ions were analysed by Yang *et al.* (2016). They found that heavy metal cations (Cr^{3+}, Pb^{2+}, Cu^{2+}, Cd^{2+}, Ag^{+}) destabilized GO suspension more aggressively than common cations (Ca^{2+}, Mg^{2+}, Na^{+}, K^{+}), and heavy metal cations can easily cross the electric double layer, bind with the GO surface, and change the surface potential, which is a more facile way for GO aggregation. The kinetic aggregation follows the order of $Cr^{3+} \gg Pb^{2+} > Cu^{2+} > Cd^{2+} > Ca^{2+} > Mg^{2+} \gg Ag^{+} > K^{+} > Na^{+}$ for the destabilizing ability of cations (Yang *et al.*, 2016). The destabilizing capability of metal cations is basically dependent on their adsorption affinity with GO, which is determined by their electronegativity and hydration shell thickness. According to their investigation, an integrative process of adsorption–transformation–aggregation takes place when GO comes in contact with heavy metal ions. The agglomeration decreases the absorption capacity of GO, so to overcome this problem, doping or functionalization with iron oxide has been reported by researchers. The iron oxide helps to prevent the agglomeration of GO in water and helps to provide more available sites for interaction with impurities. Also, the amphiphilic nature of GO helps it to remove pollutants from water easily, where most of the adsorption occurs due to the electrostatic interactions between the functional groups on GO and pollutants. The presence of greater pia–pia conjugation makes graphene more effective than GO in the removal of organic pollutants. Another approach to treat organic pollutants containing water is photodegradation. Unlike pure photodegradants, graphene-based photodegradants have unique electronic properties resulting in an ultrafast (picosecond) electron transfer process from the excited semiconductors to the graphene sheet. Scientists have modified CBN by incorporating other nanomaterials to enhance its adsorption capacities and photocatalytic abilities. This process has already been discussed previously.

8.4 REMOVAL OF HAZARDOUS MATERIALS USING CARBON-BASED NANOCOMPOSITES

The wastewater characterization has shown different classes of pollution, such as pharmaceuticals active compounds, textile dyes, and heavy metals. All such pollutants have a higher potential to damage, even at low concentrations. Among many removal technologies, adsorption has emerged as an effective technology as it is easy to operate, has low-cost adsorbent materials, and has no intermediate availability after treatment. Recently, carbonaceous materials, and especially biochar that is obtained by pyrolysis of many environmental wastes, are being utilized for pollutant removal. Two major applications of such carbonaceous materials such as adsorbent and catalysts are documented in this section.

8.4.1 Adsorption using carbon-based nanocomposites

Many such adsorbents such as algae, fungi, bacteria, bark, sludge, peat, and polysaccharides (cellulose, starch, alginates, etc.) have been extensively explored in the adsorptive removal of pollutants from wastewater (Anastopoulos & Kyzas, 2014; Bhatnagar *et al.*, 2010). All these biosorbents showed very good efficiencies towards specific pollutants due to their distinct physicochemical properties, while zero or little data was reported for mixed wastewater remediation (Qin *et al.*, 2020). For example, bentonite cannot remove non-iconic pollutants (Turabik, 2008). Chitosan is ineffective against cationic pollutants until it is chemically modified (Arya & Philip, 2016). Moreover, clay, zeolite, and alumina showed good removal efficiencies for fluoride but lower fluoride concentration in solution leads to the desorption of adsorbed fluoride (Kumar & Philip, 2006). The adsorption using such biosorbents is generally completed in two steps: (a) physical adsorption over the cell surfaces (Won *et al.*, 2009), ion exchange (Javanbakht *et al.*, 2014), where this process is driven by the chemical gradient of biological cell membranes; (b) accumulation inside the cell or precipitation over surfaces (Nies, 1999). Similarly, Chen *et al.* (2018) prepared biochar from two types of waste (rice straw and swine manure), modified with acid, and tested for the removal of the pharmaceutical pollutants tetracycline. They demonstrated that the modified biochar showed a 25% improvement in sorption than that of pristine biochar. Lonappan *et al.* (2016) developed biochar from pine wood, pig manure, and cardboard to investigate the removal of textile dyes. They found that all biochar showed very high potential, and the capacity was directly proportional to the ash content of the biochar. Yang *et al.* (2018) synthesized bamboo-derived biochar and checked its potential for oil adsorption. The results were outstanding in terms of capacity as well as the regeneration of biochar after multiple cycles. Moreover, several studies using commercial activated carbon were reported for contamination remediation where the capacity via adsorption was very much comparable. Similarly, Alhashimi and Aktas (2017) and Mall *et al.* (2007) reported better adsorption performance of biochar that is derived from waste biomass over commercial carbonaceous material. For instance, Li *et al.* (2016) and Ronix *et al.* (2017) derived hydrochar via HTC using coffee husk and bamboo waste and tested the same for the removal of methylene blue and Congo red textile dyes. Both the adsorbents showed high capacity up to 97 mg/g, which was double that of reported for biochar derived via pyrolysis. Similarly, Liu and Zhang (2009) and Regmi *et al.* (2012) have synthesized biochar and activated it for the removal of heavy metals (Pb, Cu, and Cd). They found activation of hydrochar resulted in 100% removal while non-activated also showed promising outcomes.

8.4.2 Catalysis using carbon-based nanocomposites

In recent years, waste-derived biochar-based photocatalysts, such as TiO_2-coconut shell (Zhang & Lu, 2018), TiO_2-reed straw (Zhang *et al.*, 2017), TiO_2-corn cob (Kim & Kan, 2016), TiO_2-bamboo (Men *et al.*, 2019), among others have been tested in numerous studies. Carbon-based nanomaterial derived from waste plastic also show good catalytic ability. Kumar *et al.* reported the use of waste plastic-derived graphitic carbon to reduce waste burdens in a sustainable and cost-effective manner. The adsorption capacity of this synthesized material was very low, that is, 7.41 and 4.93 mg/g for brilliant green and eosin yellow dyes, respectively. They demonstrated coupling of this material with peroxymonosulphate-promoted dye degradation and complete dye degradation, with a 61% reduction in TOC and a 95% reduction in toxicity using a dye concentration of 10 mg/L (Kumar *et al.*, 2022), as shown in Figure 8.2.

The major mechanism involved in using biochar composites is discussed in various literatures. For instance, the larger surface area can facilitate better absorption, due to which the pollutants could be prone to attack by short-lived radical species (Zou *et al.*, 2019). Kumar *et al.* (2017) showed methylparaben degraded as soon it was adsorbed on the surface. In addition, carbonaceous material

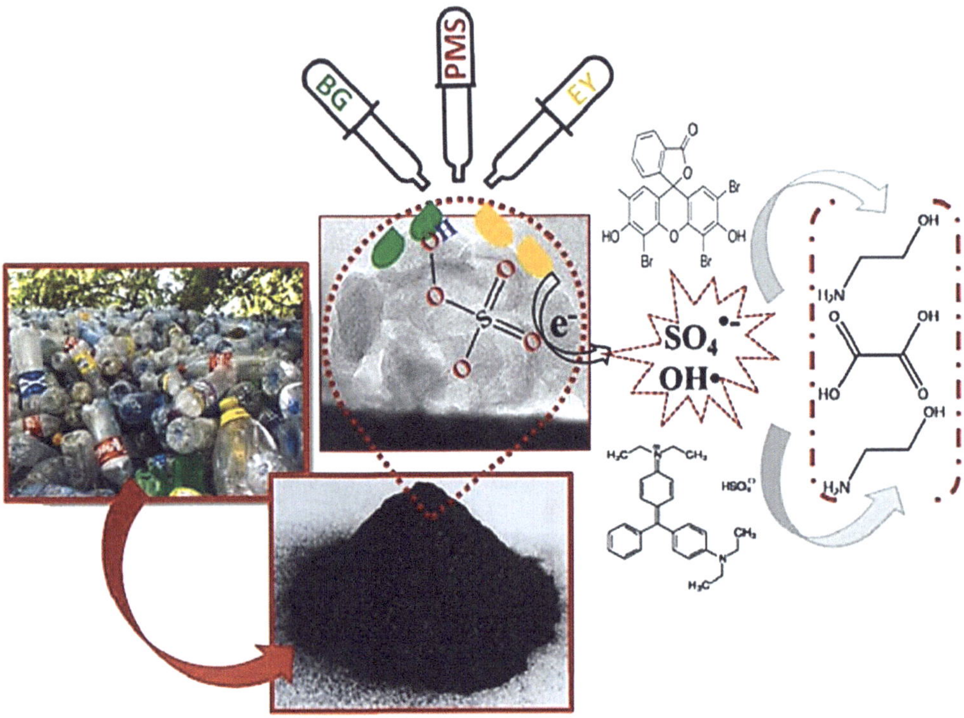

Figure 8.2 Degradation of organic dye molecules in presence of waste plastic-derived carbon-based nanomaterial.

helps in narrowing the band gap to improve the photo-absorption by forming a bridge (Di Valentin *et al.*, 2005), it acts as a photosensitizer, charge carrier (Djellabi *et al.*, 2019), and electron transport system (Zhu *et al.*, 2018). Also, such materials showed great potential in suppressing the recombination of electron–hole pairs by double conductivity. Peng *et al.* (2019) explained the retardation of electron–hole pairs recombination and enhanced ammonia removal. Shi *et al.* (2016) reported that biochar reduced photo corrosion. Biochar was synthesized by pinewood sawdust and used in a pyrite-Fenton degradation of 2, 4-dichlorophenoxyacetic acid, and the rate of degradation was significantly increased by ~2.39 times (Zhu *et al.*, 2020). For instance, Chen *et al.* (2020a, 2020b) prepared a composite of $Fe_2O_3 + TiO_2$-biochar for H_2O_2 activation and achieved MB degradation of more than 65% (higher than that of Fe_2O_3–TiO_2) with reusability up to five cycles. Liu *et al.* (2021) have carbonized cornstalk and modified the biochar with Fe and Zn and achieved higher surface area for the electrode, resulting in nitrobenzene electro degradation in a lesser time than that of the pure metal electrode. Similar outcomes demonstrated that electrocatalytic performance is highly dependent on electronic properties, surface area, and nano-structure of electrode materials (Chen *et al.*, 2020a, 2020b, 2022). Wang *et al.* (2021) synthesized magnetic biochar supported by Pd and used as electrocatalyst hydrochlorination of pentachlorophenol. Similarly, Yao *et al.* (2020) developed a composite electrode using loofah sponge-derived biochar doped with Pd and N for the electroreduction of bromate. Both studies revealed that the dopant enhances the adsorption of pollutants for effective reduction/oxidation over electrodes. Beside these noble metal-doped composites, Zhou *et al.* (2021) prepared composite cathodes using waste tea leaves and TiO_2. Their study showed the positive effect of the porosity of biochar in providing several electron transport channels and enhanced the radical species generation for phenol degradation. The efficiencies of various carbon-based catalysis are tabulated in Table 8.1.

Table 8.1 Adsorption and catalysis efficiency of various carbon-based materials for pollutants abatement.

Adsorption			
Materials	**Experimental Conditions**	**Pollutants/Results**	**Remarks**
Ferromanganese biochar made by vinasse	$C_0 = 100$ mg/L Dose = 0.1 g/L Eqm. Time = 48 h	Levofloxacin 181 mg/g	• Reusability was up to 5 cycles • π–π interaction was dominating in adsorption
Zero valent-graphene composite	$C_0 = 25$ mg/L Dose = 0.25 g/L Eqm. Time = 6.6 h	Tetracycline 660 mg/g	• Self-cleaning potential prolonged • π–π interaction was dominating in adsorption
CNTs/$CoFe_2O_4$ composites	$C_0 = 2$ mg/L Dose = 0.1 g/L Eqm. Time = 12 h	Sulphamethoxazole 354 mg/g	• Hydrophobic interaction and π–π interaction mechanism • Composites showed high magnetization, reusability, and stability
GO-supported activated magnetic biochar	$C_0 = 6$ mg/L Dose = 0.1 g/L Eqm. Time = 2 h	17β-Estradiol 46.22 mg/g	• Hydrogen bond, electrostatic, and π–π interaction mechanism • Doubled the adsorption capacity by composites
Mesoporous silica-magnetic GO	$C_0 = 10$ mg/L Dose = 0.57 g/L Eqm. Time = 48 h	Sulphamethoxazole 15.46 mg/g	• More basic pH favours the sorption • π–π interaction was dominating in adsorption

8.5 FUTURE PERSPECTIVE OF CARBON-BASED NANOCOMPOSITES FOR ENVIRONMENTAL APPLICATIONS

The use of carbon-based nanocomposites (CBNC) for environmental applications is a relatively new field of study that presents promising potential for the future. Their composite structures facilitate the added functionalities or the improvement of their physical and chemical properties, thus providing opportunities for use in water filtration. The future applications of CBNC for environmental purposes may involve the production of durable coatings for surfaces, the removal of hazardous pollutants from water, as well as for controlling environmental microorganisms. Additionally, researchers are exploring the use of nanocomposites in combination with other technologies, such as nanofiltration and membrane separation, to further enhance their efficacy. The scalability of the processes and production of these materials will also need to be improved. In the future, these materials could offer a more comprehensive solution to diverse water filtration needs. CBNC are expected to be the future of sustainable methods for the removal of hazardous materials.

There are currently several challenges in using CBNC for environmental applications, including:

Synthesis: CBNC often require complex and expensive synthesis methods, which can limit their scalability and applicability. Their synthesis involves multiple steps that are essential for obtaining the desired nanocomposite material. The most common challenge is achieving uniform particle distribution and precise control of the reaction parameters, such as temperature, pressure, concentration, and stirring rate. In addition, achieving defect-free nanostructures and controlling desorption, oxidation, and aggregation of nanomaterials are all essential steps in the synthesis that require precise experimentation and control. Lastly, developing strategies for functionalizing the nanocomposite materials and ensuring the proper integration of nanocomposite materials with other materials are also major challenges.

Dispersion: Some of the main challenges include achieving effective and uniform distribution of the nanocomposites in the matrix material, which requires maximizing the interfacial adhesion between the nanocomposite and the matrix. Additionally, certain CBNC tend to agglomerate, causing individual primary particles to clump together, thereby drastically reducing the material's performance. Agglomeration can be minimized by using surfactants, but this carries the additional challenge of incorporating them in a way that does not negatively affect the desired properties of the embedded nanocomposites. Finally, there is an issue of toxicity, as some of the nanocomposites may present toxicological risks, making it important to add mitigation methods like protective coatings or chemical modifications.

Biodegradation: Biodegradation of carbon-based nanocomposite materials is a challenge due to their complex structure, high surface area to volume ratio, and variable composition. Additionally, the low solubility of many of these nanomaterials hinders their ability to be decomposed by microorganisms.

Stability: CBNC contain finely dispersed nanoscale particles, which can significantly increase the chemical reactivity of the material due to their large surface area. The structural stability of these nanocomposites is also a challenge due to the Van Der Waals forces between particles that make them prone to agglomeration or demixing. The nanoscale size of their particles and large surfactant groups attached to the particles can increase their vulnerability to oxidation, hydrolysis, and other types of chemical reactions. This can lead to changes in their physical and mechanical properties and a decrease in their overall lifetime. In some cases, the stability of CBNC may be compromised in certain environments, such as high pH or high-temperature conditions.

Toxicology: CBNC may have toxic effects on aquatic and terrestrial living organisms, and appropriate precautions must be taken during handling and use.

Regulations: CBNC may not be regulated as a distinct class of materials, making it difficult to establish guidelines for their safe and responsible use.

Selectivity: CBNC are not very selective towards the adsorption of specific pollutants, which can lead to low efficiency and high costs for cleaning up a particular pollutant.

These challenges can restrict their use in many applications and limit their potential. Thus, further research is needed to address the chemical and structural stability issues of CBNC in order to leverage their unique mechanical, electrical, and optical properties in various industrial settings.

8.6 CONCLUSION

This chapter discussed the synthesis of CBN and the development of their composites for water treatment. Carbon is one of the most abundant elements on the earth. In the past two decades, carbon-based materials such as graphene-based compounds, CNT, and carbon fibres have been applied in various fields of research including, but not limited to, environmental applications, like water treatment. The potential for overcoming the drawbacks of conventional treatment methods is greatly increased using CBM and its nanocomposites in water treatment systems. The high adsorption capability and photocatalytic ability of CBN can strongly bind emerging chemical contaminants, which often have a weak or negligible affinity for traditional sorbents. The key obstacle to commercialization is the large-scale and cost-effective manufacture of these CBNCs. Also, getting a uniform product is a challenge associated with the synthesis of CBNCs. Significant advancements in terms of cost and mass production of these materials are required for the success of this technology. On the contrary, the effect of these materials on the safety of humans and ecological systems cannot be neglected, so the toxicological aspect should be studied in the future, and the viability and effectiveness of these CBNCs can be tested in real water samples.

REFERENCES

Akhtar N., Syakir Ishak M. I., Bhawani S. A. and Umar K. (2021). Various natural and anthropogenic factors responsible for water quality degradation: a review. *Water*, **13**(19), 2660, https://doi.org/10.3390/w13192660

Alhashimi H. A. and Aktas C. B. (2017). Life cycle environmental and economic performance of biochar compared with activated carbon: a meta-analysis. *Resources Conservation and Recycling*, **118**, 13–26, https://doi.org/10.1016/j.resconrec.2016.11.016

Anastopoulos I. and Kyzas G. Z. (2014). Agricultural peels for dye adsorption: a review of recent literature. *Journal of Molecular Liquids*, **200**, 381–389, https://doi.org/10.1016/j.molliq.2014.11.006

Arya V. and Philip L. (2016). Adsorption of pharmaceuticals in water using Fe_3O_4 coated polymer clay composite. *Microporous and Mesoporous Materials*, **232**, 273–280, https://doi.org/10.1016/j.micromeso.2016.06.033

Avila M., Burks T., Akhtar F., Göthelid M., Lansåker P. C., Toprak M. S., Muhammed M. and Uheida A. (2014). Surface functionalised nanofibers for the removal of chromium (VI) from aqueous solutions. *Chemical Engineering Journal*, **245**, 201–209, https://doi.org/10.1016/j.cej.2014.02.034

Bhatnagar A., Vilar V. J. P., Botelho C. M. S. and Boaventura R. A. R. (2010). Coconut-based biosorbents for water treatment – a review of the recent literature. *Advances in Colloid and Interface Science*, **160**(1–2), 1–15, https://doi.org/10.1016/j.cis.2010.06.011

Chen T., Luo L., Deng S., Shi G., Zhang S., Zhang Y., Deng O., Wang L., Zhang J. and Wei L. (2018). Sorption of tetracycline on H_3PO_4 modified biochar derived from rice straw and swine manure. *Bioresource Technology*, **267**, 431–437, https://doi.org/10.1016/j.biortech.2018.07.074

Chen X. L., Li F., Chen H., Wang H. and Li G. (2020a). Fe_2O_3/TiO_2 functionalized biochar as a heterogeneous catalyst for dyes degradation in water under Fenton processes. *Journal of Environmental Chemical Engineering*, **8**(4), 103905, https://doi.org/10.1016/j.jece.2020.103905

Chen Z., Duan X., Wei W., Wang S. and Ni B. J. (2020b). Iridium-based nanomaterials for electrochemical water splitting. *Nano Energy*, **78**, 105270, https://doi.org/10.1016/j.nanoen.2020.105270

Chen Z., Wei W., Chen H. and Ni B. (2022). Recent advances in waste-derived functional materials for wastewater remediation. *Eco-Environment & Health*, **1**(2), 86–104, https://doi.org/10.1016/j.eehl.2022.05.001

Damiri F., Andra S., Kommineni N., Balu S. K., Bulusu R., Boseila A. A., Akamo D. O., Ahmad Z., Khan F. S., Rahman M. H. and Berrada M. (2022). Recent advances in adsorptive nanocomposite membranes for heavy metals ion removal from contaminated water: a comprehensive review. *Materials*, **15**(15), 5392, https://doi.org/10.3390/ma15155392

Di Valentin C., Pacchioni G. and Selloni A. (2005). Theory of carbon doping of titanium dioxide. *Chemistry of Materials*, **17**(26), 6656–6665, https://doi.org/10.1021/cm051921h

Djellabi R., Yang B., Xiao K., Gong Y., Cao D., Sharif H. M. A., Zhao X., Zhu C. and Zhang J. (2019). Unravelling the mechanistic role of TiOC bonding bridge at titania/lignocellulosic biomass interface for Cr (VI) photoreduction under visible light. *Journal of Colloid and Interface Science*, **553**, 409–417, https://doi.org/10.1016/j.jcis.2019.06.052

Javanbakht V., Alavi S. A. and Zilouei H. (2014). Mechanisms of heavy metal removal using microorganisms as biosorbent. *Water Science and Technology*, **69**(9), 1775–1787, https://doi.org/10.2166/wst.2013.718

Jong K. P. D. E. and Geus J. W. (2007). Carbon nanofibers: catalytic synthesis and applications. *Catalysis Review*, **42**, 4940.

Kim J. R. and Kan E. (2016). Heterogeneous photocatalytic degradation of sulfamethoxazole in water using a biochar-supported TiO_2 photocatalyst. *Journal of Environmental Management*, **180**, 94–101, https://doi.org/10.1016/j.jenvman.2016.05.016

Kumar M. and Philip L. (2006). Adsorption and desorption characteristics of hydrophobic pesticide endosulfan in four Indian soils. *Chemosphere*, **62**(7), 1064–1077, https://doi.org/10.1016/j.chemosphere.2005.05.009

Kumar A., Sharma G., Naushad M., Kumar A., Kalia S., Guo C. and Mola G. T. (2017). Facile hetero-assembly of superparamagnetic $Fe_3O_4/BiVO_4$ stacked on biochar for solar photo-degradation of methyl paraben and pesticide removal from soil. *Journal of Photochemistry and Photobiology A: Chemistry*, **337**, 118–131, https://doi.org/10.1016/j.jphotochem.2017.01.010

Kumar S., Tewari C., Sahoo N. G. and Philip L. (2022). Mechanistic insights into carbo-catalyzed persulfate treatment for simultaneous degradation of cationic and anionic dye in multicomponent mixture using plastic waste-derived carbon. *Journal of Hazardous Materials*, **435**(January), 128956, https://doi.org/10.1016/j.jhazmat.2022.128956

Li Y., Meas A., Shan S., Yang R. and Gai X. (2016). Production and optimization of bamboo hydrochars for adsorption of Congo red and 2-naphthol. *Bioresource Technology*, **207**, 379–386, https://doi.org/10.1016/j.biortech.2016.02.012

Liu Z. and Zhang F. S. (2009). Removal of lead from water using biochars prepared from hydrothermal liquefaction of biomass. *Journal of Hazardous Materials*, **167**(1–3), 933–939, https://doi.org/10.1016/j.jhazmat.2009.01.085

Liu Q., Jiang S., Su X., Zhang X., Cao W. and Xu Y. (2021). Role of the biochar modified with $ZnCl_2$ and $FeCl_3$ on the electrochemical degradation of nitrobenzene. *Chemosphere*, **275**, 129966, https://doi.org/10.1016/j.chemosphere.2021.129966

Lonappan L., Rouissi T., Das R. K., Brar S. K., Ramirez A. A., Verma M., Surampalli R. Y. and Valero J. R. (2016). Adsorption of methylene blue on biochar microparticles derived from different waste materials. *Waste Management*, **49**, 537–544, https://doi.org/10.1016/j.wasman.2016.01.015

Mall I. D., Srivastava V. C. and Agarwal N. K. (2007). Adsorptive removal of auramine-O: kinetic and equilibrium study. *Journal of Hazardous Materials*, **143**(1–2), 386–395, https://doi.org/10.1016/j.jhazmat.2006.09.059

Men Q., Wang T., Ma C., Yang L., Liu Y., Huo P. and Yan Y. (2019). In-situ preparation of CdSe quantum dots/porous channel biochar for improving photocatalytic activity for degradation of tetracycline. *Journal of the Taiwan Institute of Chemical Engineers*, **99**, 180–192, https://doi.org/10.1016/j.jtice.2019.03.019

Moosa A. and Abed M. (2021). Graphene preparation and graphite exfoliation. *Turkish Journal of Chemistry*, **45**(3), 493–519, https://doi.org/10.3906/kim-2101-19

Nies D. H. (1999). Microbial heavy-metal resistance. *Applied Microbiology and Biotechnology*, **51**(6), 730–750, https://doi.org/10.1007/s002530051457

Pandey S., Karakoti M., Dhali S., Karki N., SanthiBhushan B., Tewari C., Rana S., Srivastava A., Melkani A. B. and Sahoo N. G. (2019). Bulk synthesis of graphene nanosheets from plastic waste: an invincible method of solid waste management for better tomorrow. *Waste Management*, **88**, 48–55, https://doi.org/10.1016/j.wasman.2019.03.023

Peng X., Wang M., Hu F., Qiu F., Dai H. and Cao Z. (2019). Facile fabrication of hollow biochar carbon-doped TiO_2/CuO composites for the photocatalytic degradation of ammonia nitrogen from aqueous solution. *Journal of Alloys and Compounds*, **770**, 1055–1063, https://doi.org/10.1016/j.jallcom.2018.08.207

Poveda R. L. and Gupta N. (2016). Carbon Nanofiber Reinforced Polymer Composites. Springer, Berlin, Germany.

Qin H., Hu T., Zhai Y., Lu N. and Aliyeva J. (2020). The improved methods of heavy metals removal by biosorbents: a review. *Environmental Pollution*, **258**, 113777, https://doi.org/10.1016/j.envpol.2019.113777

Regmi P., Moscoso J. L. G., Kumar S., Cao X., Mao J. and Schafran G. (2012). Removal of copper and cadmium from aqueous solution using switchgrass biochar produced via hydrothermal carbonization process. *Journal of Environmental Management*, **109**, 61–69, https://doi.org/10.1016/j.jenvman.2012.04.047

Ronix A., Pezoti O., Souza L. S., Souza I. P., Bedin K. C., Souza P. S., Silva T. L., Melo S. A., Cazetta A. L. and Almeida V. C. (2017). Hydrothermal carbonization of coffee husk: optimization of experimental parameters and adsorption of methylene blue dye. *Journal of Environmental Chemical Engineering*, **5**(5), 4841–4849, https://doi.org/10.1016/j.jece.2017.08.035

Shi M., Wei W., Jiang Z., Han H., Gao J. and Xie J. (2016). Biomass-derived multifunctional TiO_2/carbonaceous aerogel composite as a highly efficient photocatalyst. *RSC Advances*, **6**(30), 25255–25266, https://doi.org/10.1039/C5RA28116D

Shoukat R. and Imran M. (2022). Carbon nanotubes/nanofibers (CNTs/CNFs): a review on state of the art synthesis methods. *Microsystem Technologies*, **28**(4), 885–901, https://doi.org/10.1007/s00542-022-05263-2

Tewari C., Tatrari G., Kumar S., Pandey S., Rana A., Pal M. and Sahoo N. G. (2022). Green and cost-effective synthesis of 2D and 3D graphene-based nanomaterials from *Drepanostachyum falcatum* for bio-imaging and water purification applications. *Chemical Engineering Journal Advances*, **10**, 100265, https://doi.org/10.1016/j.ceja.2022.100265

Turabik M. (2008). Adsorption of basic dyes from single and binary component systems onto bentonite: simultaneous analysis of basic red 46 and basic yellow 28 by first order derivative spectrophotometric analysis method. *Journal of Hazardous Materials*, **158**(1), 52–64, https://doi.org/10.1016/j.jhazmat.2008.01.033

Wang Y., Cui C., Zhang G., Xin Y. and Wang S. (2021). Electrocatalytic hydrodechlorination of pentachlorophenol on Pd-supported magnetic biochar particle electrodes. *Separation and Purification Technology*, **258**, 118017, https://doi.org/10.1016/j.seppur.2020.118017

Won S. W., Yun H. J. and Yun Y. S. (2009). Effect of pH on the binding mechanisms in biosorption of reactive orange 16 by *Corynebacterium glutamicum*. *Journal of Colloid and Interface Science*, **331**(1), 83–89, https://doi.org/10.1016/j.jcis.2008.11.005

Yang K., Chen B., Zhu X. and Xing B. (2016). Aggregation, adsorption, and morphological transformation of graphene oxide in aqueous solutions containing different metal cations. *Environmental Science & Technology*, **50**(20), 11066–11075, https://doi.org/10.1021/acs.est.6b04235

Yang E., Yao C., Liu Y., Zhang C., Jia L., Li D., Fu Z., Sun D., Kirk S. R. and Yin D. (2018). Bamboo-derived porous biochar for efficient adsorption removal of dibenzothiophene from model fuel. *Fuel*, **211**, 121–129, https://doi.org/10.1016/j.fuel.2017.07.099

Yao F., Yang Q., Yan M., Li X., Chen F., Zhong Y., Yin H., Chen S., Fu J., Wang D. and Li X. (2020). Synergistic adsorption and electrocatalytic reduction of bromate by Pd/N-doped loofah sponge-derived biochar electrode. *Journal of Hazardous Materials*, **386**, 121651, https://doi.org/10.1016/j.jhazmat.2019.121651

Yoshimoto H., Shin Y. M., Terai H. and Vacanti J. P. (2003). A biodegradable nanofiber scaffold by electrospinning and its potential for bone tissue engineering. *Biomaterials*, **24**(12), 2077–2082, https://doi.org/10.1016/S0142-9612(02)00635-X

Zhang S. and Lu X. (2018). Treatment of wastewater containing reactive brilliant blue KN-R using TiO_2/BC composite as heterogeneous photocatalyst and adsorbent. *Chemosphere*, **206**, 777–783, https://doi.org/10.1016/j.chemosphere.2018.05.073

Zhang H., Wang Z., Li R., Guo J., Li Y., Zhu J. and Xie X. (2017). TiO_2 supported on reed straw biochar as an adsorptive and photocatalytic composite for the efficient degradation of sulfamethoxazole in aqueous matrices. *Chemosphere*, **185**, 351–360, https://doi.org/10.1016/j.chemosphere.2017.07.025

Zhou F. L., Gong R. H. and Porat I. (2009a). Mass production of nanofibre assemblies by electrostatic spinning. *Polymer International*, **58**(4), 331–342, https://doi.org/10.1002/pi.2521

Zhou Z., Lai C., Zhang L., Qian Y., Hou H., Reneker D. H. and Fong H. (2009b). Development of carbon nanofibers from aligned electrospun polyacrylonitrile nanofiber bundles and characterization of their microstructural, electrical, and mechanical properties. *Polymer*, **50**(13), 2999–3006, https://doi.org/10.1016/j.polymer.2009.04.058

Zhou P., Wan J., Wang X., Chen J., Gong Y., Xu K. and Liu C. (2021). Preparation and electrochemical property of TiO_2/porous carbon composite cathode derived from waste tea leaves for electrocatalytic degradation of phenol. *Journal of Applied Electrochemistry*, **51**, 653–667, https://doi.org/10.1007/s10800-020-01527-9

Zhu Z., Fan W., Liu Z., Yu Y., Dong H., Huo P. and Yan Y. (2018). Fabrication of the metal-free biochar-based graphitic carbon nitride for improved 2-mercaptobenzothiazole degradation activity. *Journal of Photochemistry and Photobiology A: Chemistry*, **358**, 284–293, https://doi.org/10.1016/j.jphotochem.2018.03.027

Zhu X., Li J., Xie B., Feng D. and Li Y. (2020). Accelerating effects of biochar for pyrite-catalyzed Fenton-like oxidation of herbicide 2, 4-D. *Chemical Engineering Journal*, **391**, 123605, https://doi.org/10.1016/j.cej.2019.123605

Zou W., Gao B., Ok Y. S. and Dong L. (2019). Integrated adsorption and photocatalytic degradation of volatile organic compounds (VOCs) using carbon-based nanocomposites: a critical review. *Chemosphere*, **218**, 845–859, https://doi.org/10.1016/j.chemosphere.2018.11.175

Zussman E., Chen X., Ding W., Calabri L., Dikin D. A., Quintana J. P. and Ruoff R. S. (2005). Mechanical and structural characterization of electrospun PAN-derived carbon nanofibers. *Carbon*, **43**(10), 2175–2185, https://doi.org/10.1016/j.carbon.2005.03.031

doi: 10.2166/9781789063714_0097

Chapter 9

Bio-polymer-reinforced nanocomposites for water and wastewater treatment: applications and future prospects

*M. S. V. Naga Jyothi, P. Amesh and Shihabudheen M. Maliyekkal**

Department of Civil and Environmental Engineering, Indian Institute of Technology, Tirupati, India
**Corresponding author: shihab@iittp.ac.in*

ABSTRACT

Providing clean water to every citizen is a mammoth task due to the increasing pollution and the shrinking and non-uniform distribution of freshwater reserves. It is affecting and threatening environmental well-being and economic growth. The treatment of water and wastewater is essential, along with other measures, to achieve desired water quality. Nanoscale materials-enabled water and wastewater treatment are attractive due to their enhanced ability to scavenge or degrade pollutants. However, the development of practical and sustainable nanomaterials plays a crucial role in the success of the process. Biopolymer-reinforced nanocomposites (BPNC) are gaining interest as practical adsorbents and catalysts for water and wastewater treatment due to their versatility, the ability to contain a wide range of nanomaterials, abundant availability, eco-friendliness, and low cost. BPNCs can be tailored into various forms, including granules with different shapes and sizes, hydrogels, membranes, and coated substrates. Grafting or immobilization of bio-polymer onto nanomaterials prevents aggregation and shapes functionality, selectivity, physical stability, and controlled pore size. This chapter briefly reviews various BPNCs reported for removing contaminants in water and wastewater. The synthesis protocols, challenges, and the associated prospects of BPNCs in the field application are discussed.

Keywords: nanocomposites, biopolymer, treatment, immobilization, water and wastewater

9.1 INTRODUCTION

A nanocomposite is a heterogeneous material formed by combining two or more materials of distinct physical and chemical properties where at least one of the constituents must be less than 100 nm. Combining materials of distinct properties allows one to utilize the most vital individual features, and the new functionality arises from the synergetic effect of the combination of materials. There are several types of nanocomposites, and they are generally classified based on the dispersed phase and dispersed matrix used in the composites. Polymers are widely used as dispersing medium due to their tuneable surface functional groups, viable regeneration, and skeleton strength. Polymer nanocomposites are used in air and water purification, packaging, automotive, electronics, and aerospace applications.

The theory of polymer nanocomposite synthesis started in the early 1990s when the Toyota research group patented a specific method for synthesizing nylon-6 clay hybrid for timing belt cover (Usuki, 2016). In the first step, the montmorillonite clay is functionalized and intercalated with 12-aminolauric acid. The ammonium ions bond to the clay silicate layers, and the carboxyl group initiates the polymerization of caprolactone (a monomer for nylon 6). The heat polymerization in the interlayers exfoliates and disperses the montmorillonite sheets. The process enhanced the mechanical and thermal resistance of the composites as compared to a nylon-6 polymer. Later, nanocomposites made of synthetic polymers, such as polyethene, polyvinyl chloride, polypropylene, polyacrylonitrile, or polyethene terephthalate, with excellent mechanical and chemical stability and long service life, have been employed in various fields, including water and wastewater purification. However, the monomers, such as acrylonitrile, vinyl chloride, and formaldehyde, that are used for polymerization, are highly toxic and eco-unfriendly. Besides, energy-intensive synthesis processes and the disposal of spent synthetic polymer are significant obstacles in their use for field-level water and wastewater purifications.

The search for safer and environmentally friendly polymers resulted in biopolymers instead of synthetic polymers. Biopolymers are also preferable in producing nanocomposites owing to their abundant availability, high economic value, renewability, and surface functionality. Currently, biopolymer-derived nanocomposites (BPNCs) have been used in environmental remediation (adsorbents, photocatalysts, membranes), catalysis, packaging, biomedical (drug delivery, porous scaffolds, tissue engineering), and sensors applications. These composites can reconcile optical, magnetic, and conductivity properties by incorporating a comprehensive variety of nanofillers such as graphene-based, metal-oxides, and organic molecules.

This chapter outlines the synthesis of BPNCs and their applications in water and wastewater treatment, emphasizing our research findings and contemporary research in India. Different types of biopolymers based on source and synthesis are discussed in brief. Finally, the chapter discusses the challenges and future perspectives of BPNCs in field application.

9.2 BIOPOLYMERS AND BIOPOLYMER NANOCOMPOSITES

Biopolymers are long-chain molecules composed of repeating chemical blocks called monomers derived from renewable or natural origin. These are often categorized based on their carbon source (fossil, bio, or synthetic) and biodegradability. The broad classification of various biopolymers is given in Figure 9.1. Popular naturally derived biopolymers such as chitosan, cellulose, alginate, and

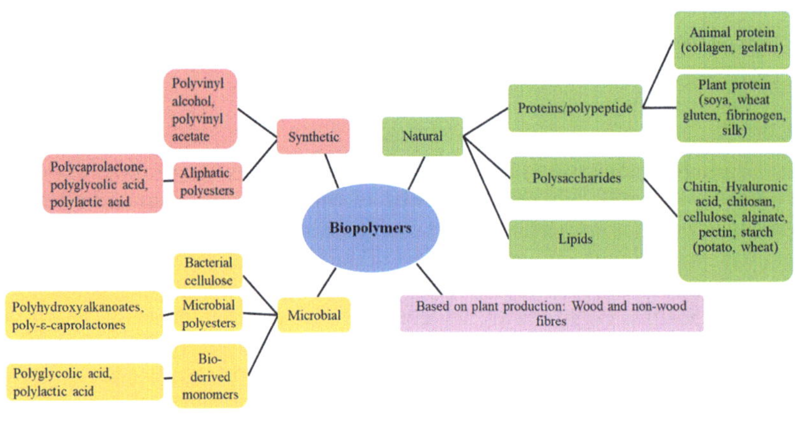

Figure 9.1 Broad classification of biopolymers. (*Source*: Adapted from Sharma *et al.*, 2018 with the permission of Elsevier).

pectin are produced by the chemical treatment of animal or plant biomass. Cellulose and chitosan are abundant cationic biopolymers and are insoluble in water under normal conditions. The underlying stability of these biopolymers comes from the hydrogen bonding network constructed among the crystalline-ordered polysaccharide chains. The bonding helps develop adsorbents in the form of films, beads, membranes, and foams (Shen & Gnanakaran, 2009). These polymers are employed for removing various aqueous pollutants, oil/water separation, and catalysis, but they have low mechanical strength and chemical stability. Besides, toxic crosslinking agents and low hydraulic stability limit their application in water and wastewater treatment. BPNCs can overcome the shortcomings of virgin polymers and improve their prospects for water and wastewater purification.

Diverse BPNCs with different physical and chemical properties are available now. Typically, they are classified based on the nanofiller dimension (0-, 1-, 2-, and 3-D), type of nanofiller (metal and metal-oxide, clay-based, carbon-based, polymers, or natural fibre), physical form (hydrogels, films, or granules), and method of synthesis. These BPNCs find application in environmental remediation, solar vapour generation, energy storage, soft robotics, drug delivery, and catalysis.

9.3 SYNTHESIS OF BIOPOLYMER NANOCOMPOSITE

The synthesis of biopolymer nanocomposites (BPNCs) can be done ex-situ or in-situ. Selecting the proper synthesis process plays a crucial role in the final product's physical, mechanical, and chemical properties. Figure 9.2 shows the different synthesis protocols of BPNCs. Ex-situ synthesis involves the synthesis of nanofiller or nanoparticles (NPs) separately and dispersing it into the polymer matrix. The NPs are physically entrapped in the biopolymer matrix through casting and solvent evaporation (films or membranes) or neutralization (Sarkar *et al.*, 2012). In the casting and solvent evaporation method, the polymer composite is mixed with solvent and NPs and cast as a membrane or spin-coated surface. The solvent evaporates when heated, leaving a composite layer, as shown in Figure 9.2a. It is reported that partial NP/polymer interaction and fast solvent evaporation are suitable conditions for uniform dispersion of NPs in the matrix (Cheng & Grest, 2016). Even though the

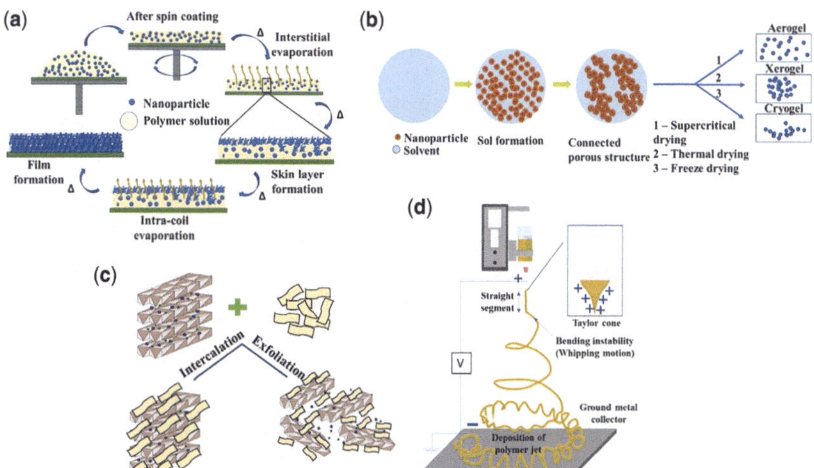

Figure 9.2 Schematics of BPNC synthesis processes: (a) solvent evaporation (*Source*: Adapted from Huang *et al.*, 2021 with the permission of Elsevier). (b) Sol–gel (*Source*: Adapted from Bokov *et al.*, 2021 open source subjected to International CC BY 4.0 creative common attribution license). (c) Intercalation and exfoliation, (d) electrospinning (*Source*: Adapted from Lee *et al.*, 2018; Reneker & Yarin, 2008 with the permission of Elsevier).

synthesis is facile, blending and homogeneous dispersion of nanofillers into the matrix in bulk scale is challenging. The neutralization and precipitation allow the preparation of nanocomposite in different forms. The process comprises polymer dissolution, NPs addition, and neutralization or ionotropic gelation. Polymer dissolution involves solvent diffusion and polymer chain disentanglement steps. In ionotropic gelation, polymers can form different structures by electrostatically crosslinking with counter ions forming a three-dimensional (3D)-mesh network of beads, fibres, or films. The gelation is induced by controlling pH and crosslinking species such as trisodium citrate, tripolyphosphate, glutaraldehyde, and epichlorohydrin (Pedroso-Santana & Fleitas-Salazar, 2020). In this process, the ratio of polymer/linker, chemical properties of the polymer, amount of filler added, and process conditions significantly affect the reaction efficiency and size and polydispersity index of the final product (Hoang *et al.*, 2022). Besides, adding nanoparticles beyond the limit may hinder gelation and form micro-sized particles. Even though aggregation of NPs during mixing can be avoided by the functionalization of NPs by surfactants, silane coupling, or grafted ligands, the homogeneous dispersion of synthesized NPs in host polymer still poses a significant challenge. Alternatively, in-situ synthesis is followed, where the polymer phase acts as a nano-reactor to synthesize NPs from a precursor.

An in-situ synthesis is a bottom-up approach consisting of NP generation, surface functionalization, and integration with the host polymer matrix. The functional groups of polymers act as nucleation centres for the growth of NPs with controlled particle size and inhibit aggregation (Naga Jyothi *et al.*, 2022). The methods provide better synthesis control and allow uniform nanoparticle dispersion in the polymer matrix. The sol–gel process is a low-cost in-situ approach that consists of hydrolysis of precursors, condensation to form a polymeric network, and drying (Bokov *et al.*, 2021). The homogeneous dispersion of NPs in the polymer matrix is achieved through hydrogen bonding between polymer (hydroxyl and amine groups) with formed sol–gel in the ageing stage. For carbon-based fillers, introducing silica groups to the polymer by the sol–gel process is recommended for pollutant removal. Layered silicates like clay or double-layer hydroxide-based BPNCs are synthesized by polymer intercalation/exfoliation. The polymer chain intercalates by displacing the solvent within the silicate. In the case of monomers, polymerization can be started by either heat or radiation through the selection of a suitable initiator (Mukhopadhyay *et al.*, 2020). On solvent removal, a layered structure is formed. This method is beneficial for water-soluble polymers where excessive use of organic solvent is avoided. The swelling of these nanocomposites is influenced by the geometry of the filler, aspect ratio, and molecular interaction (Wilson *et al.*, 2012). An increase in filler content increases the adsorption capacity. Electrospun/spray nanofibres are extensively studied for the removal of pollutants due to tunable surface area, porosity, facile modification, and scalability (Zhang *et al.*, 2021). The electrospun fibres are widely used for membrane separation. Besides, highly porous and less agglomerated one-dimensional metal-oxide self-standing fibres can be fabricated by calcination involving nucleation of nanoparticles and degradation of the sacrificial polymer, extending its applications in oil–water separation, catalysis, adsorption, and degradation. The polymer fibre formation involves charging of polymer, Taylor cone formation, whipping motion, and deposition. Factors affecting the process include solution viscosity, flow rate, ambient condition, the distance between the nozzle and collector, applied voltage, and molecular weight of the polymer. Readers may refer Haider *et al.* (2018) and Immich *et al.* (2017) for more details. Typically, the adsorption capacities of nanofibres reduce after regeneration, which can be due to surface modification. Hence, surface functionalization is to be performed to prevent porosity change after adsorption (Haider *et al.*, 2018).

9.4 APPLICATIONS OF BPNCS FOR WATER AND WASTEWATER REMEDIATION

BPNCs are employed in various fields, including pharmaceuticals, tissue engineering, food packaging, cosmetics, agriculture, construction, and automobiles. The following section briefly describes the application of BPNCs in water and wastewater treatment.

9.4.1 BPNCs as adsorbent

BPNCs are widely reported as adsorbents for removing organic and inorganic pollutants. They are dominant alternatives to conventional adsorbents due to their surface functionality, facile synthesis, tunable particle size, and ease of use. Various metal-oxide nanoparticles are dispersed in a chitosan matrix through an in-situ sol–gel process at atmospheric pressure and temperature to remove lead, fluoride, mercury, and arsenic (Saha *et al.*, 2015; Sankar *et al.*, 2013). The method allows the in-situ synthesis of nanoparticles and granulation of the polymer-supported nanoparticles into various shapes and sizes, as shown in Figure 9.3e. Cellulose–nanoscale–manganese oxide composite was prepared through an in-situ impregnation method to remove lead from water (Maliyekkal *et al.*, 2010). The adsorption capacity is high at a low dosage of manganese oxide on cellulose, possibly due to better dispersion of NP fibres. Saha *et al.* (2015) have developed a versatile method to granulate metal-oxide nanoparticles like AlOOH using chitosan polymers. They observed that the settling ability of the nanocomposite gel and the stability of granulated composites were influenced by synthesis pH and drying temperature, respectively, as shown in Figure 9.3d. They reported that the functional groups on the chitosan matrix act as nucleation centres for the in-situ formation of metal-oxides and provide stability to the polymer matrix. Sankar *et al.* (2013) developed a point-of-use household water purification system (Figure 9.3a and b) for the removal of heavy metals and microorganisms (*Escherichia coli* (*E. coli*)) using chitosan-based nanocomposites. Using similar materials, the AMRIT filter was developed for arsenic removal at a treatment cost of 2.5 paise/L and installed at several places in India. Naga Jyothi *et al.* (2022) studied the stability of bimetal oxide dispersed in a chitosan

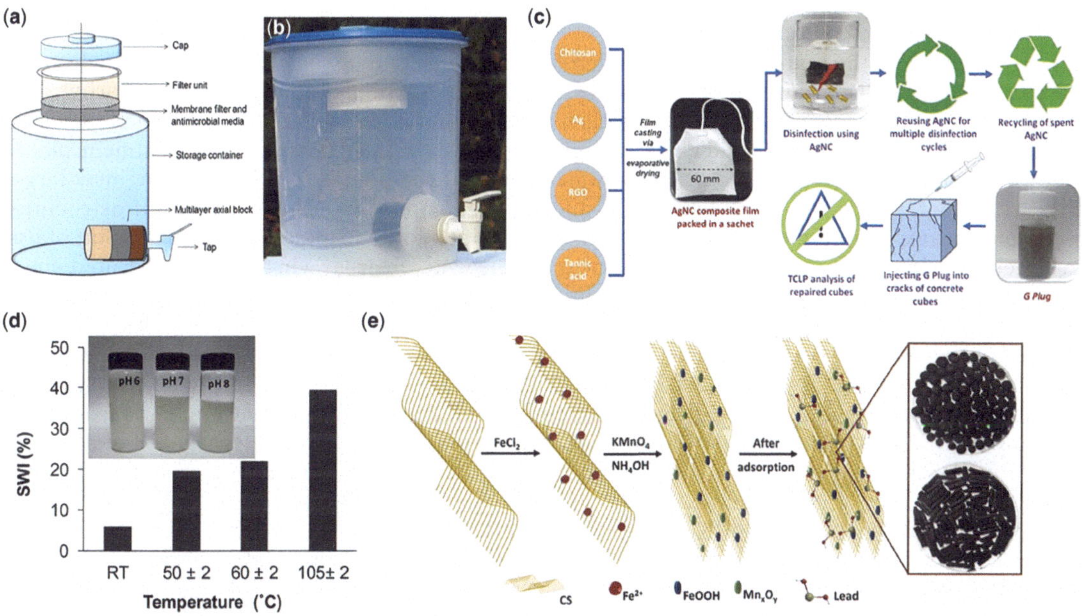

Figure 9.3 (a) and (b) Schematic diagram and actual photograph of chitosan-AlOOH based water purification device (*Source*: Reprinted from Sankar *et al.*, 2013). (c) Schematic of experiment plan followed for disinfection and recycling potential of silver-loaded waste-derived polymer film (*Source*: Reprinted from Kannan and Maliyekkal, 2023). (d) Swelling index of AlOOH granule at varying temperature. Inset is the digital image of composite gel synthesized at various pH (*Source*: Reprinted from Saha *et al.*, 2015). (e) Pathway showing FeMnG formation and adsorption mechanism (*Source*: Reprinted from Naga Jyothi *et al.*, 2022). All images are reprinted with the permission of Elsevier.

matrix. The authors found that the composite is chemically stable at typical groundwater and surface water pH and temperature condition with negligible leaching of Fe and Mn. These materials are safe for use in drinking water purification. Kumar *et al.* (2019) confirmed the viability of chitosan/poly(vinyl alcohol)/zinc-oxide nanocomposite film for human contact, as evidenced by the MTT assay using the NIH3T3 cells after the successful removal of acid black – 1 (AB 1) dye from water.

9.4.2 BPNCs as photocatalysts

Organic pollutants such as dyes, pesticides, pharmaceuticals, endocrine-disrupting compounds, and personal care products can harm biotic life (Kusuma *et al.*, 2021). Photocatalysis is a simple, low-cost, and less energy-intensive advanced oxidation process for degrading organic pollutants into harmless products like carbon dioxide and water. Traditionally used metal-oxides-based photocatalysts like titanium dioxide (TiO_2) are susceptible to agglomeration and catalyst fouling (Yang *et al.*, 2021). Using BPNCs for photocatalysis is gaining interest due to their synergistic electrical, optical, and catalytic properties (Melinte *et al.*, 2019). Nylon (N)6-TiO_2-coated cellulose paper showed synergistic adsorption and photodegradation of estrogen (E1, E2, and E3) compounds under sunlight. Mary *et al.* (2020) reported that adding chitosan to the binary semiconductor metal-oxide NPs (cerium-oxide and bismuth-oxide) reduced electron–hole pair recombination by lowering the band gap. Hasanin *et al.* (2021) reported that amine-functionalized carboxymethyl cellulose loaded with TiO_2 for UV light degradation of 2,4-dichlorophenol, and the degradation rate is higher than the conventional TiO_2 catalyst. Besides, the amine groups on the functionalized composites enabled *E. coli* and *Candida* disinfection.

9.4.3 BPNCs in disinfection of water

Drinking water is the major source of microbial hazards and associated waterborne diseases worldwide, especially in developing economies like India. National water quality monitoring data and other published reports support the evidence of the poor quality of drinking water in the country (Hirani & Dimble, 2019). This indicates that existing state-run centralized water treatment plants are not adequate to meet the safe drinking water needs of the population. Establishing new large-scale treatment systems is also challenging due to large capital investment, poor infrastructure, recontamination, and difficulty in operation and maintenance. Studies show that point-of-use treatment systems are sustainable ways for providing safe water in areas where access to treated piped water is limited (Kannan *et al.*, 2021). According to studies, a well-designed point-of-use household intervention contributes to a 30–40% reduction in diarrhoeal diseases (Kannan & Maliyekkal, 2023). However, designing a sustainable and easy-to-use system is essential to improve its acceptability and achieve large-scale use. Disinfectants loaded BPNCs are attractive due to their ability to hold disinfectant effectively and release disinfectants to water at a controlled and sustained rate (Kannan & Maliyekkal, 2023). Sankar *et al.* (2013) conducted extensive studies on silver-loaded aluminium oxyhydroxide–chitosan granules to remove bacteria and viruses by leaching silver from the granular nanocomposite. The granules have shown sustained and controlled release of silver within the permissible drinking water limit. Kannan *et al.* (2021) developed an affordable, easy-to-use silver-loaded waste-derived polymer to remove *E. coli*. The chitosan–graphene matrix acts as a substrate for holding silver nanoparticles and releasing silver ions at a controlled and sustained rate. The system protects silver nanoparticles interacting with natural organic matter and background ions like chloride and improves the service life. The sustainability indicators show that the composite is environmentally friendly and sustainable. Kannan *et al.* (2021) also reported that silver nanoparticles embedded in a chitosan–graphene matrix showed enhanced disinfecting ability in the presence of bicarbonate ions, a commonly found anion in drinking water. This attributes to the collapse of the proton motive force of the bacterial cell, leading to the enhanced uptake of silver ions and reactive oxygen species to kill the bacteria.

Table 9.1 Various BPNCs employed in water and wastewater treatment.

Pollutant Class	Adsorbent	Physical Parameters of BPNC	Adsorption Capacity (mg/g)/Performance	References
Heavy metals	Cellulose–nanoscale–manganese oxide	Fibres, 5–10 nm size NPs on surface	80.1 (Pb^{2+})	Maliyekkal *et al.* (2010)
	FeMnG	0.3–0.6 mm; 156.37 m^2/g	108.12 (Pb^{2+})	Naga Jyothi *et al.* (2022)
	Iron oxide incorporated alginate beads	126 m^2/g	29.14 (Cr^{4+})	Periyasamy *et al.* (2018)
Dyes	Chitosan/PVA/ZnO nanocomposite film	–	86% in 90 min (AB – 1)	Kumar *et al.* (2019)
Disinfection	Silver-loaded aluminium oxyhydroxide–chitosan granules	–	120 g composite provides safe drinking water for a family of 5 for 1 year (assume a daily consumption of 10 L) at $2/family/year	Sankar *et al.* (2013)
	AgNC film	–	4.8–10.3 g of film for 30 min produced 1000 L of water at a cost of <5 paise/L	Kannan *et al.* (2021)
Pharmaceuticals	Fe_3O_4-coated polymer clay composite	94.81 m^2/g	15.6, 39.1 and 24.8 (atenolol, ciprofloxacin, and gemfibrozil)	Arya and Philip (2016)
	$MnFe_2O_4$ NPs embedded chitosan-diphenylureaformaldehyde resin	442.6 m^2/g	168.42 (tetracycline)	Ahamad *et al.* (2019)
	N6-TiO_2-loaded cellulose paper		99.5% of E1 and E2 and 98.5% of E3 after 180 min under sunlight	Mafra *et al.* (2021)
Herbicide	TiO_2–chitosan photocatalyst (UV active)	Beads	86% (50 mg/L of 2,4-dichlorophenoxyacetic acid). Rs. 1323 for 2 L	Balakrishnan *et al.* (2021)
Nutrient removal	GNOF	1–2 mm	NH_4^+ removal for 267 days in nine phases and cycle time from 2 to 0.5 days in SBR mode	Desireddy *et al.* (2018)
Field scale or point-of-use systems				
Disinfection	Silver/copper NP paper filters for bacteria removal in Limpopo, South Africa	Metal content - 0.1–6.5%	0.22–0.37 USD/m^2. 6.5 cm^2 size paper produces 90 L treated water	Dankovich *et al.* (2016)
Heavy metal	Iron oxyhydroxide chitosan nanocomposite	Granular	<2.5 paise/L for arsenic and iron removal	Sankar *et al.* (2013)
Nutrient removal	Iron-oxide NP-loaded chitosan composite	Beads	52.3% (PO_4^{-3}), 0.059 mg/g. Pilot study at Seoho stream, Republic of Korea	Kim *et al.* (2018)

9.4.4 Recycling and disposal of spent materials

Safe handling and disposal of spent nanocomposite materials are essential to prevent them from the waste stream and reduce the environmental impact. Typically, the filter parts and materials used in water purification are discarded into the open environment. No systematic mechanism exists to collect the used filter parts or materials and recycle or dispose of them safely. This can significantly threaten the environment (Kannan & Maliyekkal, 2023). Therefore, developing fewer toxic materials that produce less waste is essential to improving the overall sustainability of the process. In this context, BPNCs are attractive due to their low toxicity and the ability to immobilize nanoparticles and prevent them from releasing into the environment. Besides, several spent BPNCs can be reused as catalysts, desiccants, secondary adsorption of another pollutant, the release of nutrients, or the reinforcement of cement-based composites. Nutrient-recovered adsorbents are used as fertilizers for crops. Wang *et al.* (2016) reported the reusability of wheat straw-derived modified cellulose adsorbent as a slow-release fertilizer in the soil after the uptake of ammonia and phosphorous. Ballav *et al.* (2018) reported the reuse of mercury-adsorbed polypyrrole@L-cysteine as a catalyst for converting phenylacetylene to acetophenone. Biopolymers are used as reinforcement in cementitious materials due to their low toxicity, low cost, abundant availability, low density, and toughness (Ardanuy *et al.*, 2015). Utilizing this benefit, Kannan and Maliyekkal (2023) obtained a novel strategy to recycle the exhausted chitosan–silver nanoparticle composite film into a value-added concrete micro-crack sealant, as shown in Figure 9.3c. The toxicity characteristic leaching procedure (TCLP) test results also confirmed that the leaching of Ag^{+} from the repaired mortar cubes is within permissible limits, signifying that the disposal is safe. A comparison of various BPNCs employed in water and wastewater treatment systems is presented in Table 9.1.

9.5 CONCLUSION, CHALLENGES, AND THE WAY FORWARD

Greater biocompatibility, abundant availability, economic viability, tunable size, ease of separation, and presence of large functional groups make biopolymers an excellent choice for dispersing nanoparticles and forming nanocomposites. BPNCs are currently used in various fields, such as biomedical, pharmaceutical, cosmetics, automobiles, construction, and environmental remediation. However, the selection of synthesis routes, polymer and filler types, and post-synthesis processing governs the properties of the composites and their application. In-situ synthesis is proven more efficient in dispersing nanoparticles and producing composites with uniform properties than ex-situ synthesis. Synthesis techniques such as casting/solvent evaporation and electrospinning are suitable for deriving two-dimensional forms like films and membranes. However, sol–gel, neutralization, and precipitation methods typically derive nanoparticles and 3D structures like granules and beads. Nanocomposites made of chitosan, cellulose, and alginate, as dispersing phases, are extensively used to remove pollutants in water and wastewater. However, cost, acceptability, durability, regeneration and reuse potential, long-term hydraulic stability, and bulk production are the key factors that govern the application of BPNCs in water and wastewater purification. Several parameters, such as dispersion, stability, and interactions of a nanofiller in the polymer host matrix, crystallization ability of the nanofiller, surface charge, and polymer chain flexibility, must be considered to produce the nanocomposite with the desired properties. The use of bio-derived surfactants may be encouraged for biodegradability, enhanced dispersion of NPs in a polymer matrix, and functionality for simultaneous removal of organic and inorganic pollutants. The leaching of nanoparticles from the matrix and ecotoxicity must be evaluated before employing the material for water and wastewater treatment. The recyclability of spent materials should be explored, and more research should be carried out on a practical and sustainable way to utilize spent materials. Research findings show that the spent BPNCs have potential applications in making building materials like ceramics, bricks, and concrete and repairing materials like micro-crack sealants. The socio-economic and environmental impact of the process, service, and disposal of the composites should be conducted to determine the overall sustainability of using BPNCs in water and wastewater purification.

REFERENCES

Ahamad T., Ruksana Chaudhary A. A., Naushad M. and Alshehri S. M. (2019). Fabrication of $MnFe_2O_4$ nanoparticles embedded chitosan-diphenylureaformaldehyde resin for the removal of tetracycline from aqueous solution. *International Journal of Biological Macromolecules*, **134**, 180–188, https://doi.org/10.1016/j.ijbiomac.2019.04.204

Ardanuy M., Claramunt J. and Toledo Filho R. D. (2015). Cellulosic fiber reinforced cement-based composites: a review of recent research. *Construction and Building Materials*, **79**, 115–128, https://doi.org/10.1016/j.conbuildmat.2015.01.035

Arya V. and Philip L. (2016). Adsorption of pharmaceuticals in water using Fe_3O_4 coated polymer clay composite. *Microporous and Mesoporous Materials*, **232**, 273–280, https://doi.org/10.1016/j.micromeso.2016.06.033

Balakrishnan A., Gopalram K. and Appunni S. (2021). Photocatalytic degradation of 2,4-dichlorophenoxyacetic acid by TiO_2 modified catalyst: kinetics and operating cost analysis. *Environmental Science and Pollution Research*, **28**(25), 33331–33343, https://doi.org/10.1007/s11356-021-12928-4

Ballav N., Das R., Giri S., Muliwa A. M., Pillay K. and Maity A. (2018). l-Cysteine doped polypyrrole (PPy@L-Cyst): a super adsorbent for the rapid removal of Hg + 2 and efficient catalytic activity of the spent adsorbent for reuse. *Chemical Engineering Journal*, **345**, 621–630, https://doi.org/10.1016/j.cej.2018.01.093

Bokov D., Turki Jalil A., Chupradit S., Suksatan W., Javed Ansari M., Shewael I. H., Valiev G. H. and Kianfar E. (2021). Nanomaterial by sol-gel method: synthesis and application. *Advances in Materials Science and Engineering*, **2021**, 1–21, https://doi.org/10.1155/2021/5102014

Cheng S. and Grest G. S. (2016). Dispersing nanoparticles in a polymer film via solvent evaporation. *ACS Macro Letters*, **5**(6), 694–698, https://doi.org/10.1021/acsmacrolett.6b00263

Dankovich T. A., Levine J. S., Potgieter N., Dillingham R. and Smith J. A. (2016). Inactivation of bacteria from contaminated streams in Limpopo, South Africa by silver- or copper-nanoparticle paper filters. *Environmental Science: Water Research & Technology*, **2**(1), 85–96, https://doi.org/10.1039/C5EW00188A

Desireddy S., Sabumon P. C. and Maliyekkal S. M. (2018). Anoxic ammonia removal using granulated nanoscale oxyhydroxides of Fe (GNOF) in an SBR. *Journal of Environmental Chemical Engineering*, **6**(4), 4273–4281, https://doi.org/10.1016/j.jece.2018.05.033

Haider A., Haider S. and Kang I. K. (2018). A comprehensive review summarizing the effect of electrospinning parameters and potential applications of nanofibers in biomedical and biotechnology. *Arabian Journal of Chemistry*, **11**(8), 1165–1188, https://doi.org/10.1016/j.arabjc.2015.11.015

Hasanin M., Abdelhameed R. M., Dacrory S., Abou-Yousef H. and Kamel S. (2021). Photocatalytic degradation of pesticide intermediate using green eco-friendly amino functionalized cellulose nanocomposites. *Materials Science and Engineering: B*, **270**, 115231, https://doi.org/10.1016/j.mseb.2021.115231

Hirani P. and Dimble V. (2019). Water pollution is killing millions of Indians. Here's how technology and reliable data can change that. *World Economic Forum*. Accessed 5 December 2022. https://www.weforum.org/agenda/2019/10/water-pollution-in-india-data-tech-solution/

Hoang N. H., Le Thanh T., Sangpueak R., Treekoon J., Saengchan C., Thepbandit W., Papathoti N. K., Kamkaew A. and Buensanteai N. (2022). Chitosan nanoparticles-based ionic gelation method: a promising candidate for plant disease management. *Polymers 2022*, **14**(4), 662, https://doi.org/10.3390/POLYM14040662

Huang Y. J., Yeh J. W. and Chang-Mou Yang A. (2021). 'High-entropy polymers': a new route of polymer mixing with suppressed phase separation. *Materialia*, **15**, 100978, https://doi.org/10.1016/j.mtla.2020.100978

Immich A. P. S., Tornero J. A., Casas F. C., Arias M. J. L., Immich A. P. S., Tornero J. A., Casas F. C. and Arias M. J. L. (2017). Electrospun PLLA membranes for caffeine delivery: diffusional approach. *Journal of Biomedical Science and Engineering*, **10**(12), 563–574, https://doi.org/10.4236/jbise.2017.1012042

Kannan U. and Maliyekkal S. M. (2023). A resource-efficient and portable nanotechnology-enabled disinfection system: performance studies and a novel strategy to recycle spent material. *Process Safety and Environmental Protection*, **171**, 532–540, https://doi.org/10.1016/j.psep.2023.01.043

Kannan U., Sabumon P. C. and Maliyekkal S. M. (2021). Development of an eco-friendly and reusable point-of-use disinfection system. *Process Safety and Environmental Protection*, **148**, 104–113, https://doi.org/10.1016/j.psep.2020.09.055

Kim J. H., Kim S. B., Lee S. H. and Choi J. W. (2018). Laboratory and pilot-scale field experiments for application of iron oxide nanoparticle-loaded chitosan composites to phosphate removal from natural water. *Environmental Technology (United Kingdom)*, **39**(6), 770–779, https://doi.org/10.1080/09593330.2017.1310937

Kumar S., Krishnakumar B., Sobral A. J. F. N. and Koh J. (2019). Bio-based (chitosan/PVA/ZnO) nanocomposites film: thermally stable and photoluminescence material for removal of organic dye. *Carbohydrate Polymers*, **205**, 559–564, https://doi.org/10.1016/j.carbpol.2018.10.108

Kusuma T. D., Naga Jyothi M. S. V., Rao C. P. and Maliyekkal S. M. (2021). Advanced oxidation processes: a promising route for abatement of emerging contaminants in water. In: Energy, Environment, and Sustainability, A. K. Singh, S. P. Rathinam, K. Gupta and T. Agarwal (eds.), Springer Nature, pp. 275–305, https://doi.org/10.1007/978-981-16-3256-3_12

Lee J. K. Y., Chen N., Peng S., Li L., Tian L., Thakor N. and Ramakrishna S. (2018). Polymer-based composites by electrospinning: preparation & functionalization with nanocarbons. *Progress in Polymer Science*, **86**, 40–84, https://doi.org/10.1016/j.progpolymsci.2018.07.002

Mafra G., Brognoli R., Carasek E., López-Lorente Á. I., Luque R., Lucena R. and Cárdenas S. (2021). Photocatalytic cellulose-paper: deepening in the sustainable and synergic combination of sorption and photodegradation. *ACS Omega*, **6**(14), 9577–9586, https://doi.org/10.1021/acsomega.1c00128

Maliyekkal S. M., Lisha K. P. and Pradeep T. (2010). A novel cellulose–manganese oxide hybrid material by in situ soft chemical synthesis and its application for the removal of Pb(II) from water. *Journal of Hazardous Materials*, **181**(1–3), 986–995, https://doi.org/10.1016/j.jhazmat.2010.05.112

Mary N. U., Umapathy M. J. and Sivasamy A. (2020). Biomaterial supported binary semiconductor metal oxide nanocomposite for water remediation under solar irradiation. *Optik*, **208**, 164219, https://doi.org/10.1016/j.ijleo.2020.164219

Melinte V., Stroea L. and Chibac-Scutaru A. L. (2019). Polymer nanocomposites for photocatalytic applications. *Catalysts*, **9**(12), 986, https://doi.org/10.3390/catal9120986

Mukhopadhyay R., Bhaduri D., Sarkar B., Rusmin R., Hou D., Khanam R., Sarkar S., Kumar Biswas J., Vithanage M., Bhatnagar A. and Ok Y. S. (2020). Clay–polymer nanocomposites: progress and challenges for use in sustainable water treatment. *Journal of Hazardous Materials*, **383**, 121125, https://doi.org/10.1016/j.jhazmat.2019.121125

Naga Jyothi M. S. V., Harafan A., Sen Gupta S., Neethu N., Singhal G., Ramaiah B. J. and Maliyekkal S. M. (2022). Chitosan immobilised granular FeOOH-MnxOy bimetal-oxides nanocomposite for the adsorptive removal of lead from water. *Journal of Environmental Chemical Engineering*, **10**(2), 107353, https://doi.org/10.1016/j.jece.2022.107353

Pedroso-Santana S. and Fleitas-Salazar N. (2020). Ionotropic gelation method in the synthesis of nanoparticles/microparticles for biomedical purposes. *Polymer International*, **69**(5), 443–447, https://doi.org/10.1002/pi.5970

Periyasamy S., Gopalakannan V. and Viswanathan N. (2018). Hydrothermal assisted magnetic nano-hydroxyapatite encapsulated alginate beads for efficient Cr(VI) uptake from water. *Journal of Environmental Chemical Engineering*, **6**(1), 1443–1454, https://doi.org/10.1016/j.jece.2018.01.007

Reneker D. H. and Yarin A. L. (2008). Electrospinning jets and polymer nanofibers. *Polymer*, **49**(10), 2387–2425, https://doi.org/10.1016/j.polymer.2008.02.002

Saha G., Maliyekkal S. M., Sabumon P. C. and Pradeep T. (2015). A low cost approach to synthesize sand like AlOOH nanoarchitecture (SANA) and its application in defluoridation of water. *Journal of Environmental Chemical Engineering*, **3**(2), 1303–1311, https://doi.org/10.1016/j.jece.2014.11.030

Sankar M. U., Aigal S., Maliyekkal S. M., Chaudhary A., Kumar A. A., Chaudhari K. and Pradeep T. (2013). Biopolymer-reinforced synthetic granular nanocomposites for affordable point-of-use water purification. *Proceedings of the National Academy of Sciences*, **110**(21), 8459–8464, https://doi.org/10.1073/pnas.1220222110

Sarkar S., Guibal E., Quignard F. and SenGupta A. K. (2012). Polymer-supported metals and metal oxide nanoparticles: synthesis, characterization, and applications. *Journal of Nanoparticle Research*, **14**(2), 1–24, https://doi.org/10.1007/s11051-011-0715-2

Sharma B., Malik P. and Jain P. (2018). Biopolymer reinforced nanocomposites: a comprehensive review. *Materials Today Communications*, **16**, 353–363, https://doi.org/10.1016/j.mtcomm.2018.07.004

Shen T. and Gnanakaran S. (2009). The stability of cellulose: a statistical perspective from a coarse-grained model of hydrogen-bond networks. *Biophysical Journal*, **96**(8), 3032, https://doi.org/10.1016/j.bpj.2008.12.3953

Usuki A. (2016). Nylon 6-clay hybrid: from invention to practical use. *R&D Review of Toyota CRDL*, **47**(1), 45–55, http://www.tytlabs.com/review/

Wang X., Lü S., Gao C., Feng C., Xu X., Bai X., Gao N., Yang J., Liu M. and Wu L. (2016). Recovery of ammonium and phosphate from wastewater by wheat straw-based amphoteric adsorbent and reusing as a multifunctional slow-release compound fertilizer. *ACS Sustainable Chemistry and Engineering*, **4**(4), 2068–2079, https://doi.org/10.1021/acssuschemeng.5b01494

Wilson R., George S. M., Maria H. J., Plivelic T. S., Anil Kumar S. and Thomas S. (2012). Clay intercalation and its influence on the morphology and transport properties of EVA/clay nanocomposites. *Journal of Physical Chemistry C*, **116**(37), 20002–20014, https://doi.org/10.1021/JP302177Y

Yang X., Sun H., Li G., An T. and Choi W. (2021). Fouling of TiO_2 induced by natural organic matters during photocatalytic water treatment: mechanisms and regeneration strategy. *Applied Catalysis B: Environmental*, **294**, 120252, https://doi.org/10.1016/j.apcatb.2021.120252

Zhang Y., Wang F. and Wang Y. (2021). Recent developments of electrospun nanofibrous materials as novel adsorbents for water treatment. *Materials Today Communications*, **27**, 102272, https://doi.org/10.1016/j.mtcomm.2021.102272

doi: 10.2166/9781789063714_0109

Chapter 10

A holistic approach to assess the toxic behaviour of emerging nanomaterials in aquatic system

Nikita Shraogi[1,2], *Rahul Verma*[1,2], *Joel Saji*[1], *Amrita Singh*[1,2], *Durgesh P. Mourya*[1,2], *Sneha Verma*[1,2], *Alok K. Pandey*[1,2] *and Satyakam Patnaik*[1,2]*

[1]*CSIR-Indian Institute of Toxicology Research (CSIR-IITR), Lucknow, India*
[2]*Academy of Scientific and Innovative Research (AcSIR), Ghaziabad 201002, India*
**Corresponding author: satyakampatnaik@iitr.res.in*

ABSTRACT

With the advancement of technology, the use of new and improved materials starting from macro- to micro- and nano-range in varied fields has increased exponentially. Although the scientific community has been interested in nanomaterials due to their unique properties, the use of these innovative materials has given rise to concern regarding their potential toxicity to living beings as well as the environment. This chapter elaborates on how emerging nanomaterials behave in relation to dynamic microenvironments at the nano–bio–eco interface level and how this affects their toxicity, fate, and exposure potential. In addition, a brief account of the exposure pathways and different models used for toxicity evaluation is also discussed, vis-à-vis the ongoing research work on toxicity, the role of contributing factors, and the probable alternatives developed to attenuate related hazards, challenges, and future outlook.

Keywords: emerging nanomaterials, nanomaterials exposure, aquatic toxicity, bioaccumulation, genotoxicity, environmental fate, metabolites, greener alternatives

10.1 INTRODUCTION

The field of nanotechnology is evolving into a multifaceted stream of scientific and technological studies with tremendous potential for transforming fundamental sciences due to their unique properties (Sudha *et al.*, 2018). In general, nanomaterials have at least one dimension in the range of 1–100 nm. At this scale, the surface area of materials is greatly increased, enabling them to gain some unique properties not available in bulk. The use of nanotechnology is extensive, whether in medicine, agriculture, pollution remediation, sensors, cosmetics robotics, etc. However, most of these materials end up contaminating water bodies at the end of their life cycle. Their distinctive physicochemical characteristics, high penetration ability, large surface area, and chemical activity make them not just desirable for varied applications but also conceivably hazardous to the environment and life forms.

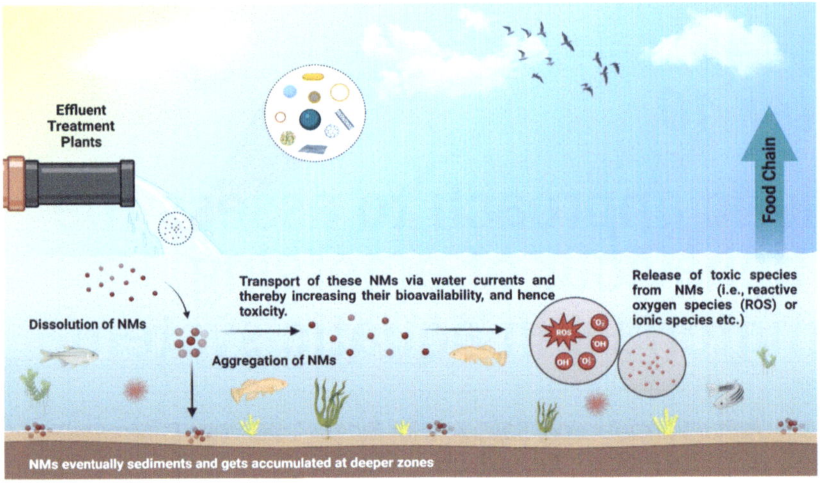

Figure 10.1 Graphical representation of the pathways by NMs entering aquatic bio-environment and thereby affecting the food chain.

Aquatic ecosystems host diverse organisms, such as plankton, fishes, crustaceans, and amphibians. Each of these groups contains within them distinct species which interact with each other at different levels of the food chain (Ishii *et al.*, 2020). Due to their small size, charge, and functionalization, nanomaterials can easily make their way into the biological system (Singh *et al.*, 2018) and tend to get accumulated in the organism depending on their lipophilicity and/or if their log K_o/w >4.5. The toxicity caused to these aquatic organisms can be evaluated as a measure of change in behaviour, morphology, physiology, reproductive capacity, survival, and teratogenic abnormalities. Higher organisms, including human beings, can then get exposed to these nanomaterials through the food chain as shown in Figure 10.1. The wide range of *in-vitro* studies discussed in this chapter provides simple, cost-effective, and efficient approaches which are widely utilized to replace the use of animals in toxicity testing. Although toxicity studies in humans are much more complex and require both *in-vitro* and *in-vivo* toxicity assessments, a section in this chapter is dedicated to understanding the exposure pathways, blood-contact properties, nanoparticles biodistribution, and immune response as well as toxicokinetic studies that help to determine the safe or permissible limit of nanomaterials. Nonetheless, various toxicity-enhancing factors concerning nanomaterials could be controlled to make them safe and yet highly efficient. These safe-by-design greener alternatives and sustainable usage of nanomaterials are also highlighted. A discussion of the challenges and future outlooks as perceived by the author is also included.

10.2 POTENTIAL TOXICITY OF EMERGING NANOMATERIALS IN AQUATIC ECOSYSTEMS

This section briefly outlines the latest in nanotechnology vis-à-vis emerging nanomaterials or nanocomposites, their usages, and their impact on the aquatic ecosystem.

10.2.1 Nanomaterials

Nanomaterials offer numerous opportunities and strategies to address current challenges and their foreseeable negative impact on aquatic ecosystems are outlined in Figure 10.2. In general, nanoparticles, nanotubes, or nanofibre polymers with a particle size range of 4–100 nm are used in materials synthesis (Mehwish *et al.*, 2014). In addition to the aforementioned unique properties that

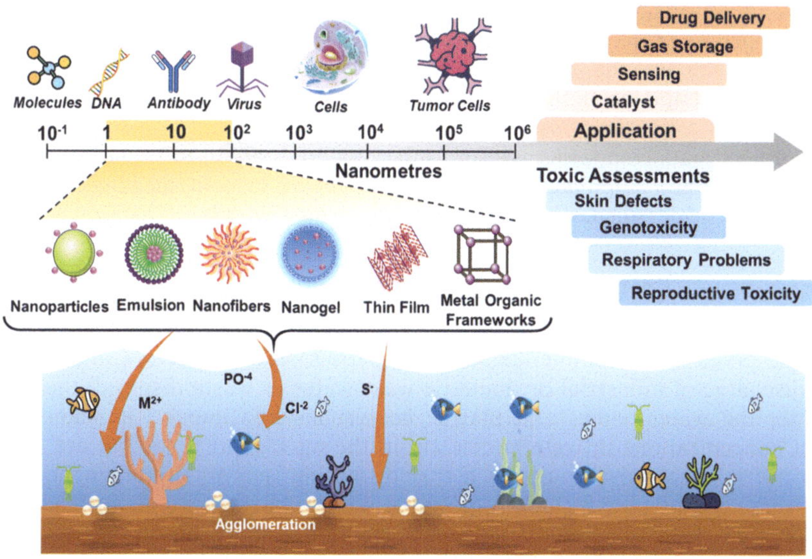

Figure 10.2 Summary of the toxicological data gathered from different models of aquatic microorganisms and vertebrates that are discussed in this chapter.

make them highly reactive, they can also be tuned by incorporating appropriate functional groups to enhance their reactivity, affinity, and selectivity towards drugs, pesticides, and chemically active species, biomolecules, etc., which ultimately improve the performance of the nanomaterials. The most commonly used nanomaterials for water treatment are metallic and metalloid nanomaterials, zeolites, graphene or carbon-based, fullerenes, metal-organic-frameworks (MOFs), covalent-organic-frameworks (COFs), etc. Even though the nanoparticle's aforementioned characteristics make them highly valuable for a variety of applications, it is believed that these characteristics are important contributors to the toxicity of nanomaterials in an aquatic environment (Sajid *et al.*, 2015).

10.2.2 Graphene-based materials

Graphene is an allotrope of carbon with a hexagonal lattice structure and sp^2 hybridized carbon atoms. The unaltered basal plane sites of graphene contain hydrophobic free surface electrons. Due to agglomeration, aggregation, sedimentation, and dissolution, graphene is present in the environment at a high level. Numerous studies have proven that graphene impacts a variety of aquatic creatures, including plankton, crustaceans, fishes, and amphibians, regardless of its type (Volkov *et al.*, 2017).

10.2.3 Metal-organic-frameworks

Metal-organic frameworks (MOFs) are a rapidly expanding class of crystalline hybrid porous materials with ultrahigh surface areas and access to multiple binding sites. MOFs' structural and modular design enables a variety of uses in gas storage, separation, catalysis, sensing, and drug delivery. Most of the literature on MOFs discusses thermally and chemically stable MOFs that are stable up to 500°C. Although these stable materials are not directly exposed to aquatic systems, they may enter agricultural runoff, local, industrial, or waste drainage systems. The question is how they will behave in aquatic systems. Alternatively, if the MOFs are not so stable, the leaching of metals into the cellular compartments of aquatic organisms may result in harmful accumulations, but this will depend on the nature and concentration of the metal (Colombo *et al.*, 2011).

10.2.4 Other nanocomposites

The enhanced properties of the nanocomposites are determined by the individual properties of each component material, the amounts present, and the nanocomposites' overall geometry. This, in turn, render excellent thermal and mechanical stability to the nanocomposites, making them versatile, easy for chemical functionalization, and possessing a large interphase region (Ama & Ray, 2020). In addition to the aforementioned nanocomposites, various other materials are used as nanomaterials, including alumina, anatase, cadmium sulphide, cobalt ferrite, copper oxide, gold, maghemite, iron, iron oxide, iron hydroxide, nickel oxide, silica, stannous oxide, zinc oxide, zinc sulphide, zirconia, and some alloys.

10.3 FATE AND TOXIC EFFECT OF NANOMATERIALS IN AQUATIC SYSTEMS

Questions regarding the safety of engineered nanomaterials and nanoparticles have now been raised due to their widespread use in commercial items, industries, strategic sectors, and the clinical world. Through the discharge of industrial or domestic wastewater or waste disposal, these nanoparticles potentially enter the water bodies and marine ecosystems and impact the same. The predicted environmental concentrations in waterbodies are in the ng/L (ppt) range but are expected to rise further. The concern is that most nanoparticle toxicity mechanisms are unknown; however, it is being reported that they may involve instability of outer-membrane integrity, disturbance of membrane potential, cytotoxicity, genotoxicity, disruption of energy transduction, and production of reactive oxygen species in living organisms. Due to their tiny size, nanoparticles can enter biological systems in unusual ways (Mukherjee *et al.*, 2021). The following sections focus on how these engineered nanomaterials impart toxicity to the aquatic ecosystem through trophic levels.

10.3.1 Plankton

Phytoplankton, often referred to as contamination indicators, have crucial implications for marine ecosystems since even a slight drop-in growth rate can significantly impact the food chain ecosystem (Miller *et al.*, 2010).

10.3.2 Crustaceans and fish

Several studies have concluded increased mortality of marine crustaceans (Callaghan & MacCormack, 2017).

10.3.3 Amphibians

Due to their bi-phasic life cycle, amphibians are sensitive to water and soil pollution, making them effective bio-monitors in ecotoxicological investigations and environmental risk assessment. Various studies have reported that amphibian exposure to nanomaterials can cause physiological, morphological, embryological, and molecular alterations (Bour *et al.*, 2017).

10.4 METHODS OF TOXICITY EVALUATION IN AQUATIC ORGANISMS

Aquatic organisms are the primary entities of nanomaterials exposure. The following section discusses the different methods through which nanomaterial toxicity is typically evaluated in aquatic organisms.

10.4.1 Behavioural studies

Behavioural studies are a potentially significant tool for ecotoxicological studies because they are 10–1000 times more sensitive than conventional LC_{50} studies (Ašmonaitė *et al.*, 2016). Behavioural methods are non-destructive, and results are inferred by considering the three R's principle and ethical guidelines. Method behavioural analysis following an encounter with toxic substances include analysing various parameters such as swimming speed, alteration in circadian rhythms, social interaction, predator avoidance, spatial learning and memory, habituation to dark or noise stimulus, and novel tank test (Cachat *et al.*, 2010).

10.4.2 Physiological studies

Various researchers have argued the importance of studying the effect on haematology, plasma biochemistry, hormonal changes, and histopathology of aquatic organisms for assessing the toxic effects of nanomaterials on aquatic organisms. Remya *et al.* (2015) conducted a study that demonstrated iron oxide nanoparticles altered some physiological parameters.

10.4.3 Reproduction studies

Long-term exposure to nanomaterials can potentially deteriorate reproductive capability and also have a detrimental effect on the survival and development of offsprings. Chronic toxicity testing by administering a calculated dose of nanomaterial to the organism and observing the reduction in egg production or defect in folliculogenesis, the number of sperm produced, average offspring in each brood, the mortality rate of embryos, histopathological examination of testis and ovary are conducted to gain an understanding of how nanomaterials affect the reproductive system of aquatic fauna (Teng *et al.*, 2022).

10.4.4 Mortality studies

Mortality is considered an acceptable end-point of any toxicological study. It is determined both at an acute and chronic level. Following the OECD (Organisation for Economic Co-operation and Development) protocol, acute toxicity studies are generally conducted for 24, 48, and 72 h for *Daphnia magna* (a commonly used model organism for aquatic studies). EC_{50} and LC_{50} are calculated based on immobilization and mortality over this period. Based on acute toxicity results, chronic toxicity studies for *D. magna* are conducted following a 21-day exposure (Zhu *et al.*, 2010).

10.4.5 Transgenerational studies

The effects on the development, physiology, and endocrinology of an offspring without any direct contact with the toxicant, but due to the exposure of parents in their early life to the same are known as transgenerational effects. Varieties of nanomaterials pass the placental barrier and pose greater concern for transgenerational exposure in aquatic mammals. In transgenerational studies, various parameters of toxicity are observed over two or more generations, such as LC_{50}/LD_{50} and EC_{50} determination, gene expression, fertility, etc., to determine if effects are transferred through the germ line (Kalichak *et al.*, 2019).

10.4.6 Bioaccumulation studies

Having a comprehensive understanding of nanomaterial biodistribution in organs and tissues is crucial to understanding the mechanisms and potential toxicity they possess. To assess the bioaccumulation potential of any material, end-points such as bioconcentration factor (BCF), bioaccumulation factor (BAF), and biomagnification factor (BMF) are typically employed. BCF of 5000 or more and BMF greater than 1 indicate high bioaccumulation or biomagnification potential, respectively (Zheng & Nowack, 2022).

10.4.7 Exposure to humans through the aquatic environment

Nanomaterials recently have taken over all aspects of our life; be it agriculture, medicine, water remediation, cosmetics, food, etc. The release of nanomaterials to water bodies is generally a result of emission during their production, runoff from agricultural fields, disposal from industries, or direct release of certain nanomaterials meant for water treatment. Nanomaterials may then find their way to human bodies either through drinking water or the food chain (Tiede *et al.*, 2016).

10.5 TOXICITY ASSESSMENT OF NANOMATERIALS

10.5.1 In vitro toxicity assessment

The initial process in screening for material biocompatibility is to conduct cytotoxicity studies utilizing cell culture-based methodologies. Despite their lack of specificity, they are sensitive, dependable, convenient, and repeatable. Any substance must be evaluated *in vitro* before proceeding to *in vivo* procedures. Besides, efforts are constantly being put forth for alternate animal models by suitable *in vitro* methods. *In vitro* testing is cost-effective, quicker, and bears no ethical issues. The different cell lines used in the *in vitro* experiments are chosen by the possible route of exposure of the test substance in the system. Immortalized malignant cells are most commonly used, although stem cells or primary cells are also being used more frequently (Joris *et al.*, 2013). There are different cell viability and proliferation assays to determine cytotoxicity. Table 10.1 summarizes in-vitro toxicity assays reportedly used for nanoparticle toxicity determination.

10.5.2 In-vivo toxicity assessment

10.5.2.1 Exposure pathways

Nanomaterials may enter the human body through different exposure pathways, such as inhalation, ingestion, and skin penetration. Skin, the outermost layer of the body, serves as an effective barrier against nanoparticles; nevertheless, damaged skin, as well as the presence of sweat ducts and hair follicles, makes it vulnerable to penetration by small-sized nanomaterials (Nafisi & Maibach, 2018). Inhalation of water aerosols becomes another route of exposure to nanomaterials in humans (Remya *et al.*, 2016). Direct ingestion of water, vegetables, or aquatic organisms such as fish and shellfish contaminated with nanomaterials becomes another important route of exposure to the human body (Turan *et al.*, 2019).

10.5.2.2 Blood contact

Nanomaterials (<500 nm) can easily cross the epithelium to reach blood vessels. Blood proteins are negatively charged and are attracted to positively charged nanoparticles. A ‘protein corona’ forms on the surface of NPs triggering different immune responses as cytokines are released by the corona (De Matteis, 2017). Studies by Kongseng *et al.* (2016) and Nadesh *et al.* (2013) demonstrate that nanoparticle size and surface modifiers play an important role in the toxicity they exhibit in the biological system.

10.5.2.3 Immune system response

The physio-chemical characteristics of nanoparticles become an important factor in immune system responses. The nanoparticles are either ignored, eliminated silently, or generate an immune response based on these characteristics. Most nanoparticles, coated with biopolymers or proteins such as poly(ethylene glycol), poly(vinyl pyrrolidone), albumin, retinol, and serum mixtures, are sensed by the immune system as self, and hence do not trigger a response. Similarly, nanoparticles smaller than 10 nm diameter undergo rapid renal clearance, while nanoparticles of comparatively larger diameter are prone to be phagocytosed or get encapsulated (Ernst *et al.*, 2021).

10.5.2.4 Biodistribution and toxicokinetics

Different nanoparticles with varying sizes, surface functionalities, and charges are translocated to various organs within the biological system and exert their toxic effects upon them. To obtain a dose that generates the desired response without causing toxicity in any non-target organ, it is often necessary to conduct a biodistribution and toxicokinetic study (Jain *et al.*, 2018).

Table 10.1 In vitro toxicity/internalization in cell line system.

Assays/ Method	Type of Method	Target Molecule/ Specificity	Organism/ Cell Line	Nanoparticle	Results	References
MTT	In-vitro absorbance	Dehydrogenase (enzyme)	Gill cell line (*L. rohita*)	Silver (Ag)	EC_{50} (21.54 μg/mL)	Taju *et al.* (2014)
NRU	In-vitro colourimetric	Lysosomal activity	PLHC (topminnow fish cell line)	Uncoated zinc oxide	IC_{50} (86.07 μg/mL)	Bermejo-Nogales *et al.* (2017)
LDH	In-vitro colourimetric	Membrane stability	WAG cell line (Gills of Wallago Attu)	Titanium oxide Zinc oxide	IC_{50} (35.06 mg/L) 5.57 mg/L	Dubey *et al.* (2015)
DCFDA/DHE	In-vitro fluorometric	ROS generation	Intestine cell line of rainbow trout RTgutGC	Silver	No significant change	Chanda *et al.* (2021)
Trypan blue	Dye-exclusion test	Live/Dead assessment	Zebra fish liver cell line	Titanium oxide	Decreased viability	Siqueira *et al.* (2021)
Alamar blue		Metabolic impairment	Fish cell line from six species	Silicon oxide	Reduced cell viability 50% at greater than 100 μg/mL	Vo *et al.* (2014)
PI uptake	Flow cytometry	Live/dead assessment	HACAT cell line	Cobalt oxide nanoparticle	Decrease in the viability of cell line	Mauro *et al.* (2015)
Annexin V	Flow cytometry	Apoptotic evaluation	A549 cell line	Silver nanoparticle	Increase in apoptotic cell%	Foldbjerg *et al.* (2011)
Chromosomal aberration	Microscopy	Chromosomal abnormality and damages	Medaka fish cells	Silver nanoparticle	Damage metaphase and increase in the aberration %	Wise *et al.* (2010)
Micronucleus assay	Microscopy/ flow cytometer	Nuclear damage	Rainbow trout cell line	Titanium oxide nanoparticle	Induction of cytogenetic damage	Vevers and Jha (2008)
HPRT assay	Microscopy	Mutations of the hypoxanthine-guanine phosphoribosyl transferase (HPRT) gene	V-79 cell line	Titanium oxide nanoparticle	Increase in mutant frequency	Jain *et al.* (2017)
Ames test	Number of revertants in culture plate (his+ and his−)	Short-term bacterial reverse mutation assay	NA	SiO_2	Found non-toxic at test concentration	Kwon *et al.* (2014)
Comet assay	Alkaline comet assay for	All types of DNA damage	V79 cell line	Air particulate matters	Significant DNA damage	Dubey *et al.* (2022)
	Neutral comet assay	Double-strands break	Catfish erythrocytes	Silver nanoparticle	Severe DNA damage	Naguib *et al.* (2022)
	Fpg modified comet assay	Oxidative and alkylated DNA damage	PLHC-1 fish cell line	NA	Increase in multiple fold	Kienzler *et al.* (2012)

10.6 FACTORS CONTRIBUTING TOWARDS TOXICITY ENHANCEMENT

The three D's (dosage, dimension, and durability) are thought to be the most essential characteristics in determining the toxic consequences of nanoparticles (Oberdörster, 2002). Recent reports, however, demonstrate varying connections between several physicochemical features of nanoparticles and the related health impacts, raising some doubts about which factors are more relevant in determining their toxicity: aggregation, mass, quantity, size, bulk, surface chemistry, or all of the above. The following section highlights the most relevant factors of nanomaterial toxicity.

10.6.1 Dose-dependent toxicity

The amount or quantity of material that reaches a biological system is characterized as a dose. The dosage is proportional to the period of contact multiplied by the concentration of the NPs in the relevant medium. In general, the negative health impacts of nanoparticles do not appear to be related to nanoparticle mass doses. Surface area is the metric that appears to be related to the effects (Oberdoester, 2010).

10.6.2 Size-dependent toxicity

Toxicological investigations have shown that tiny nanoparticles (100 nm) have a negative impact on lung health, generating more inflammation than bigger particles generated from the same substance. According to studies, the same particles with varying sizes and the same crystalline structure caused a chronically high inflammatory response in the lungs when compared to bigger particles (Donaldson & Stone, 2003). This led to the conclusion that the inflammatory impact of nanoparticles may be shaped by their size and surface area.

10.6.3 Surface coating and functionalization-dependent toxicity

Particle surface toxicity is important because it comes in contact with cells and biological substances. Surfactants can significantly alter the physicochemical features of nanomaterials, such as magnetic, electric, optical, and chemical reactivity, hence influencing their lethality. Surface coatings can make noxious particles non-toxic while making innocuous particles extremely poisonous. On nanoparticle surfaces, the presence of oxygen, ozone, oxygen radicals, and transition metals causes the formation of reactive oxygen species and the production of inflammation. Despite being absorbed, spherical gold nanoparticles with varied surface coatings are not harmful to human cells (Connor *et al.*, 2005).

10.7 GREENER ALTERNATIVES TOWARDS REDUCTION OF NON-TARGET TOXICITY

Nanomaterial synthesis demands special attention, and as such, green synthetic routes involve various microbial synthesis routes, plant extracts, polymeric precursor-assisted methods, and many other methods, including microwave treatment, sonochemical have gained interest of the researchers during the past decades as a measure to follow the principle of sustainable development. Biocompatible polymers are used as reducing, capping, and stabilizing agents and logically control the diameter of nanomaterials (Kar *et al.*, 2022). Multiple bacterial, fungal, and algal species have been investigated in numerous rigorous efforts towards synthesizing various nanomaterials. These methods of synthesis are not only inexpensive but also less arduous, less time-consuming, simple, and, most significantly, non-toxic. Plants are home to a variety of polyphenols, which act both as reducing as well as capping agents and convert metal ions to nanoparticles in a single step (Bala *et al.*, 2015).

10.8 CHALLENGES, FUTURE OUTLOOK, AND CONCLUSION

Nanomaterials are emerging as a new class of water pollutants that can be hazardous to aquatic and human lives. However, these are still the subject of preliminary ecotoxicological research. Research on nanomaterials toxicity is still evolving because studies are largely limited to controlled tests. Since the properties of nanomaterials change depending on their environment, such as the presence of organic matter, water salinity, pH, temperature, and so on, it is hard to evaluate the effect of nanomaterials in a free environment. Moreover, studies on ecological destiny, transport, and health consequences are essential for defining exposure scenarios. Similarly, robust methods and procedures specifying the actual biological conditions should be developed, taking into consideration the fate, behaviour, and potential effects on the aggregation state of nanomaterials in the aquatic environment. Additionally, the aggregation of nanomaterials in water can cause other organic or inorganic substances to get attached and thereby alter bioavailability, posing new toxicological challenges. Altogether, it could be concluded that although nanomaterials are widely used in almost every aspect of human life, there exist several reports about the negative effects of these materials on other biological systems. These reports need to be addressed. The toxicity of nanomaterials on aquatic animals is primarily influenced by the size of the nanoparticles and the concentration or dosages of exposure time, in addition to physicochemical features.

REFERENCES

Ama O. M. and Ray S. S. (2020). Nanostructured Metal-Oxide Electrode Materials for Water Purification. Springer, https://doi.org/10.1007/978-3-030-43346-8

Ašmonaitė G., Boyer S., de Souza K. B., Wassmur B. and Sturve J. (2016). Behavioural toxicity assessment of silver ions and nanoparticles on zebrafish using a locomotion profiling approach. *Aquatic Toxicology*, **173**, 143–153, https://doi.org/10.1016/j.aquatox.2016.01.013

Bala N., Saha S., Chakraborty M., Maiti M., Das S., Basu R. and Nandy P. (2015). Green synthesis of zinc oxide nanoparticles using *Hibiscus subdariffa* leaf extract: effect of temperature on synthesis, anti-bacterial activity and anti-diabetic activity. *RSC Advances*, **5**(7), 4993–5003, https://doi.org/10.1039/C4RA12784F

Bermejo-Nogales A., Fernández-Cruz M. and Navas J. (2017). Fish cell lines as a tool for the ecotoxicity assessment and ranking of engineered nanomaterials. *Regulatory Toxicology and Pharmacology*, **90**, 297–307, https://doi.org/10.1016/j.yrtph.2017.09.029

Bour A., Mouchet F., Cadarsi S., Silvestre J., Baqué D., Gauthier L. and Pinelli E. (2017). CeO_2 nanoparticle fate in environmental conditions and toxicity on a freshwater predator species: a microcosm study. *Environmental Science and Pollution Research*, **24**(20), 17081–17089, https://doi.org/10.1007/s11356-017-9346-1

Cachat J., Stewart A., Grossman L., Gaikwad S., Kadri F., Chung K. M., Wu N., Wong K., Roy S. and Suciu C. (2010). Measuring behavioural and endocrine responses to novelty stress in adult zebrafish. *Nature Protocols*, **5**(11), 1786–1799, https://doi.org/10.1038/nprot.2010.140

Callaghan N. I. and MacCormack T. J. (2017). Ecophysiological perspectives on engineered nanomaterial toxicity in fish and crustaceans. *Comparative Biochemistry and Physiology Part C: Toxicology & Pharmacology*, **193**, 30–41, https://doi.org/10.1016/j.cbpc.2016.12.007

Chanda D., Dudefoi W., Anadu J. and Minghetti M. (2021). Evaluation of the effect of silver and silver nanoparticles on the function of selenoproteins using an in-vitro model of the fish intestine: the cell line RTgutGC. *Ecotoxicology and Environmental Safety*, **211**, 111930, https://doi.org/10.1016/j.ecoenv.2021.111930

Colombo V., Galli S., Choi H. J., Han G. D., Maspero A., Palmisano G., Masciocchi N. and Long J. R. (2011). High thermal and chemical stability in pyrazolate-bridged metal–organic frameworks with exposed metal sites. *Chemical Science*, **2**(7), 1311–1319, https://doi.org/10.1039/c1sc00136a

Connor E. E., Mwamuka J., Gole A., Murphy C. J. and Wyatt M. D. (2005). Gold nanoparticles are taken up by human cells but do not cause acute cytotoxicity. *Small*, **1**(3), 325–327, https://doi.org/10.1002/smll.200400093

De Matteis V. (2017). Exposure to inorganic nanoparticles: routes of entry, immune response, biodistribution and in vitro/in vivo toxicity evaluation. *Toxics*, 5(4), 29, https://doi.org/10.3390/toxics5040029

Donaldson K. and Stone V. (2003). Current hypotheses on the mechanisms of toxicity of ultrafine particles. *Annali dell'Istituto Superiore di Sanitã*, **39**(3), 405–410.

Dubey A., Goswami M., Yadav K. and Chaudhary D. (2015). Oxidative stress and nano-toxicity induced by TiO_2 and ZnO on WAG cell line. *PloS One*, **10**(5), e0127493, https://doi.org/10.1371/journal.pone.0127493

Dubey K., Maurya R., Mourya D. and Pandey A. K. (2022). Physicochemical characterization and oxidative potential of size fractionated particulate matter: uptake, genotoxicity and mutagenicity in V-79 cells. *Ecotoxicology and Environmental Safety*, **247**, 114205, https://doi.org/10.1016/j.ecoenv.2022.114205

Ernst L. M., Casals E., Italiani P., Boraschi D. and Puntes V. (2021). The interactions between nanoparticles and the innate immune system from a nanotechnologist perspective. *Nanomaterials*, **11**(11), 2991, https://doi.org/10.3390/nano11112991

Foldbjerg R., Dang D. A. and Autrup H. (2011). Cytotoxicity and genotoxicity of silver nanoparticles in the human lung cancer cell line, A549. *Archives of Toxicology*, **85**(7), 743–750, https://doi.org/10.1007/s00204-010-0545-5

Ishii Y., Shin-Ichiro S. M. and Hayashi S. (2020). Different factors determine 137Cs concentration factors of freshwater fish and aquatic organisms in lake and river ecosystems. *Journal of Environmental Radioactivity*, **213**, 106102, https://doi.org/10.1016/j.jenvrad.2019.106102

Jain A. K., Senapati V. A., Singh D., Dubey K., Maurya R. and Pandey A. K. (2017). Impact of anatase titanium dioxide nanoparticles on mutagenic and genotoxic response in Chinese hamster lung fibroblast cells (V-79): the role of cellular uptake. *Food and Chemical Toxicology*, **105**, 127–139, https://doi.org/10.1016/j.fct.2017.04.005

Jain A. K., Singh D., Dubey K., Maurya R., Mittal S. and Pandey A. K. (2018). Models and methods for in vitro toxicity. In: In vitro toxicology. Elsevier.

Joris F., Manshian B. B., Peynshaert K., De Smedt S. C., Braeckmans K. and Soenen S. J. (2013). Assessing nanoparticle toxicity in cell-based assays: influence of cell culture parameters and optimized models for bridging the in vitro–in vivo gap. *Chemical Society Reviews*, **42**(21), 8339–8359, https://doi.org/10.1039/c3cs60145e

Kalichak F., de Alcantara Barcellos H. H., Idalencio R., Koakoski G., Soares S. M., Pompermaier A., Rossini M. and Barcellos L. J. G. (2019). Persistent and transgenerational effects of risperidone in zebrafish. *Environmental Science and Pollution Research*, **26**(25), 26293–26303, https://doi.org/10.1007/s11356-019-05890-9

Kar A. K., Singh A., Singh D., Shraogi N., Verma R., Saji J., Jagdale P., Ghosh D. and Patnaik S. (2022). Biopolymeric composite hydrogel loaded with silver NPs and epigallocatechin gallate (EGCG) effectively manages ROS for rapid wound healing in type II diabetic wounds. *International Journal of Biological Macromolecules*, **218**, 506–518, https://doi.org/10.1016/j.ijbiomac.2022.06.196

Kienzler A., Tronchère X., Devaux A. and Bony S. (2012). Assessment of RTG-W1, RTL-W1 and PLHC-1 fish cell lines for genotoxicity testing of environmental pollutants by means of an Fpg-modified comet assay. *Toxicology in Vitro*, **26**(3), 500–510, https://doi.org/10.1016/j.tiv.2012.01.001

Kongseng S., Yoovathaworn K., Wongprasert K., Chunhabundit R., Sukwong P. and Pissuwan D. (2016). Cytotoxic and inflammatory responses of TiO_2 nanoparticles on human peripheral blood mononuclear cells. *Journal of Applied Toxicology*, **36**(10), 1364–1373, https://doi.org/10.1002/jat.3342

Kwon J. Y., Kim H. L., Lee J. Y., Ju Y. H., Kim J. S., Kang S. H., Kim Y.-R., Lee J.-K., Jeong J. and Kim M.-K. (2014). Undetectable levels of genotoxicity of SiO_2 nanoparticles in in vitro and in vivo tests. *International Journal of Nanomedicine*, **9**(Suppl 2), 173.

Mauro M., Crosera M., Pelin M., Florio C., Bellomo F., Adami G., Apostoli P., De Palma G., Bovenzi M. and Campanini M. (2015). Cobalt oxide nanoparticles: behavior towards intact and impaired human skin and keratinocytes toxicity. *International Journal of Environmental Research and Public Health*, **12**(7), 8263–8280, https://doi.org/10.3390/ijerph120708263

Mehwish N., Kausar A. and Siddiq M. (2014). Advances in polymer-based nanostructured membranes for water treatment. *Polymer-Plastics Technology and Engineering*, **53**(12), 1290–1316, https://doi.org/10.1080/03602559.2014.909465

Miller R. J., Lenihan H. S., Muller E. B., Tseng N., Hanna S. K. and Keller A. A. (2010). Impacts of metal oxide nanoparticles on marine phytoplankton. *Environmental Science & Technology*, **44**(19), 7329–7334, https://doi.org/10.1021/es100247x

Mukherjee S., Gautam A., Pal K., Karmakar P., Ray M. and Ray S. (2021). Copper oxide nanoparticle and copper sulfate induced impairment of innate immune parameters in a common Indian sponge. *Journal of Hazardous Materials Letters*, **2**, 100036, https://doi.org/10.1016/j.hazl.2021.100036

Nadesh R., Narayanan D., Sreerekha P. R., Vadakumpully S., Mony U., Koyakkutty M., Nair S. V. and Menon D. (2013). Hematotoxicological analysis of surface-modified and-unmodified chitosan nanoparticles. *Journal of Biomedical Materials Research Part A*, 101(10), 2957–2966, https://doi.org/10.1002/jbm.a.34591

Nafisi S. and Maibach H. I. (2018). Skin penetration of nanoparticles. In: Emerging Nanotechnologies in Immunology, edited by R. Shegokar and E. B. Souto (eds), Elsevier. https://doi.org/10.1016/B978-0-323-40016-9.00003-8

Naguib M., Mekkawy I. A., Mahmoud U. M. and Sayed A. E.-D. H. (2022). Genotoxic evaluation of silver nanoparticles in catfish *Clarias gariepinus* erythrocytes; DNA strand breakage using comet assay. *Scientific African*, **16**, e01260, https://doi.org/10.1016/j.sciaf.2022.e01260

Oberdoester G. (2010). Nanotoxicology: an emerging discipline evolving from studies of ultrafine particles (vol 113, pg 823, 2005). *Environmental Health Perspectives*, **118**(9), A380–A380.

Oberdörster, G. (2002). Toxicokinetics and effects of fibrous and nonfibrous particles. *Inhalation Toxicology*, **14**(1), 29–56, https://doi.org/10.1080/089583701753338622

Remya A. S., Ramesh M., Saravanan M., Poopal R. K., Bharathi S. and Nataraj D. (2015). Iron oxide nanoparticles to an Indian major carp, *Labeo rohita*: impacts on hematology, iono regulation and gill Na+/K+ ATPase activity. *Journal of King Saud University-Science*, **27**(2), 151–160, https://doi.org/10.1016/j.jksus.2014.11.002

Remya N., Syama S., Sabareeswaran A. and Mohanan P. (2016). Toxicity, toxicokinetics and biodistribution of dextran stabilized Iron oxide Nanoparticles for biomedical applications. *International Journal of Pharmaceutics*, **511**(1), 586–598, https://doi.org/10.1016/j.ijpharm.2016.06.119

Sajid M., Ilyas M., Basheer C., Tariq M., Daud M., Baig N. and Shehzad F. (2015). Impact of nanoparticles on human and environment: review of toxicity factors, exposures, control strategies, and future prospects. *Environmental Science and Pollution Research*, **22**(6), 4122–4143, https://doi.org/10.1007/s11356-014-3994-1

Singh P., Pandit S., Mokkapati V., Garg A., Ravikumar V. and Mijakovic I. (2018). Gold nanoparticles in diagnostics and therapeutics for human cancer. *International Journal of Molecular Sciences*, **19**(7), 1979, https://doi.org/10.3390/ijms19071979

Siqueira P., do Carmo T. L. L., Bonomo M. M., dos Santos F. A. and Fernandes M. N. (2021). Proliferative response avoids mutagenic effects of titanium dioxide (TiO_2) nanoparticles in a zebrafish hepatocyte cell line. *Journal of Hazardous Materials Advances*, **4**, 100036, https://doi.org/10.1016/j.hazadv.2021.100036

Sudha P. N., Sangeetha K., Vijayalakshmi K. and Barhoum A. (2018). Nanomaterials history, classification, unique properties, production and market. In: Emerging Applications of Nanoparticles and Architecture Nanostructures, A. Barhoum and A. S. H. Makhlouf (eds), Elsevier, pp. 341–384, https://doi.org/10.1016/C2016-0-01906-X

Taju G., Majeed S. A., Nambi K. and Hameed A. S. (2014). In vitro assay for the toxicity of silver nanoparticles using heart and gill cell lines of *Catla catla* and gill cell line of *Labeo rohita*. *Comparative Biochemistry and Physiology Part C: Toxicology & Pharmacology*, **161**, 41–52, https://doi.org/10.1016/j.cbpc.2014.01.007

Teng M., Zhao X., Zhao L., Shi D., Li Y., Sun J., Zhao W., Zhou L., Wang X. and Giesy J. P. (2022). Zebrafish (*Danio rerio*) reproduction is affected by life-cycle exposure to differently charged polystyrene nanoplastics with sex-specific responses. *ACS ES&T Water*, **2**(12), 2558–2566, https://doi.org/10.1021/acsestwater.2c00348

Tiede K., Hanssen S. F., Westerhoff P., Fern G. J., Hankin S. M., Aitken R. J., Chaudhry Q. and Boxall A. B. (2016). How important is drinking water exposure for the risks of engineered nanoparticles to consumers? *Nanotoxicology*, **10**(1), 102–110.

Turan N. B., Erkan H. S., Engin G. O. and Bilgili M. S. (2019). Nanoparticles in the aquatic environment: usage, properties, transformation and toxicity—a review. *Process Safety and Environmental Protection*, **130**, 238–249, https://doi.org/10.1016/j.psep.2019.08.014

Vevers W. F. and Jha A. N. (2008). Genotoxic and cytotoxic potential of titanium dioxide (TiO_2) nanoparticles on fish cells in vitro. *Ecotoxicology*, **17**(5), 410–420, https://doi.org/10.1007/s10646-008-0226-9

Vo N. T., Bufalino M. R., Hartlen K. D., Kitaev V. and Lee L. E. (2014). Cytotoxicity evaluation of silica nanoparticles using fish cell lines. *In Vitro Cellular & Developmental Biology-Animal*, **50**(5), 427–438, https://doi.org/10.1007/s11626-013-9720-3

Volkov Y., McIntyre J. and Prina-Mello A. (2017). Graphene toxicity as a double-edged sword of risks and exploitable opportunities: a critical analysis of the most recent trends and developments. *2D Materials*, **4**(2), 022001, https://doi.org/10.1088/2053-1583/aa5476

Wise J. P., Sr, Goodale B. C., Wise S. S., Craig G. A., Pongan A. F., Walter R. B., Thompson W. D., Ng A.-K., Aboueissa A.-M. and Mitani H. (2010). Silver nanospheres are cytotoxic and genotoxic to fish cells. *Aquatic Toxicology*, **97**(1), 34–41, https://doi.org/10.1016/j.aquatox.2009.11.016

Zheng Y. and Nowack B. (2022). Meta-analysis of bioaccumulation data for nondissolvable engineered nanomaterials in freshwater aquatic organisms. *Environmental Toxicology and Chemistry*, **41**(5), 1202–1214, https://doi.org/10.1002/etc.5312

Zhu X., Chang Y. and Chen Y. (2010). Toxicity and bioaccumulation of TiO_2 nanoparticle aggregates in *Daphnia magna*. *Chemosphere*, **78**(3), 209–215, https://doi.org/10.1016/j.chemosphere.2009.11.013

Section 3

New Technologies for Water and Wastewater Treatment

INTRODUCTION

The global population is increasing steadily and is expected to be about 8.2 billion by 2025. To meet the requirements of the swelling consumer society, industrial activities and associated waste generation are increasing rapidly. Water consumption and used water generation have also gone up significantly. However, especially in the Global South, water resources are dwindling both in terms of quality and quantity. Only a fraction of the wastewater generated is getting treated for various reasons. A water-secure ecosystem needs to be prepared to address the impending challenges linked to water stress.

Though many technologies are available for water and wastewater treatment, there is a need to develop more efficient and affordable technologies, to meet the growing demands in developing countries. The increasing occurrence of various emerging pollutants in water environments is driving the development of new technologies capable of abating such pollutants. As the used water quantity is increasing and most of it is going untreated in emerging economies, it is advisable to adopt low-cost nature-based technologies instead of capital and energy-intensive technologies. Recovery and reuse of water, nutrients, carbon, etc., from wastewater are essential to achieve circularity. In the context of circularity in waste management, one must also investigate such circular systems' sustainability.

The seven chapters of Section 3 present the new technologies developed for water and wastewater treatment in recent years. The chapter on **New technologies for drinking water** focuses on the application of nanotechnologies to produce sustainable and affordable clean water. Recalcitrant and hazardous pollutants pose a severe challenge to conventional water treatment technologies. The chapter on **Pulse power technology for the removal of emerging contaminants from water and wastewater** gives an overview of the possible application of pulse power plasma technology (PPT) for efficient and rapid degradation of a wide range of emerging water pollutants, including dyes, pesticides, toxic solvents, pharmaceuticals, and so on for the disinfection of water and wastewater. Nature has a significant capability to abate many of the pollutants. The chapter on **Application of engineered natural treatment systems for the pollution abatement** focuses on emerging contaminants in various wastewater (greywater, blackwater, and domestic wastewater) and their fate in the CWs. It also attempts to discuss the factors affecting the removal of pharmaceuticals and personnel care products (PPCPs), such as flow configurations of CWs, choice of substrate materials and plant species, and operating conditions. Carbon is a unique material with extensive use in various fields of science and engineering. Purification of water and wastewater is an early application of carbon, and now it has become an integral part of most household and community-based water treatment units. The chapter on **Carbon-based filters for water and wastewater treatment** provides a review of the evolution and application of carbon-based materials in water and wastewater treatment. The synthesis, removal mechanism, and factors affecting purification by carbon filters, along with the challenges and prospects of carbon-based filtration technology, are also briefly discussed in this chapter.

The focus in the wastewater management sector has considerably shifted from conventional treatment processes to resource and nutrient recovery approaches in recent years to promote a circular economy. The recovery of nutrients such as nitrogen, phosphorus, and carbon from wastewater is a sustainable approach to wastewater management and increases ecological and economic sustainability. The chapter on **Nutrient recovery from wastewater for circular economy** provides a comprehensive overview of existing conventional technologies used to recover nutrients from different nutrient-rich wastewater generated from domestic, industrial, and agricultural sources from anaerobic digestate. The chapter on **Water pollution abatement using waste-derived materials: a sustainable approach** provides a holistic approach to solid waste management by generating value-added materials and their potential application for water pollution abatement. High-performance water and wastewater treatment has become feasible with the development of innovative and advanced technologies, with potential impact on health and the environment. At the same time, they can help protect the environment and meet specific social needs in a more sustainable way. Hence, there is a need to select sustainable technologies. The chapter on **Technology evaluation for sustainability** discusses the need of evaluating the sustainability of such technology and processes, ways to conduct such studies, and examples of sustainable technologies employed in India. In summary, this section focuses on the important aspects and concepts that form the basis for emerging technologies for water and wastewater treatment.

doi: 10.2166/9781789063714_0123

Chapter 11

New technologies for drinking water

*Ankit Nagar, Md Rabiul Islam and Thalappil Pradeep**

DST Unit of Nanoscience (DST UNS), and Thematic Unit of Excellence (TUE), Department of Chemistry, Indian Institute of Technology Madras, Chennai, India
**Corresponding author: pradeep@iitm.ac.in*

ABSTRACT

The global population is expected to be about 8.2 billion by 2025. This suggests a dire need to prepare for the ongoing and impending challenges linked with water stress to create a water-secure ecosystem. Densely populated countries such as India and China are enhancing the production of semiconductor, petrochemical, automotive, chemical, and pharmaceutical products, which are materially enriching. Simultaneously, our quickly diminishing resources are being further stressed by unprocessed industrial waste. Therefore, to build a sustainable and hygienic livelihood for all, it is imperative to be able to remove conventional and emerging pollutants from water in an affordable and energy-efficient manner. Nanotechnology-based solutions have made substantial contributions towards offering sustainable solutions in the clean water space. Although nanotechnology-based water purification has been an active research area for decades, it is yet to occupy a significant market share because of irregularities in the context of sensitivity, efficient operation in the field, and higher cost in comparison to traditional technologies. This chapter will highlight emerging technologies for drinking water which have also reached society in the form of easily operable devices, such as nanomaterial-incorporated handpumps, desalination devices, and atmospheric water generators.

Keywords: water harvesting, water purification, membranes, capacitive deionization, commercialization

11.1 INTRODUCTION

Due to the growing population, climate change, decline in water quality, and inadequate management, there is a significant imbalance between the demand and availability of freshwater globally. There are three categories of water pollutants: organic, inorganic, and biological (Khalil *et al.*, 2016). Many of them are hazardous to humans and ecology. As an example, arsenic is among the most toxic elements, as it is prevalent in excess of its acceptable limits in numerous places. Heavy metals like mercury, uranium, lead, chromium, copper, nickel, cadmium, and zinc are also extremely hazardous. In addition, high levels of phosphates, selenides, chromates, nitrates, sulphates, and oxalates are harmful and may affect the taste of water. Toxic organic pollutants include detergents, hydrocarbons, phenols, pesticides, organochlorine chemicals, and phthalates. Emerging contaminants include non-biodegradable pharmaceutical and personal care products (PPCPs) (Carballa *et al.*, 2007; Eregowda & Mohapatra, 2020; Kunduru *et al.*, 2017). Their sources are domestic and hospital waters containing pharmaceuticals, their packaging, and detergent additives. PPCPs in water are found in the concentration range of 0.001 to 1 ppb (Kunduru *et al.*, 2017). Consequently, typical treatment technologies employed in wastewater treatment facilities are ineffective against such contaminants. The combustion of fossil fuels also generates polycyclic aromatic hydrocarbons (PAHs), which are another category of water contaminants. Due to their limited solubility and high affinity for particles, their concentration in water is low. The PAHs found in high amounts in drinking water are pyrene, phenanthrene, fluranthene, and anthracene (Mojiri *et al.*, 2019). Often, faeces are a source of microbial contamination of water. Water treatment plants, manufacturing facilities, and hospitals are the root drivers of such contamination. It has been demonstrated that urban activities increase the concentration of infections. Natural organic matter (NOM) is among the most important factors to comprehend water quality and select the treatment method in such circumstances.

The rapidly rising global population suggests a dire need to prepare for the ongoing and impending challenges linked with water stress to create a water-secure ecosystem. Densely populated countries like India and China are enhancing the production of semiconductor, petrochemical, automotive, chemical, and pharmaceutical products, which are materially enriching. Simultaneously, our quickly diminishing resources are being further stressed by unprocessed industrial waste. Therefore, to build a sustainable and hygienic livelihood for all, it is imperative to be able to remove conventional and emerging pollutants from water in an affordable and energy-efficient manner.

To ensure good public health and hygiene, it is important to remove both traditional and emerging toxins from water at a reasonable cost and with low energy input, ideally with no negative environmental impact. Conventional purification techniques include adsorption, filtration, ultraviolet irradiation, desalination, disinfection, and coagulation, whereas emerging nanotechnology-enabled purification techniques involve the use of carbon nanocomposites, metal or metal oxide nanoparticles (NPs), nanoporous ceramics, and other nanomaterials.

Although nanotechnology-based water purification has been an active research area for decades, it is yet to occupy a significant market share because of irregularities in the context of sensitivity, efficient operation in the field, and higher cost in comparison to traditional technologies. This chapter will highlight emerging technologies for purifying drinking water which have also reached the society in the form of easily operable devices, such as nanomaterial-incorporated handpumps, desalination devices, and atmospheric water generators (AWGs).

11.2 ADSORPTION-BASED PURIFICATION TECHNOLOGIES

The removal capacity of adsorbents is dependent on their density, porosity, external and internal surface area, pore-size distribution, surface structure, functionalization, and process variables such as temperature, pH, and contact time (Teodosiu *et al.*, 2018). Engineered nanomaterials (ENMs) have been used to generate diverse organic and inorganic adsorbents for the efficient remediation of environmental pollutants with varying molecular sizes, hydrophobicity, and speciation behaviour (Qu *et al.*, 2013). These ENMs consist of nanoscale metal and metal oxides of Fe, Ti, Ag, Zr, and Zn; carbonaceous materials such as carbon nanotubes and their derivatives, fullerenes, graphenic materials, and activated carbon; and magnetic nanoparticles involving Fe, Ni, Co, their oxides and alloys, and core–shell nanoparticles (Teodosiu *et al.*, 2018; van der Hoek *et al.*, 2014).

Composite materials are prepared by combining a matrix (the continuous phase) with reinforced materials (dispersed). Nanocomposites consist of nanoparticles integrated with a variety of functionalized materials, such as multiwall CNTs (MWCNTs), activated carbon, graphene oxide, polymeric or biopolymer fibres, clays, zeolites, and so on (Peng *et al.*, 2020). Their unique characteristics, such as high durability, high stiffness, high strength, corrosion resistance, low density, and heat resistance prove them advantageous over other materials. For instance, Al_2O_3/MWCNTs composite has been identified to remove Cd^{2+} and trichloroethylene from polluted water, whilst TiO_2-zeolite nanocomposites are known to remove industrial dyes from wastewater (Kamali *et al.*, 2019; Saini, 2015). AMRIT, which stands for Anion and Metal Removal by Indian Technology, is supplying arsenic-free water to more than 1.2 million people per day; it contains innovative nanostructured materials with high adsorption capacities for arsenite and arsenate ions in water in the field (Nagar & Pradeep, 2020). The composite's inherent structure with metastable 2-line ferrihydrite enclosed in chitosan biopolymer cages enables the formation of efficient adsorption sites, which are responsible for the composite's unprecedentedly high capacity (Kumar *et al.*, 2017). Similarly, nanochemistry is utilized by filters comprised of noble metal nanoparticle-coated metal oxides to remove pesticides from drinking water (Nair & Pradeep, 2003; Nair *et al.*, 2003). This technique had already reached more than 7.5 million people till 2016, the most recent year for which implementation data was obtained. It also led to a decrease in pesticide levels in certain regions of India from over 20 times above the safety standard to significantly below the prescribed limit (Sreeprasad *et al.*, 2011). In 2013, Sankar *et al.* developed an antibacterial composition comprised of a composite of aluminium oxyhydroxide and chitosan containing silver particles with a diameter of 10–20 nm (Sankar *et al.*, 2013). These solutions do not require energy and are affordable even in the poorest regions of the globe. The sustainability of sorbents was improved by constructing composites with reinforced micro- and nanocellulose, including active nanoscale adsorption sites that demonstrated greater absorption and excellent mechanical stability (Mukherjee *et al.*, 2018, 2019).

11.3 MEMBRANES

In comparison to traditional purification technologies such as coagulation and flocculation, membrane filtration is comparatively more robust and diverse, and provides crucial benefits in water and wastewater treatment applications. For instance, nano-TiO_2 filtration membranes do not undergo fouling and provide high filtration efficiency at high water flux (DelaiáSun, 2010). Nanoporous zeolites, nanoporous polymers, and attapulgite clays have a binding capacity 100,000 times more than that of typical activated carbon (Singer *et al.*, 2005). Pure silica particles coated with an active material are significantly more cost-effective at removing harmful compounds and microorganisms (Majewski & Chan, 2008). Conventional water treatment plants have been able to effectively remove biological pollutants with CNT-based technology, hence resolving the issue of maintaining the biostability of treated water in the distribution lines (Upadhyayula *et al.*, 2009). These membranes have been identified as promising replacements for RO membranes in the context of desalination. The large-scale fabrication of CNT membranes is becoming more cost-effective, and their entry into the desalination market is anticipated within the next five to six years, as scaling-related challenges are presently being addressed (Technology and Action for Rural Advancement (TARA), 2012). Numerous methods have been investigated as potential substitutes for traditional RO membranes in desalination and water recycling applications. Nanocomposite membranes have 1–3 times higher water permeability but the same rejection as commercial RO membranes, and the former are multifunctional with antibacterial and photoreactive properties (Technology and Action for Rural Advancement (TARA), 2012). The nanotechnology-based filtration devices are inexpensive, transportable, user-friendly, extremely effective, and have exceptionally large surface areas. Commercial applications of nanomaterials in the water industry include MoS_2 nanosheets for disinfection, CNF films for antimicrobial activity, silver nanobrushes for humidity harvesting, and ferrihydrite for heavy metal removal (Nagar & Pradeep, 2020). It is anticipated that the manufacturing of nanomaterials and their composites will increase, due to the growing need to generate clean water in a time- and energy-efficient manner.

11.4 CAPACITIVE DEIONIZATION

Capacitive deionization (CDI) is an emerging technology that involves the adsorption and desorption of ions on an electrode's surface by applying a DC low potential energy (1.2–1.8 V) across a pair of porous carbon electrodes. It is thus both cost-efficient and energy-efficient compared to other existing desalination methods. When a flowing water stream is passed across a CDI system, anions and cations are transported towards oppositely charged electrodes and get adsorbed on them. In this manner, deionized and 'potable' water is generated from saline or brackish water. The adsorbed ions are subsequently removed from the electrode by reversing the polarity, thereby regenerating the electrode surface for further reuse (Porada *et al.*, 2013). Thus, clean water can be generated by repeating the adsorption and desorption cycles continuously.

Numerous carbonaceous materials and their composites have been utilized as CDI electrodes due to their high surface area, suitable pore size, and high salt adsorption capacities (Porada *et al.*, 2013). Different graphenic materials, including activated carbon (AC), activated carbon nanofibre (ACF), graphene-like nanoflakes, reduced graphene oxide (rGO), carbon nanotubes (CNT), graphene–CNT composites, reduced graphene oxide-ACF, 3D-graphenic materials, sponge-templated graphene, graphene–Fe_3O_4, graphene chitosan-Mn_3O_4, rGO-activated carbon (AC) composites, functionalized graphene nanocomposite, and so on, have been used as CDI electrodes (Figure 11.1) (Islam *et al.*, 2021; Porada *et al.*, 2013). The adsorption capacities of the graphenic composites CO_2-activated rGO, graphene/carbon nanotubes, sulphonic functional graphite nanosheets, MgAl-Ox/graphene nanohybrids, SO_3H/NH_2 functionalized-graphene/AC, graphene sponges, and 3D-graphenic material were found to be 6.26, 1.4, 8.6, 13.6, 10.3, 14.9, and 14.7 mg/g, respectively, when given an input of 500 ppm NaCl solution (Gupta *et al.*, 2019; Islam *et al.*, 2021; Porada *et al.*, 2013). Emerging CDI electrode materials (i.e., activated carbon, ordered mesoporous carbon, carbon aerogel, other carbon

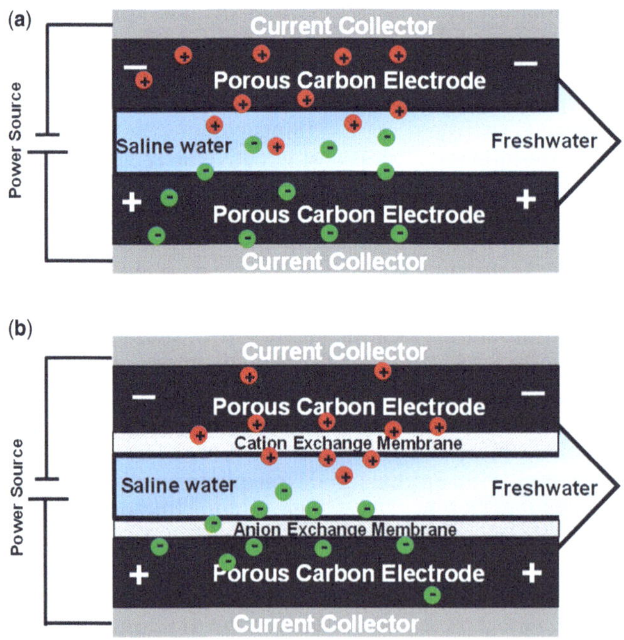

Figure 11.1 A schematic showing the working principle of (a) CDI and (b) MCDI during the electro-adsorption process. (*Source*: Modified and redrawn from Porada *et al.*, 2013).

derivatives, carbon nanotubes, graphene and graphene-based composites, and other two-dimensional materials such as MXenes, MOFs, COFs, MoS_2, etc.) have recently been studied in laboratories all over the world.

The main objective of CDI technology is to implement energy- and cost-efficient processes for water purification. Thus, the major tasks for the advancement of CDI technology are to improve the electrode architecture and synthesize suitable new electrode materials. The desalination capacity of the electrodes depends upon their material and the initial salt concentration. Mostly, carbon materials and other modified carbon materials integrated with novel materials have been used in electrode fabrication to enhance performance (Gupta *et al.*, 2019; Islam *et al.*, 2021). Electrode materials should have high electrical conductivity, high electrochemical stability over different pH, high specific surface area, and suitable voltage ranges (~1.2–1.8 V). Additionally, the pore size, total pore volume, and pore connectivity of the active electrode materials are the main parameters that affect electro-adsorption capacity. Furthermore, other electrode parameters, such as exceptional wetting behaviour, high bio-inertness to prevent bio-fouling, electrode stacks, input water flow rate, electrode characteristics, the lifetime of the electrodes, the process's cost-efficiency, and suitable porosity of the electrode materials, are significant factors to be considered while designing an efficient CDI technology (Gupta *et al.*, 2019; Porada *et al.*, 2013).

CDI has been extensively explored for brackish water desalination because it is considered to be energetically competitive with reverse osmosis (RO). The quanta of energy consumed by different CDI are less than those of a comparable RO in identical conditions. Energy consumed by MCDI could be lower than that of RO. It has been reported that the energy requirement of MCDI is 2–3 times lower than RO's for brackish water desalination (under the same conditions, i.e. flux (J_w) = 10.0 L/m²/h, salt rejection (R_{salt}) = 80%, water recovery (WR) = 80%, and a feed salinity of 34.22 mM (2 g/L)). It was also observed that under the same conditions, the energy consumption of MCDI (~0.4 kWh/m³) is less than that of CDI (~2.5 kWh/m³) (Porada *et al.*, 2020). Furthermore, it was seen that for these desalination conditions (flux (J_w) = 11.9 L/m²/h, salt rejection (R_{salt}) = 80%, water recovery (WR) = 93.5%, and a feed salinity of 40 mM), the energy consumption of RO (~0.5 kWh/m³) is higher than of MCDI (~0.4 kWh/m³) (Porada *et al.*, 2020). The cost of desalination using our CDI prototype was calculated to be approximately 3–4 paisa (US$0.00040 to 0.00054) per litre.

With regard to the commercialization of CDI technologies, Voltea was the first US-based recognized CDI Company founded in 2006. Ionic Engineering Technology Pvt. Ltd. in India, INNODI and NGen from India, Idropan Dell'Orto in Italy, and PowerTech Water in the United States have recently entered the CDI market. Atlantis Technologies claims that their product removes up to 99% of impurities and can reduce water salinity up to 100,000 ppm with reduced maintenance. Idropan Dell'Orto, founded by Tullio Servida, has products named Plimmer 4G Alpha and Delta, which claim to reduce total dissolved solids concentration (TDS) of 2000 μS by 70%–85% in a single pass and 85%–95% in a double pass. They have international partners including ECOWATER in the United States, EUROWATER in Denmark, AquaSphere in India, and others in Australia, Tunisia, Saudi Arabia, and Russia. INNODI is a new start-up from IIT Madras, which was established in 2017, and launched its first domestic CDI product in the market in 2017. Moreover, IIT Tirupati incubated a CDI company, NGen, which was established in 2022.

11.5 ATMOSPHERIC WATER HARVESTING

Atmospheric humidity is an abundant source of freshwater that can be easily harvested using a sustainable energy source such as solar, wind, or geothermal energy. This section provides a comprehensive and critical overview of current research and prospects in the area of atmospheric water harvesting (AWH). From an application standpoint, the greatest challenge is to build a humidity harvester that can produce sufficient freshwater across a wide range of weather conditions with minimal energy footprint. Therefore, several harvesting techniques, such as radiative cooling, solar distillation, and sorption-based water collection, are examined and explored in terms of their capture materials, system designs, and thermodynamic cycles. In addition, we comprehensively evaluate the performance of recently reported atmospheric water harvesters. We also explore four critical issues that limit cost-effectiveness and offer prospective solutions.

Over the past two decades, AWH technologies have made considerable advancements. However, broad-spectrum research on atmospheric water harvesters is limited and system integrations have been insufficiently studied. It is anticipated that additional research will address these challenges in order to facilitate efforts to translate decades of research on AWH into concrete benefits in our daily lives.

Despite the considerable value of the potentially recoverable freshwater globally, there are currently very few AWH devices in commercial operation. Any feasible AWH method must satisfy the following five key criteria: It must be efficient, inexpensive, scalable, operational in a broad range of weather conditions, and reliable enough to function for an entire year. At present, none of the commercially available AWGs meet all five of these criteria. As per thermodynamics, this is primarily due to the process's energy inefficiency.

The practise of collecting water from fog dates back millennia (Beysens & Mulumouk, 2000). Humans appear to have considered dew as a source of potable water since the beginning of time. Research on modern AWH technologies has continued, and a variety of atmospheric water-collection techniques, which are primarily utilized in arid and semi-arid regions, have been investigated. AWH technologies can be split into three distinct groups based on the type of airborne water: artificial rain collection (Bruintjes, 1999; DeFelice & Axisa, 2017; Wang *et al.*, 2016), fog water collection (Fessehaye *et al.*, 2014; Klemm *et al.*, 2012), and dew water collection (Khalil *et al.*, 2016). Weather manipulation, also referred to as cloud seeding, may produce large precipitation, but only in the troposphere, where clouds with high water content congregate. There is no proof that the same operation can be accomplished at ground level in a routine and controlled manner (Bruintjes, 1999; DeFelice & Axisa, 2017). In certain arid places, fog collecting, as opposed to weather modification, is a proven technology for obtaining considerable quantities of drinkable water. By blowing humid ambient air over a cooled surface, dew can be collected, and liquid water is produced if the surface temperature is below the dewpoint temperature of the air (Khalil *et al.*, 2016).

Fog harvesting technologies are viable and accessible, and can alleviate the dearth of potable water that often plagues arid coastal regions. Ordinarily, fog water is collected by installing a rectangular mesh orthogonal to the direction of wind flow, which captures fog droplets. When exposed to a foggy atmosphere, wind-borne water droplets push against the mesh and become trapped. The droplets grow by coalescence until they become large enough to drop by gravity, and a gutter carries the water to a storage tank. The greatest issue for fog collection today is its low efficiency, which is defined as the ratio of water that enters the collector's gutter to the liquid water flux perpendicular to the collector's mesh. The poor efficiency is primarily due to droplet pinning, which has been overcome at lab scale using superhydrophobic surfaces. However, such modified surfaces are yet to prove their durability by undergoing field tests.

While such 'passive' fog harvesters were effective, the amount of water they produced was insufficient in moderate-humidity weather. Therefore, active water harvesting was investigated, in which AWGs utilized the Peltier effect or a refrigeration cycle to lower the surrounding air's temperature below the dew point to facilitate condensation (Tu *et al.*, 2018). This permitted the collection of atmospheric water in water-scarce locations to overcome the limitations of long-distance freshwater transport and heavily polluted groundwater and surface water bodies. While several commercial devices are already available to consumers, research is ongoing to develop energy-efficient condensing surfaces for use in regions with relative humidity (RH) below 40%.

Adsorption-based harvesting has emerged as a viable alternative to systems that rely on condensation. In this technique, sorption of water vapour occurs during the night, followed by desorption in a closed chamber using sunlight in the daytime (Figure 11.2). During desorption, as soon as the enclosure

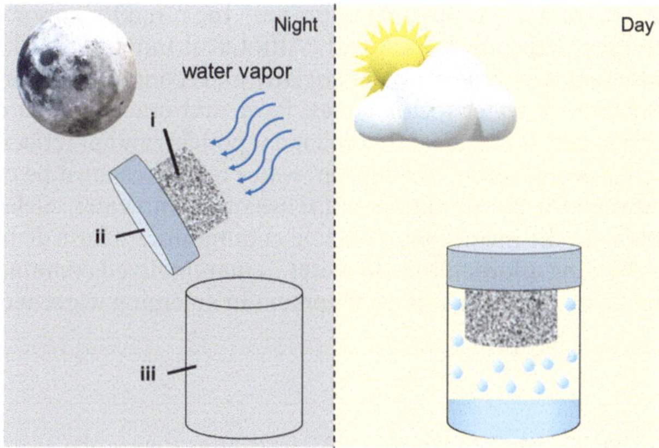

Figure 11.2 An illustration depicting the working of a sorption-based water harvesting device, wherein the material (i) integrated with the cap, (ii) of a container, (iii) is allowed to adsorb humidity at night (left panel), and desorb the humidity in the closed transparent container in the morning.

saturates with humidity, the desorbed vapour begins to condense on the chamber walls. Although typical desiccants such as zeolites and silica gel absorb humidity, their poor sorption kinetics, low absorption capacity, and high-energy requirement for regeneration prevent their practical application. Several intriguing alternatives, such as metal-organic frameworks (MOFs), have variable pore size and a high surface area. A kilogram of $Zr_6(OH)_4(fumarate)_6$, for instance, has the capacity to gather 2.8 L/day at 20% RH, thereby encouraging additional research on the development of molecular materials for water harvesting in deserts (Kim *et al.*, 2017). The incorporation of these MOF NPs into thermosensitive polymers, such as poly(N-isopropylacrylamide), has permitted harvesting at temperatures closer to the ambient temperature (Yilmaz *et al.*, 2020). Improving the absorption capacity by adding hydrophilic ligands into nanopores, minimizing the energy required for desorption, accelerating sorption kinetics, and maintaining cyclic stability for several hundreds of sorption cycles are current research priorities in the subject area.

11.6 EMERGING TECHNOLOGIES FOR WATER PURIFICATION

Carbon has been the primary component of water treatment materials since prehistoric times. In 3750 BC, Egyptians and Sumerians utilized wood charcoal to purify water (Patrick, 1995). Activated carbon has been a crucial component in water-cleaning applications since the 1940s. Carbon nanotubes (Liu *et al.*, 2013), carbon nanofibres (Shen *et al.*, 2016; Zhang *et al.*, 2022), and graphene-based materials (Jiang *et al.*, 2016; Li *et al.*, 2019), have been used in water treatment techniques like membrane-based separation, disinfection, and adsorption. Recent years have seen an influx of novel materials into water treatment, and their usage in combination with carbon will present fresh opportunities. The sensing and treatment applications of materials science for clean water technologies remain fascinating.

It is possible to incorporate developing nanomaterials-based purification methodologies into commercial water filtration systems and products present in the market. Several nanomaterials, such as carbon nanotubes, cellulose-based nanomaterials, aquaporins, and MXenes, have shown exceptional performance at the laboratory scale and have been effectively mass produced and incorporated into products as well (Nagar & Pradeep, 2020). CNTs, for instance, are among the most commonly used nanomaterials for purifying applications (Smith & Rodrigues, 2015). CNT-based purification is effective at eliminating inorganic, organic, and biological water pollutants. Nanomaterials-based membranes are currently manufactured by multinational corporations such as GEA Group ('GEA

Group,' n.d.), Applied Membranes ('Applied Membranes, Inc.,' n.d.), Inopor ('Inopor,' n.d.), Koch Separation Solutions ('Inopor,' n.d.), and Alfa Laval ('Alfa Laval India Pvt. Ltd.,' n.d.).

A future with clean water that is environmentally, socially, and economically sustainable would include the development and execution of water technologies, laws, and practices that ensure accessibility to clean and safe water for everyone. It is crucial to implement policies and practices that result in reduced water footprints and efficient water reuse. In addition, water resources must be coherently managed by establishing a balance between the consumption and reuse of groundwater, surface water, and recycled water. This is only possible with the involvement of local communities whose demands and perspectives are taken into account. Providing affordable clean water to marginalized communities can only happen with continuous investments on research and development in emerging water technologies.

REFERENCES

Alfa Laval India Pvt. Ltd. (n.d.). Retrieved 24 January 2023, from https://www.alfalaval.in/

Applied Membranes, Inc. (n.d.). Retrieved 24 January 2023, from https://www.appliedmembranes.com/

Beysens D. and Milimouk I. (2000). The case for alternative fresh water sources. *Pour les resources alternatives en eau, Secheresse*, **11**(4), 1–16, http://rexresearch.com/dewharvest/Secheresse.pdf

Bruintjes R. T. (1999). A review of cloud seeding experiments to enhance precipitation and some new prospects. *Bulletin of the American Meteorological Society*, **80**(5), 805–820, https://doi.org/10.1175/1520-0477(1999)080<0805:AROCSE>2.0.CO;2

Carballa M., Omil F., Ternes T. and Lema J. M. (2007). Fate of pharmaceutical and personal care products (PPCPs) during anaerobic digestion of sewage sludge. *Water Research*, **41**(10), 2139–2150, https://doi.org/10.1016/j.watres.2007.02.012

DeFelice T. P. and Axisa D. (2017). Modern and prospective technologies for weather modification activities: developing a framework for integrating autonomous unmanned aircraft systems. *Atmospheric Research*, **193**, 173–183, https://doi.org/10.1016/j.atmosres.2017.04.024

DelaiáSun D. (2010). Hierarchically multifunctional TiO_2 nano-thorn membrane for water purification. *Chemical Communications*, **46**(35), 6542–6544, https://doi.org/10.1039/c0cc01143f

Eregowda T. and Mohapatra S. (2020). Fate of micropollutants in engineered and natural environment. In: Resilience, Response, and Risk in Water Systems, M. Kumar, F. Munoz-Arriola, H. Furumai and T. Chaminda (eds.), Springer, Singapore, pp. 283–301.

Fessehaye M., Abdul-Wahab S. A., Savage M. J., Kohler T., Gherezghiher T. and Hurni H. (2014). Fog-water collection for community use *Renewable and Sustainable Energy Reviews*, **29**, 52–62, https://doi.org/10.1016/j.rser.2013.08.063

GEA Group. (n.d.). Retrieved 24 January 2023, from https://www.gea.com/en/index.jsp

Gupta S. S., Islam M. R. and Pradeep T. (2019). Capacitive deionization (CDI): an alternative cost-efficient desalination technique. In: Advances in Water Purification Techniques, A. Satinder (ed.), Elsevier, Cambridge, MA 02139, United States, pp. 165–202.

Inopor. (n.d.). Retrieved 24 January 2023, from https://inopor.com/en/13-en.html

Islam M. R., Gupta S. S., Jana S. K., Srikrishnarka P., Mondal B., Chennu S.,... Pradeep T. (2021). A covalently integrated reduced graphene oxide-ion-exchange resin electrode for efficient capacitive deionization. *Advanced Materials Interfaces*, **8**, 2001998, https://doi.org/10.1002/admi.202001998

Jiang Y., Biswas P. and Fortner J. D. (2016). A review of recent developments in graphene-enabled membranes for water treatment. *Environmental Science: Water Research & Technology*, **2**(6), 915–922, https://doi.org/10.1039/C6EW00187D

Kamali M., Persson K. M., Costa M. E. and Capela I. (2019). Sustainability criteria for assessing nanotechnology applicability in industrial wastewater treatment: current status and future outlook. *Environment International*, **125**, 261–276, https://doi.org/10.1016/j.envint.2019.01.055

Khalil B., Adamowski J., Shabbir A., Jang C., Rojas M., Reilly K. and Ozga-Zielinski B. (2016). A review: dew water collection from radiative passive collectors to recent developments of active collectors. *Sustainable Water Resources Management*, **2**(1), 71–86, https://doi.org/10.1007/s40899-015-0038-z

Kim H., Yang S., Rao S. R., Narayanan S., Kapustin E. A., Furukawa H.,... Wang E. N. (2017). Water harvesting from air with metal-organic frameworks powered by natural sunlight. *Science (New York, N.Y.)*, **356**(6336), 430–434, https://doi.org/10.1126/science.aam8743

Klemm O., Schemenauer R. S., Lummerich A., Cereceda P., Marzol V., Corell D.,... Olivier J. (2012). Fog as a fresh-water resource: overview and perspectives. *Ambio*, **41**(3), 221–234, https://doi.org/10.1007/s13280-012-0247-8

Kumar A. A., Som A., Longo P., Sudhakar C., Bhuin R. G., Gupta S. S.,... Pradeep T. (2017). Confined metastable 2-line ferrihydrite for affordable point-of-use arsenic-free drinking water. *Advanced Materials*, **29**(7), 1604260, https://doi.org/10.1002/adma.201604260

Kunduru K. R., Nazarkovsky M., Farah S., Pawar R. P., Basu A. and Domb A. J. (2017). Nanotechnology for water purification: applications of nanotechnology methods in wastewater treatment. *Water Purification*, 33–74, https://doi.org/10.1016/B978-0-12-804300-4.00002-2

Li M., Liu Y., Zeng G., Liu N. and Liu S. (2019). Graphene and graphene-based nanocomposites used for antibiotics removal in water treatment: a review. *Chemosphere*, **226**, 360–380, https://doi.org/10.1016/j.chemosphere.2019.03.117

Liu X., Wang M., Zhang S. and Pan B. (2013). Application potential of carbon nanotubes in water treatment: a review. *Journal of Environmental Sciences*, **25**(7), 1263–1280, https://doi.org/10.1016/S1001-0742(12)60161-2

Majewski P. J. and Chan C. P. (2008). Water purification by functionalised self-assembled monolayers on silica particles. *International Journal of Nanotechnology*, **5**(2–3), 291–298, https://doi.org/10.1504/IJNT.2008.016919

Mojiri A., Zhou J. L., Ohashi A., Ozaki N. and Kindaichi T. (2019). Comprehensive review of polycyclic aromatic hydrocarbons in water sources, their effects and treatments. *Science of the Total Environment*, **696**, 133971, https://doi.org/10.1016/j.scitotenv.2019.133971

Mukherjee S., Kumar A. A., Sudhakar C., Kumar R., Ahuja T., Mondal B.,... Pradeep T. (2018). Sustainable and affordable composites built using microstructures performing better than nanostructures for arsenic removal. *ACS Sustainable Chemistry & Engineering*, **7**(3), 3222–3233, https://doi.org/10.1021/acssuschemeng.8b05157

Mukherjee S., Ramireddy H., Baidya A., Amala A. K., Sudhakar C., Mondal B.,... Pradeep T. (2019). Nanocellulose-reinforced organo-inorganic nanocomposite for synergistic and affordable defluoridation of water and an evaluation of its sustainability metrics. *ACS Sustainable Chemistry & Engineering*, **8**(1), 139–147, https://doi.org/10.1021/acssuschemeng.9b04822

Nagar A. and Pradeep T. (2020). Clean water through nanotechnology: needs, gaps, and fulfillment. *ACS Nano*, **14**, 6420–6435, https://doi.org/10.1021/acsnano.9b01730

Nair A. S. and Pradeep T. (2003). Halocarbon mineralization and catalytic destruction by metal nanoparticles. *Current Science*, **84**(12), 1560–1564.

Nair A. S., Tom R. T. and Pradeep T. (2003). Detection and extraction of endosulfan by metal nanoparticles. *Journal of Environmental Monitoring*, **5**(2), 363–365, https://doi.org/10.1039/b300107e

Patrick J. W. (ed.) (1995). Porosity in Carbons: Characterization and Applications. Halsted Press, New York.

Peng Z., Liu X., Zhang W., Zeng Z., Liu Z., Zhang C.,... Tang W. (2020). Advances in the application, toxicity and degradation of carbon nanomaterials in environment: a review. *Environment International*, **134**, 105298, https://doi.org/10.1016/j.envint.2019.105298

Porada S., Zhao R., Van Der Wal A., Presser V. and Biesheuvel P. M. (2013). Review on the science and technology of water desalination by capacitive deionization. *Progress in Materials Science*, **58**(8), 1388–1442, https://doi.org/10.1016/j.pmatsci.2013.03.005

Porada S., Zhang L. and Dykstra J. E. (2020). Energy consumption in membrane capacitive deionization and comparison with reverse osmosis. *Desalination*, **488**, 114383, https://doi.org/10.1016/j.desal.2020.114383

Qu X., Brame J., Li Q. and Alvarez P. J. J. (2013). Nanotechnology for a safe and sustainable water supply: enabling integrated water treatment and reuse. *Accounts of Chemical Research*, **46**(3), 834–843, https://doi.org/10.1021/ar300029v

Saini P. (ed.) (2015). Fundamentals of Conjugated Polymer Blends, Copolymers and Composites: Synthesis, Properties, and Applications. John Wiley & Sons, Beverly, United States.

Sankar M. U., Aigal S., Maliyekkal S. M., Chaudhary A., Anshup Kumar A. A.,... Pradeep T. (2013). Biopolymer-reinforced synthetic granular nanocomposites for affordable point-of-use water purification. *Proceedings of the National Academy of Sciences*, **110**(21), 8459–8464, https://doi.org/10.1073/pnas.1220222110

Shen Y., Li L., Xiao K. and Xi J. (2016). Constructing three-dimensional hierarchical architectures by integrating carbon nanofibers into graphite felts for water purification. *ACS Sustainable Chemistry & Engineering*, **4**(4), 2351–2358, https://doi.org/10.1021/acssuschemeng.6b00030

Singer P. A., Salamanca-Buentello F. and Daar A. S. (2005). Harnessing nanotechnology to improve global equity. *Issues in Science and Technology*, **21**(4), 57–64.

Smith S. C. and Rodrigues D. F. (2015). Carbon-based nanomaterials for removal of chemical and biological contaminants from water: a review of mechanisms and applications. *Carbon*, **91**, 122–143, https://doi.org/10.1016/j.carbon.2015.04.043

Sreeprasad T. S., Maliyekkal S. M., Lisha K. P. and Pradeep T. (2011). Reduced graphene oxide–metal/metal oxide composites: facile synthesis and application in water purification. *Journal of Hazardous Materials*, **186**(1), 921–931, https://doi.org/10.1016/j.jhazmat.2010.11.100

Technology and Action for Rural Advancement (TARA). (2012). Access to Safe Water for the Bottom of Pyramid: Strategies for Disseminating Technology Research Benefits. Department for International Development, Government of UK.

Teodosiu C., Gilca A.-F., Barjoveanu G. and Fiore S. (2018). Emerging pollutants removal through advanced drinking water treatment: a review on processes and environmental performances assessment. *Journal of Cleaner Production*, **197**, 1210–1221, https://doi.org/10.1016/j.jclepro.2018.06.247

Tu Y., Wang R., Zhang Y. and Wang J. (2018). Progress and expectation of atmospheric water harvesting. *Joule*, **2**(8), 1452–1475, https://doi.org/10.1016/j.joule.2018.07.015

Upadhyayula V. K. K., Deng S., Mitchell M. C. and Smith G. B. (2009). Application of carbon nanotube technology for removal of contaminants in drinking water: a review. *Science of the Total Environment*, **408**(1), 1–13, https://doi.org/10.1016/j.scitotenv.2009.09.027

van der Hoek J. P., Bertelkamp C., Verliefde A. R. D. and Singhal N. (2014). Drinking water treatment technologies in Europe: state of the art–challenges–research needs. *Journal of Water Supply: Research and Technology–AQUA*, **63**(2), 124–130, https://doi.org/10.2166/aqua.2013.007

Wang G., Zhong D., Li T., Wei J., Huang Y., Fu X., Li J. and Zhang Y. (2016). Sky river: discovery, concept, and implications for future research. *Scientia Sinica Technologica*, **46**(6), 649–656, https://doi.org/10.1360/N092015-00367

Yilmaz G., Meng F. L., Lu W., Abed J., Peh C. K. N., Gao M.,... Ho G. W. (2020). Autonomous atmospheric water seeping MOF matrix. *Science Advances*, **6**(42), eabc8605, https://doi.org/10.1126/sciadv.abc8605

Zhang Z., Wang C., Yao Y., Zhang H., Na J., Zhou Y.,... Yamauchi Y. (2022). Modular assembly of MOF-derived carbon nanofibers into macroarchitectures for water treatment. *Chemical Science*, **13**(32), 9159–9164, https://doi.org/10.1039/D2SC02619H

doi: 10.2166/9781789063714_0133

Chapter 12

Pulsed power technology for water and wastewater treatment

Jerin Jose[1], *Raj Kamal Singh*[2] *and Ligy Philip*[3*]

[1]*Department of Chemical Engineering, Amal Jyothi College of Engineering, Kerala, India*
[2]*SGS Dayton, New Jersey, USA*
[3]*Department of Civil Engineering, Indian Institute of Technology Madras, Chennai, India*
**Corresponding author: ligy@iitm.ac.in*

ABSTRACT

The presence of recalcitrant and hazardous water pollutants poses a serious challenge to conventional water treatment technologies and is motivating the development of advanced treatment approaches like pulsed power technology (PPT). One application of PPT is to generate plasma that can be used for the efficient and rapid degradation of a wide range of emerging water pollutants, including dyes, pesticides, toxic solvents, and pharmaceuticals, and for the disinfection of water and wastewater. Plasma-based water treatment utilizes electrical discharge plasmas formed in contact with or in the vicinity of water to degrade chemicals within the contaminated water. Plasma in these conditions can produce a diverse range of highly reactive species, including strong oxidants (•OH) and reductants (e^-_{aq}), which are suitable for the simultaneous oxidation and reduction of the organic compounds present in highly complex wastewater such as industrial effluents. PPT exhibits the advantages of fast pollutant degradation kinetics, relatively low energy input, and the absence of chemical additives, making it a promising alternative water treatment technology. The increasing demand for water treatment technologies for treating hazardous chemicals with no/minimal generation of secondary pollutants may accelerate the adoption of safe and energy-efficient technologies like PPT.

Keywords: pulse power technology, plasma, water treatment, wastewater treatment, industrial effluents, advanced oxidation processes

12.1 INTRODUCTION

Rapid industrialization and uncontrolled urbanization have resulted in the generation of enormous amount of wastewater. The discharge of untreated or partially treated effluents has led to widespread contamination of the environment. Globally, over 348 million people do not have access to pure drinking water, and it is estimated that there will be a 400% increase in water requirements for industrial applications by 2050 (Connor, 2015). Industrial effluents contain high concentrations of organic solvents, dyes, cyanides, polycyclic aromatic hydrocarbons, and so on, which have high toxicity and poor biodegradability. Some of them are carcinogenic, mutagenic, and teratogenic, hence posing a severe threat to the environment and human health. In recent years, a new category of contaminants called 'emerging contaminants' (ECs) or 'micropollutants' has been detected in the aquatic environment; it includes pharmaceuticals like antibiotics, pesticides, personal care products and so on, which are present in trace concentrations in water bodies.

Conventional technologies consisting of physicochemical and biological processes are not able to effectively degrade the wide variety of recalcitrant pollutants. In addition, such methods often require more treatment time and a large area, and are sensitive to environmental conditions. Hence, it is necessary to develop advanced technologies to deal with such pollutants. Different types of advanced oxidation processes (AOPs) have been employed for degrading complex pollutants (Benner *et al.*, 2013).

Electrical discharge-based plasma technology can be considered an advanced form of AOPs. Non-thermal plasma produces ·OH radicals, hydroperoxyl radicals, atomic hydrogen and oxygen, and various other radicals and oxidizing species, along with the generation of shock waves, UV light, and so on. Due to the combined and simultaneous action of all these chemical species and physical effects, plasma-based water treatment is capable of degrading most of the complex pollutant compounds in a short duration (Stratton *et al.*, 2015).

Among the different types of plasma discharges, pulsed power plasma is a prominent technique due to its high degradation performance and low energy requirement. In pulsed power technology (PPT), energy is accumulated in capacitors or inductors for a long period of time and is subsequently released over a very short time period. The potential and applicability of PPT have been evaluated for a range of water contaminants such as dyes, organic solvents, nutrients, ECs, pathogens, and so on. This chapter deals with the mechanisms of PPT-based water treatment, the factors affecting treatment efficiency, different plasma reactor configurations, a comparison of PPT with other treatment processes, and the integration of PPT with other technologies in the treatment scheme. The chapter ends with the challenges for its implementation.

12.2 PULSED POWER TECHNOLOGY

Pulsed power is generated by using a high-voltage power supply to maintain a strong electric field with minimum energy input and rapidly excite a medium (gas or liquid) to form plasma. Electrical pulses of fast rise time and narrow width-pulsed voltage have been a key criterion for plasma generation for water and wastewater treatment. Plasma is a quasi-neutral ionized gas composed of free electrons, ions, neutral species, free radicals, excited species, and photons. Depending on the ionization degree, plasmas can be classified as 'hot' (ionization degree close to 100%) or as 'cold' (or 'non-thermal'). In hot or thermal plasma, all species are in equilibrium and are characterized by a unique temperature, whereas cold plasmas are far from equilibrium and the electron temperature is much higher than the temperature of the heavy particles (ions and neutrals). Consequently, electrons are the main vehicle of energy for cold plasmas, and they govern the plasma chemistry through collisions with the other species (mainly neutrals). Pulsed power plasma is a type of non-thermal plasma. In pulsed power plasma, only the free electrons attain high energy, while the heavier ions remain at room temperature. The high-energy electrons then collide with the gas molecules and excite, ionize, or dissociate them.

12.2.1 Chemical and physical effects of PPT

High-voltage electrical discharges in water (electrohydraulic discharge) or in the gas phase above the water produce oxidative species such as hydroxyls, hydroperoxyls, oxygen, hydrogen peroxide, molecular oxygen, and reductive species like aqueous electrons (e^-_{aq}), hydrogen and superoxide radicals, accompanied by various physical effects.

12.2.1.1 Oxidative species

The ·OH (hydroxyl) radical is a very strong oxidant, with an oxidation potential of 2.8 V. During an electrical discharge, ·OH is generated by the dissociation, ionization and vibrational/rotational excitation of a water molecule; it can also be produced from atomic oxygen, hydrogen peroxide and ozone decomposition (Jiang *et al.*, 2014). Other than ·OH, the two main oxidizing agents generated are ozone and H_2O_2. By electrical discharge, O_2 molecules dissociate to form O atom, and the further recombination of O atoms into molecular oxygen gives rise to an ozone molecule (Malik *et al.*, 2001). H_2O_2 is a comparatively stable species, formed by the recombination of ·OH, and has less oxidizing power compared to ·OH (Joshi & Locke, 1995).

12.2.1.2 Reductive species

Along with oxidants, some strong reducing agents such as aqueous electrons, hydrogen radicals and superoxide radicals also get generated during plasma discharge. High-energy electrons get transformed to aqueous electrons inside water. H· forms by the direct interaction of electrons with water molecules in an acidic aqueous solution. Similarly, the superoxide radical ($^{\cdot}O_2^-$) is produced by the reaction of aqueous electrons with oxygen molecules. $e^-_{(aq)}$ plays an important role in the degradation of highly oxidized organic compounds via the reduction pathway.

12.2.1.3 Physical effects

When excited molecules or atoms return to a lower energy state, UV light gets generated. UV irradiation can decompose an organic molecule directly. UV light can also decompose H_2O_2 and ozone, forming ·OH, thereby indirectly decomposing pollutants. During electrical discharge, the propagation or expansion of plasma against the surrounding water generates intense shockwaves. These shockwaves cause electro-hydraulic cavitation inside water, causing the formation of ·OH and H_2O_2.

12.2.2 Mechanisms of PPT-based water treatment

Electrical discharges in water form a complex reaction mixture of highly reactive oxidizing and reducing species. These species react with organic contaminants via different chemical reactions. Lukes and Locke (2005) studied the degradation of substituted phenols in a hybrid gas–liquid electrical discharge reactor. They determined that electrophilic attacks by ·OH radicals and ozone were the main oxidation pathways for the degradation of phenol in argon and oxygen atmospheres, respectively. Hydroxylated aromatic by-products were identified during the degradation of all substituted phenols. Degradation of 4-chlorophenol was studied in different gas–liquid plasma reactors and the electrophilic attack of ·OH and ozone was found to be responsible for 4-chlorophenol's degradation. However, the rate constant of ·OH was much higher compared to that of the ozone attack (Zhang *et al.*, 2007). They also concluded that the hydroxylation of the organic molecules was the main degradation pathway. In addition, it was shown that the formation of peroxyl radicals ($^{\cdot}HO_2$) helps the fragmentation of aromatic rings and forms aliphatic products. Dobrin *et al.* (2013) studied the degradation of diclofenec by pulsed corona discharge and reported that the cleavage of the C–N bond is the first step in diclofenec degradation. This is followed by further oxidative ring opening, with the formation of shorter chain fatty acids and mineralization (Dobrin *et al.*, 2013). In summary, the attack of the ·OH radical is the main mechanism for the decomposition of organic compounds. Complex organic compounds break either by bond/ring cleavage and electrophilic addition of ·OH, or by the hydroxylation, dealkylation, and dehalogenation of aromatic compounds. Many intermediates have also been reported during the oxidative degradation in plasma processes. Figure 12.1 shows the various reactive species produced during plasma discharge, their conversions, and their involvement in the degradation of an aliphatic compound and an aromatic compound.

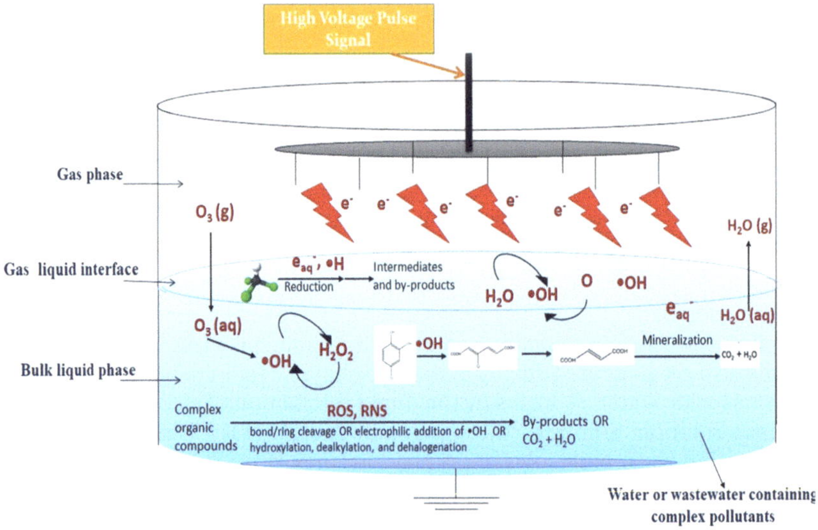

Figure 12.1 Mechanism of plasma-based water/wastewater treatment showing various reactive species.

12.2.3 Comparison of PPT with conventional AOPs

PPT can be considered a combination of various AOPs. This is because during plasma discharge, a range of reactive species such as •OH radicals, H_2O_2, O_3, $HO^{\cdot}_2$, $O_2^{\cdot-}$, aqueous electrons, H•, shock waves, UV light, and so on are produced simultaneously, all of which take part in pollutant decomposition. In plasma-based water treatment, chemical, electrical, photochemical, thermal, and cavitation processes are occurring simultaneously, which is the main difference from conventional AOPs in which only one of these processes occurs at any time. The gas-phase discharge produces reactive species, which enter the water by diffusion; the plasma discharge–water interaction also results in the generation of different chemical species, in contrast to the externally supplied chemicals in conventional AOPs. Conventional AOPs have various limitations like high energy costs, requirement for costly chemicals, toxic by-product formation, catalyst poisoning, slow kinetics, and low removal efficiency. Compared to AOPs, PPT has the advantages of: (1) no external supply of chemical oxidizing agents; (2) superior removal efficiency and faster kinetics; (3) suitable for point-of-use applications even in rural areas due to the non-necessity of consumables; (4) high energy efficiency; and (5) no requirement for high temperature or pressure as in the case of some AOPs. Gas-phase discharge reactors will be able to achieve simultaneous degradation of pollutants present in the gas phase and liquid phase of the plasma reactor, which is mainly applicable in the case of volatile pollutants like volatile organic compounds (VOCs).

12.3 FACTORS AFFECTING THE PERFORMANCE OF PPT

12.3.1 Input energy

The efficiency and rate of pollutant degradation by plasma technology predominantly depends on the energy applied to generate the plasma discharge. The input energy of PPT can be changed by varying either the input voltage or the pulse frequency. Pulses with fast rise time and narrow width will produce a strong electric field and achieve pollutant degradation with high energy efficiency. A higher input energy generates a higher concentration of reactive species and stronger physical effects.

12.3.2 Reactor configuration

Pulsed plasma can be generated using two asymmetric electrodes – one with high curvature, like a needle- or wire-shaped electrode, and the other with a small curvature, like a plate electrode. Therefore, the electrode configurations will be point-plate, wire-plate, wire-cylinder, and so on and the reactor can be either a liquid discharge reactor or gas discharge reactor (Jiang *et al.*, 2014). But the most commonly used electrode configuration in plasma-based water treatment studies is point-to-plate electrode geometry. The electrical discharges in a gas medium consume much less energy than discharges in water (Malik *et al.*, 2001).

In a liquid discharge reactor (Figure 12.2a), the reactive species directly reacts with the pollutants dissolved in water, whereas the diffusion of reactive species from gas phase to bulk liquid phase will limit the rate of degradation in a gas discharge reactor (Figure 12.2b). In a hybrid gas–liquid discharge reactor (Figure 12.2c), the high-voltage electrode is immersed in the liquid and the ground electrode is suspended in the gas phase above the liquid surface; in this, the discharge could also occur in both the gas and liquid phases, intensifying the reactive species' formation and resulting in efficient pollutant removal.

The reactor geometry should be such that it should provide the maximum area of contact between the electrical discharge and the water surface. Hence, a water feed into the discharge zone in the form of a thin sheet or spray can increase the surface-to-volume ratio, which in turn increases the mass transfer of reactive species into the water, resulting in better degradation and energy efficiency (Malik, 2010).

12.3.3 Solution pH

Solution pH influences the plasma degradation of a pollutant in two ways: (1) pH affecting the formation of various reactive species, and (2) pH affecting the speciation of the organic compounds (Singh *et al.*, 2016). Organic compounds will be in the molecular form when the solution pH is lower than the pK_a of the compound, and in the ionic form when the pH is greater than pK_a. Hence, the optimum pH varies with respect to each compound, depending on the major reactive species involved in the compound's degradation and whether its acidic or basic form is more highly reactive.

12.3.4 Gas in which discharge occurs

Pulsed plasma discharge occurs in different gas media such as air, oxygen, argon, nitrogen and so on. Discharge in an oxygen medium will enhance the generation of ozone and O-containing reactive species like •OH radicals, which increases the degradation efficiency significantly. Efficiency reduces in different gas media in the order oxygen > air > argon > nitrogen. A demerit of discharge in the presence of nitrogen gas is the dissociation of N_2 molecules and the production of HNO_2, HNO_3, and

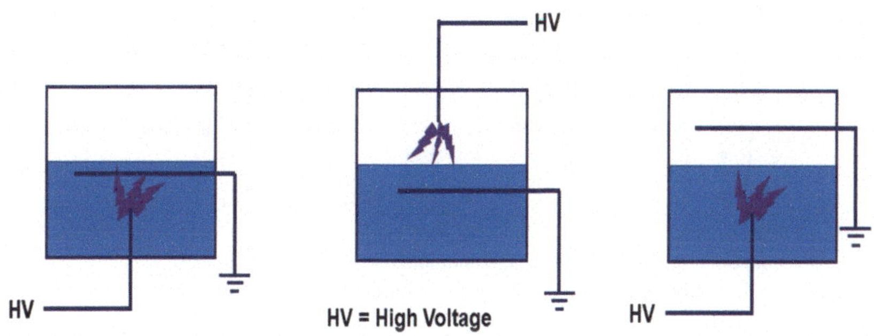

Figure 12.2 Schematics of pulsed discharge corona reactors: (a) liquid discharge reactor, (b) gas discharge reactor, and (c) hybrid gas–liquid discharge reactor.

so on, which reduces the solution pH drastically and may influence the chemical reactions occurring in the system (Kornev *et al.*, 2013).

12.3.5 Solution conductivity

The mode of electrical discharge and the concentration of free radicals generated are strongly influenced by the conductivity of the solution. The discharge current increases when conductivity increases. This leads to higher input energy to the reactor, which in turn results in the generation of more reactive species and better degradation efficiency (Lei *et al.*, 2007). In addition to that, anions like CO_3^{2-}, HCO_3^-, PO_4^{3-} and so on are scavengers of $^{\cdot}OH$ radicals, which will reduce the pollutant degradation efficiency.

12.4 APPLICATIONS OF PPT FOR WATER AND WASTEWATER TREATMENT

12.4.1 Organic pollutants

Plasma technology has been employed for the degradation of a wide range of organic pollutants present in water and wastewater. Among them, phenol is the most studied compound because of its poor biodegradability and high occurrence in many industrial effluents. The other classes of organic compounds treated using PPT include dyes (e.g. azo dyes), organic solvents, pharmaceutical compounds, and pesticides. In most of the cases, PPT was able to achieve high removal efficiency in a low reaction time. Figure 12.3 gives the results of a PPT-based (input voltage = 25 kV and frequency = 25 Hz) degradation study of three pharmaceutical compounds: diclofenac (DCF), carbamazepine (CBZ), and ciprofloxacin (CPF) in single and mixed conditions.

Generally, the intermediates produced during the plasma degradation of organic pollutants can also be completely oxidized with a prolonged reaction time. But sometimes, recalcitrant intermediates like short-chain carboxylic acids (e.g. formic acid, acetic acid, etc.) will be produced, which are resistant to plasma treatment (Jose & Philip, 2019). In some cases, the intermediates produced can be toxic, and hence toxicity studies should be carried out to evaluate the impact of the treated wastewater on the

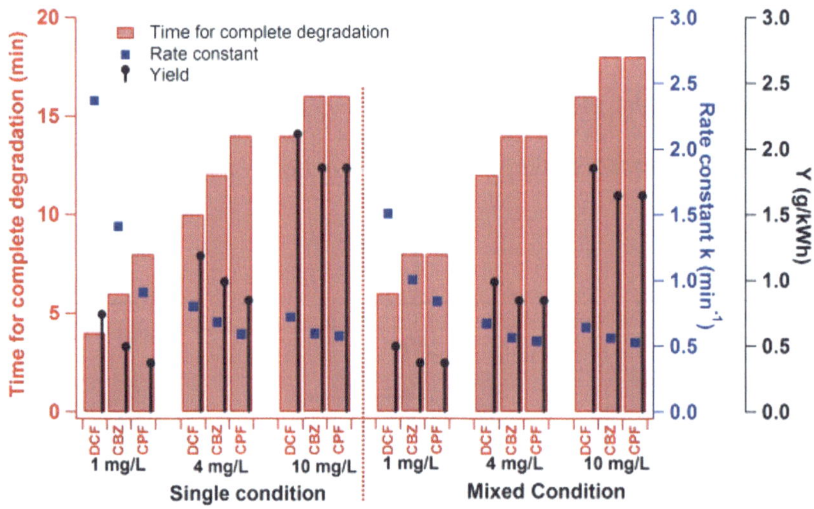

Figure 12.3 Effect of initial concentration of pharmaceutical compounds (diclofenac (DCF), carbamazepine (CBZ) and ciprofloxacin (CPF)) on degradation time, rate constant and energy yield (g/kWh) in single and mixed conditions. (*Source*: Reproduced with permission from Singh *et al.*, 2017). Copyright 2023, Elsevier.

environment. Advanced technologies like PPT can be employed for improving the biodegradability (biological oxygen demand (BOD)/chemical oxygen demand (COD) ratio) of complex industrial wastewater, which can be subsequently treated using biological methods. The integrated treatment increases the pollutant removal efficiency and reduces the treatment cost.

12.4.2 Emerging contaminants

Some studies reported the applicability of plasma for the ECs' abatement from water. The decomposition of carbamazepine, clofibric acid and iopromide in aqueous solution and landfill leachate using corona discharge over a liquid surface is investigated (Krause *et al.*, 2011). The degradation and mineralization of diclofenac in water using pulsed corona discharge were investigated and the degradation pathway was proposed by identifying the intermediates (Dobrin *et al.*, 2013). Gerrity *et al.* (2010) investigated a pilot-scale unit with a needle-to-plane geometry corona discharge reactor for the treatment of different pharmaceutical compounds. Panorel *et al.* (2012) employed a pulsed corona discharge reactor for the treatment of a solution that was showered in droplets or jets in the plasma channel. The results of these pulsed corona plasma processes used for the degradation of ECs are summarized in Table 12.1.

12.4.3 Disinfection

Plasma generates various reactive species ($^{\bullet}OH$, H_2O_2 and O_3, etc.) along with UV light and shockwaves, which significantly affect the structural integrity of the bacterial cell membrane. A rapid disinfection capability of PPT was observed at an applied pulsed voltage of 23 kV and 25 Hz frequency shown in Figure 12.4 (Singh *et al.*, 2016). Complete disinfection (7 log reduction) was obtained in 6 min of treatment time. A few plausible mechanisms involved in the plasma inactivation of microorganisms are: (a) damage of the cell wall and membrane by short-lived reactive oxygen species (ROS) ($^{\bullet}OH$, $^{\bullet}O_2^-$, $^1O_2$), resulting in leakage of the cellular constituents; (b) destruction of the genetic material of microorganisms by UV irradiation and ROS; (c) shockwave-assisted destruction of microorganisms; (d) oxidative damage to DNA, proteins, and lipids, ultimately leading to microbial cell death due to plasma-induced oxidative stress; and (e) water acidification which can change the structure of proteins (e.g. enzymes), consequently altering protein activity. Membrane damage and deterioration are direct causes of cell inactivation. The plausible reasons for subcellular disinfection mechanism through PPT are shown in Figure 12.5. These mechanisms make plasma sterilization entirely different from classical sterilization techniques. Moreover, this technique is also applicable for the inactivation of protozoans and viruses.

12.5 COMPARISON OF PPT'S ENERGY EFFICIENCY WITH THAT OF OTHER TECHNOLOGIES

In pulsed corona discharge, only the free electrons will be raised to high energy levels, leaving the ions and neutrals close to room temperature; hence the plasma generated can be classified as cold plasma. The reason for this is the short duration of high-voltage pulse signals, which provide high energy only to the electrons, which are the lightest charged particles (Malik *et al.*, 2001).

The electrical energy per order (E_{EO}) is the electrical energy in kWh required to degrade a contaminant C by one order of magnitude in 1 m^3 of contaminated water. Based on E_{EO} values, various AOPs have been classified into different groups: (1) AOPs with median E_{EO} values of <1 kWh/m^3 (e.g. O_3/H_2O_2, O_3/UV), (2) in the range of 1–100 kWh/m^3 (e.g. Photo-Fenton, plasma, and electrochemical AOPs), and (3) >100 kWh/m^3 (e.g. UV-based photocatalysis, ultrasound) (Miklos *et al.*, 2018). Hence, among all AOPs, the energy requirement of plasma technologies is in the middle range. The energy efficiency of plasma treatments can be improved by increasing the mass transfer of plasma-reactive species to water, which can be achieved by increasing the surface-to-volume ratio by making water flow in a thin layer or by spraying the solution.

Table 12.1 A comparison of different plasma systems used for degradation of ECs.

ECs	Types of Discharge	Reactor/ Electrode Configurations	Operating Conditions	Mode of Operation/ Conditions	Removal Efficiency and Treatment Time	Energy Requirement/ mg/L ECs Reduction	References
Clofibric acid, carbamazepine, iopromide	Corona discharge	Two-barrier-electrode	$V = 25–35$ kV $f = 30$ kHz	Batch, volume = 200 mL ECs conc = 0.1 mM	>98% in 30 min	0.0031–0.011 kWh/mg/L	Krause *et al.* (2011)
Diclofenac	Pulsed corona discharge	Wire-to-plane geometry	$V = 18$ kV O_2 flow rate = 1 L/min	Batch, volume = 55 mL, EC conc = 50 mg/L	>99% in 15 min	0.023 kWh/ mg/L for 90% degradation	Dobrin *et al.* (2013)
Meprobamate, phenytoin, primidone, carbamazepine, atenolol, atrazine	Pulsed corona discharge	Needle plane configuration	$V = 8$ kV $f = 500$ Hz	Batch and continuous, Volume = 150 L Water flow rate = 8 L/min ECs conc = 36–378 ng/L	>90% in 19 min	16.93–61.11 kWh/mg/L	Gerrity *et al.* (2010)
17β-estradiol, salicylic acid, indometacin, ibuprofen	Pulsed corona discharge	Needle-to-plane configuration	$V = 20$ kV $I = 400$ A $f = 840$ Hz	Batch, volume = 40 L ECs conc = 100 mg/L	70–99% in 30 min	0.000006–0.000016 kWh/mg/L	Panorel *et al.* (2012)

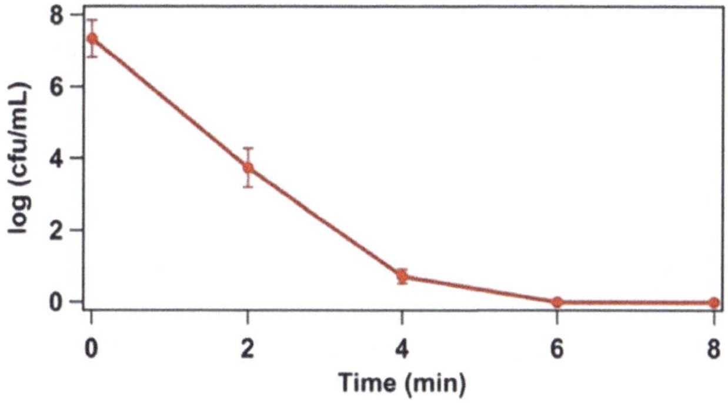

Figure 12.4 Log reduction of *Escherichia coli* in physiological saline as a function of time for power delivery of 85.2 kW, applied voltage 23 kV and frequency 25 Hz. (*Source*: Reproduced from Singh *et al.* (2016), with permission from the Royal Society of Chemistry).

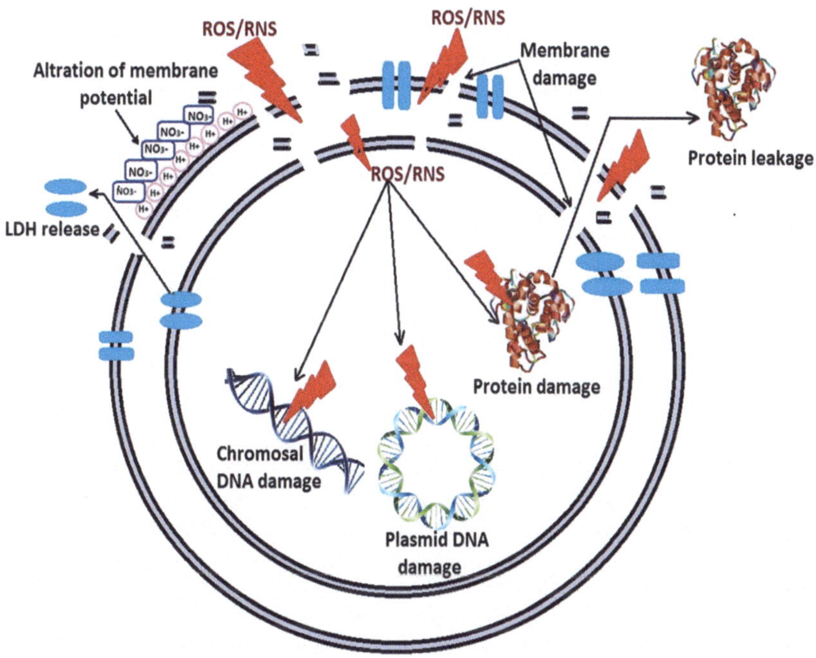

Figure 12.5 A schematic of sub-cellular level bacterial disinfection mechanism. (*Source*: Reproduced from Singh *et al.* (2016), with permission from the Royal Society of Chemistry).

12.6 IMPACTS OF PPT ON OTHER WATER QUALITY PARAMETERS

The gaseous medium in which plasma discharge takes place directly influences the formation of nitrate in the treated solution and also affects the solution pH significantly. When discharge occurs in air, it dissociates nitrogen and oxygen molecules and leads to the generation of nitrite and nitrate ions in water, and it also produces H^+ ions in water, which reduces the pH.

After plasma treatment or any AOP, if complete mineralization or 100% total organic carbon removal is not achieved, it indicates the formation of degradation intermediates during the treatment process. Such intermediates can also be toxic compounds. Hence, a toxicity analysis of the treated water should be carried out before it is sent for various reuse applications. Long-lived species such as H_2O_2 and O_3 produced in water during plasma discharge may cause harmful effects to microorganisms if the plasma-treated wastewater is further sent to a biological treatment system.

12.7 INTEGRATION OF PPT WITH OTHER TREATMENT TECHNOLOGIES

The performance of any AOP in the treatment of a particular wastewater is directly related to the concentration of pollutants or the contaminant load in it. The treatment efficacy and energy efficiency of PPT can be enhanced by coupling it with various other processes. PPT can be combined with conventional biological wastewater treatment as a pre-treatment or post-treatment technology, depending on the characteristics of the wastewater. Application of PPT as pre-treatment can enhance the biodegradability and reduce the toxicity of the wastewater, which can then be biologically treated. Generally, biological wastewater treatment is suitable for effluents having a BOD_5/COD ratio > 0.5. The sequential application of various treatment processes is suitable for industrial effluents containing a range of contaminants with varying oxidizability. Complete mineralization of target pollutants by PPT or AOP results in increased treatment costs, due to the refractory nature of highly oxidized end-products like short-chain carboxylic acids. Hence, in such cases, PPT can be combined with biological processes to make the process economical.

Since a cocktail of reactive species is produced during plasma discharge, adding other oxidants or catalysts can generate other AOPs along with plasma. For example, added Fe^{2+} can react with the H_2O_2 produced during the discharge to develop Fenton's oxidation and the UV light generated can be combined with added TiO_2 to develop photocatalysis. Since the major reactive species involved in plasma-based water treatment is the $^{\cdot}OH$ radical, it is possible to make use of slower oxidants produced during plasma discharge, such as H_2O_2, O_3, and so on, by combining them with catalysts like activated carbon, metal oxide, metal ions, and so on (Jiang *et al.*, 2014).

12.8 CHALLENGES FOR THE IMPLEMENTATION OF PLASMA-BASED WATER TECHNOLOGIES

Even though PPT has proven to be a very effective and dependable technology for water treatment at the laboratory scale, pilot-scale and field applications are scarce. The main reasons behind this are the difficulties faced in the scaling up of plasma reactors and the development of pulse power generators to supply power for treating large volumes of water or wastewater. The low throughput of a plasma reactor is related to the diffusion limitations of reactive species to react with the pollutants in a bulk solution. The reason behind the diffusion limitations of the radicals is their short lifetime. Throughput can be improved by increasing the area of contact of plasma and the water or wastewater; for this, either the water can be sprinkled into the plasma discharge zone or be allowed to flow in thin sheets through the reactor. For better control of the process, more understanding is needed for the processes occurring in the interface and the reactive species' transport in a bulk liquid solution. Highly volatile compounds diffuse into the plasma zone and directly react with the oxidizing radicals or molecules and get decomposed faster, whereas less volatile contaminants get decomposed at the gas–liquid interface and in the bulk solution. In the case of dilute wastewater, the enrichment of pollutants by membrane filtration or adsorption followed by plasma treatment of the concentrate can reduce to an extent the problems associated with scaling up PPT (Shang *et al.*, 2019).

In some cases, the reason for low degradation efficiency could be the recombination of $^{\cdot}OH$ radicals to produce less reactive species. For example, at very high conductivities, $^{\cdot}OH$ will recombine to form H_2O_2, whose oxidation potential is lesser than that of $^{\cdot}OH$, resulting in slow oxidation kinetics.

As in the case of any AOPs, the intermediates or by-products produced during pollutant decomposition by plasma should be analysed and confirmed non-toxic in nature before the discharge or reuse of the treated effluent. Plasma discharge occurring in real effluents will be different from that in synthetic wastewater. Therefore, real effluent-based studies are important to understand the achievable efficiency in field-scale applications and the scavenging effects of various interferences present in wastewater and to evaluate the toxic intermediates produced, if any.

12.9 SUMMARY

Water and wastewater treatment using PPT is a potential technology for degradation of highly recalcitrant and toxic pollutants. In contrast to other AOPs, PPT makes use of a wide range of chemical reactive species, physical effects, and electric field-driven effects simultaneously for the decomposition of contaminants. PPT can handle a wide range of organic pollutants and ECs, such as pharmaceuticals, pesticides, and personal care products, and is efficient for disinfection as well. One of the obstacles in large-scale application of PPT for water treatment is the difficulty in scaling up. Suitable reactor configurations and a clear understanding of the physical processes taking place at the plasma–liquid interface and the fluid dynamical effects may help to solve this issue. The increasing demand for advanced technologies for dealing with emerging pollutants in water, and for the decomposition of hazardous and recalcitrant chemicals present in industrial wastewater, with environmental friendliness and non-generation of secondary pollutants will accelerate the adoption of technologies like pulsed power plasma.

REFERENCES

Benner J., Helbling D. E., Kohler H. P. E., Wittebol J., Kaiser E., Prasse C., Ternes T. A., Albers C. N., Aamand J., Horemans B., Springael D., Walravens E. and Boon N. (2013). Is biological treatment a viable alternative for micropollutant removal in drinking water treatment processes? *Water Research*, **47**(16), 5955–5976, https://doi.org/10.1016/j.watres.2013.07.015

Connor R. (2015). The United Nations world water development report 2015: water for a sustainable world (Vol. 1). UNESCO publishing, place de Fontenoy, 75352 Paris 07 SP, France, Accessed 13 December 2022.

Dobrin D., Bradu C., Magureanu M., Mandache N. B. and Parvulescu V. I. (2013). Degradation of diclofenac in water using a pulsed corona discharge. *Chemical Engineering Journal*, **234**, 389–396, https://doi.org/10.1016/j.cej.2013.08.114

Gerrity D., Stanford B. D., Trenholm R. A. and Snyder S. A. (2010). An evaluation of a pilot-scale nonthermal plasma advanced oxidation process for trace organic compound degradation. *Water Research*, **44**(2), 493–504, https://doi.org/10.1016/j.watres.2009.09.029

Jiang B., Zheng J., Qiu S., Wu M., Zhang Q., Yan Z. and Xue Q. (2014). Review on electrical discharge plasma technology for wastewater remediation. *Chemical Engineering Journal*, **236**, 348–368, https://doi.org/10.1016/j.cej.2013.09.090

Jose J. and Philip L. (2019). Degradation of chlorobenzene in aqueous solution by pulsed power plasma: mechanism and effect of operational parameters. *Journal of Environmental Chemical Engineering*, **7**(6), 103476, https://doi.org/10.1016/j.jece.2019.103476

Joshi A. A. and Locke B. R. (1995). Formation of hydroxyl radicals, hydrogen peroxide and aqueous electrons by pulsed streamer corona discharge in aqueous solution. *Journal of Hazardous Materials*, **41**, 3–30, https://doi.org/10.1016/0304-3894(94)00099-3

Kornev I., Osokin G., Galanov A., Yavorovskiy N. and Preis S. (2013). Formation of nitrite- and nitrate-ions in aqueous solutions treated with pulsed electric discharges. *Ozone: Science & Engineering*, **35**(1), 22–30, https://doi.org/10.1080/01919512.2013.720898

Krause H., Schweiger B., Prinz E., Kim J. and Steinfeld U. (2011). Degradation of persistent pharmaceuticals in aqueous solutions by a positive dielectric barrier discharge treatment. *Journal of Electrostatics*, **69**, 333–338, https://doi.org/10.1016/j.elstat.2011.04.011

Lei L. C., Zhang Y., Zhang X. W., Du Y. X., Dai Q. Z. and Han S. (2007). Degradation performance of 4-chlorophenol as a typical organic pollutant by a pulsed high voltage discharge system. *Industrial & Engineering Chemistry Research*, **46**(17), 5469–5477, https://doi.org/10.1007/s40201-019-00433-3

Lukes P. and Locke B. R. (2005). Degradation of substituted phenols in a hybrid gas–liquid electrical discharge reactor. *Industrial & Engineering Chemistry Research*, **44**(9), 2921–2930, https://doi.org/10.1021/ie0491342

Malik M. A. (2010). Water purification by plasmas: which reactors are most energy efficient? *Plasma Chemistry and Plasma Processing*, **30**(1), 21–31, https://doi.org/10.1007/s11090-009-9202-2

Malik M. A., Ghaffar A. and Malik S. A. (2001). Water purification by electrical discharges. *Plasma Sources Science and Technology*, **10**(1), 82–91, https://doi.org/10.1088/0963-0252/10/1/311

Miklos D. B., Remy C., Jekel M., Linden K. G., Drewes J. E. and Hübner U. (2018). Evaluation of advanced oxidation processes for water and wastewater treatment – a critical review. *Water Research*, **139**, 118–131, https://doi.org/10.1016/j.watres.2018.03.042

Panorel I., Preis S., Kornev I., Hatakka H. and Louhi-Kultanen M. (2012). Oxidation of aqueous pharmaceuticals by pulsed corona discharge. *Environmental Technology*, **34**(7), 923–930, https://doi.org/10.1080/09593330.2012.722691

Shang K., Jie L. I. and Morent R. (2019). Hybrid electric discharge plasma technologies for water decontamination: a short review. *Plasma Science and Technology*, **21**(4), 043001, https://doi.org/10.1088/2058-6272/aafbc6

Singh R. K., Philip L. and Sarathi R. (2016). Disinfection of water using pulse power technique: A mechanistic perspective. *RSC Advances*, **6**, 11980–11990, https://doi.org/10.1039/C5RA26941E

Singh R. K., Philip L. and Sarathi R. (2017). Rapid degradation, mineralization and detoxification of pharmaceutically active compounds in aqueous solution during pulsed corona discharge treatment. *Water Research*, **121**, 20–36, https://doi.org/10.1016/j.watres.2017.05.006

Stratton G. R., Bellona C. L., Dai F., Holsen T. M. and Thagard S. M. (2015). Plasma-based water treatment: conception and application of a new general principle for reactor design. *Chemical Engineering Journal*, **273**, 543–550, https://doi.org/10.1016/j.cej.2015.03.059

Zhang Y., Zhou M., Hao X. and Lei L. (2007). Degradation mechanisms of 4-chlorophenol in a novel gas-liquid hybrid discharge reactor by pulsed high voltage system with oxygen or nitrogen bubbling. *Chemosphere*, **67**(4), 702–711, https://doi.org/10.1016/j.chemosphere.2006.10.065

doi: 10.2166/9781789063714_0145

Chapter 13

Application of engineered natural treatment systems for pollution abatement

*Manthiram Karthik Ravichandran, Mohammed Iqbal Thayyil and Ligy Philip**

Department of Civil Engineering, Indian Institute of Technology Madras, Chennai, India
**Corresponding author: ligy@iitm.ac.in*

ABSTRACT

In recent years, the availability of freshwater resources has been depleting due to increased water consumption by the growing population worldwide. In this regard, the treatment and reuse of wastewater aid in achieving water security and environmental protection. In low-income countries of the Global South, the infrastructure for water and sanitation is neither sufficient nor sustainable to meet current and future demands. This lack of resources has caused numerous communities in these countries to face difficulties in obtaining clean water, sanitation and hygiene services. The water treatment landscape in India is characterized by a lack of infrastructure and resources, with only 37% of the generated sewage getting treated. The cost of treatment is significantly high, with a large portion of the total expense going towards the infrastructure needed to collect and transport sewage to centralized treatment plants. Also, these systems demand substantial operational and maintenance costs. In addition, the traditional sewage treatment plants are not designed for the removal of emerging contaminants (ECs) like pharmaceuticals and personal care products, which are becoming increasingly prevalent in wastewater systems. In this context, engineered natural treatment systems (ENTS) are emerging as a viable wastewater treatment and reuse option. In particular, constructed wetlands (CWs) are low-cost, passive natural treatment options with minimal capital investment and technological intervention. Furthermore, their treatment potential and ecosystem services have been proven effective in removing ECs along with organic pollutants and nutrients. This makes them an ideal option for sustainable water treatment and reuse, particularly in the low-income countries of the Global South. This chapter focuses on the performance of CWs in removing a wide range of pollutants from municipal wastewater. Special attention is given to ECs in wastewater systems and their fate and removal mechanisms in the CWs. This study also summarizes some of the successful practical models for the application of ENTS.

Keywords: engineered natural treatment systems, constructed wetlands, floating treatment wetlands, emerging contaminants, fate and removal mechanism

13.1 INTRODUCTION

Ensuring the adequacy of water in terms of quantity and quality is one of the biggest challenges in the present century. Rapid population growth, urbanization, resource-intensive developmental activities, and irrigated agriculture are critical anthropogenic drivers of increasing water demand. The available freshwater resources, particularly in urban and peri-urban regions of low-income countries, cannot cater to the rising demand as they are increasingly stressed due to over-exploitation and anthropogenic pollution. Reusing wastewater after subjecting it to appropriate levels of treatment is widely accepted as a potential solution to reduce the stress on scarce water sources. However, there is a severe lacuna in wastewater treatment infrastructure in low- and middle-income countries.

According to the Central Pollution Control Board (CPCB, 2021), there was a functional capacity to treat only 37% of the generated sewage in 2020. Many sewage treatment plants (STPs) in the country are not operational or are under construction, leading to a significant amount of untreated wastewater being released into the environment. Moreover, 39% of STPs functional in India do not conform to the mandated discharge standards. Furthermore, the performance of conventional STPs is not satisfactory for the removal of nutrients and emerging contaminants (ECs) (Chen *et al.*, 2019). Hence there is an urgent need to explore and adopt decentralized sustainable treatment systems that are less energy intensive and are equipped to remove both conventional and ECs.

This chapter describes how nature-based systems for wastewater treatment, such as constructed wetlands (CWs), can be used as the solution for sustainably and economically removing pollutants from water systems. These systems utilize natural agents such as plants, soil, microorganisms, and solar energy to remove or degrade pollutants. Additionally, nature-based treatment systems promote biodiversity and carbon sequestration and can act as buffers to combat the negative impacts of storms and floods (Cross *et al.*, 2021). They have also effectively removed ECs such as antibiotics, hormones, and pesticides and personal care products (PPCPs) (Chen *et al.*, 2019). This chapter evaluates the use of CWs as engineered natural treatment systems (ENTS) to remove conventional and ECs from water. It also discusses the potential environmental risks associated with the release of treated effluent from CWs into water bodies or its reuse for irrigation purposes.

13.2 PHARMACEUTICALS AND PERSONAL CARE PRODUCTS

13.2.1 Sources and categories of PPCPs

ECs comprise a broad spectrum of chemicals, including PPCPs, insecticides, and preservatives. Pharmaceutical compounds have garnered significant attention due to their biologically active nature. ECs stem from natural and anthropogenic sources, with domestic wastewater as the primary contributor (Figure 13.1). Human and animal excretion of drugs via urine and faeces contributes substantially to the prevalence of pharmaceuticals in domestic sewage. Additionally, household and personal care products play a significant role in the presence of ECs in wastewater. Their fate in the environment is subject to various factors, including their physicochemical properties, sources, and pathways. The most common pathway for ECs is through wastewater treatment plants (WWTPs). If these facilities fail to remove ECs, they can re-enter the water cycle and have a significant environmental impact. Moreover, some ECs can transform in the environment, forming transformation products with varying toxicity profiles, which complicates the assessment of their environmental fate and effects.

13.2.2 Occurrence of PPCPs in various environmental matrices

The presence of ECs in riverine systems is a growing concern due to their potential impact on human health and the environment. Asia has the highest presence of PPCPs in riverine systems, with 48 different compounds detected, and followed by Europe, with 45 pollutants (Wilkinson *et al.*, 2022). In India, the installed wastewater treatment capacity is less than the volume of wastewater generated; leading to the discharge of partially treated or untreated sewage into water bodies (CPCB, 2021). Consequently, these contaminants have been detected in major Indian water bodies, posing a moderate-to-high risk to aquatic organisms. Furthermore, the natural attenuation of ECs is highly uncertain, and the potential for their presence in drinking water sources cannot be ignored. Therefore,

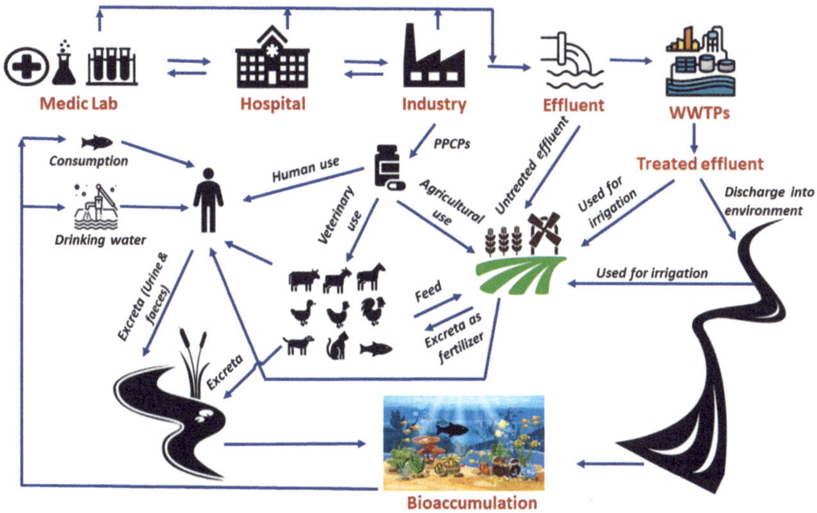

Figure 13.1 Sources and pathways of ECs.

it is imperative to monitor and manage the occurrence of ECs in riverine systems, particularly in areas where inadequate wastewater treatment facilities exist.

13.2.3 Adverse effects of PPCPs

Most of the ECs, especially PPCPs, are manufactured to target specific receptors, enzymes, or biological processes in humans and animals to achieve expected medicinal effects. Thus, when these compounds are released into the environment, their exposure to the aquatic organisms in the receiving water bodies can cause changes in histological and behavioural effects, biochemical activities, and gene regulation in those organisms. Significant detrimental effects of ECs include defects to the reproductive systems, hormonal imbalance, carcinogenicity, development of antibiotic resistance in microbes, and formation of antibiotic resistance genes. Even though there are no known negative impacts of ECs on human health, the possibility of long-term effects cannot be ignored.

13.3 ENGINEERED NATURAL TREATMENT SYSTEMS

ENTS are techniques that mimic natural processes for removing pollutants from wastewater, using biological, physical, and chemical degradation pathways found in natural systems like wetlands. Examples of popular ENTS include CWs, waste stabilization ponds (WSPs), and floating wetlands. WSPs can effectively remove organic matter and pathogens but require a longer retention time, large land area and regular desludging (Figure 13.2a). WSPs are often paired with phytoremediation techniques that use plants to extract, filter, volatize, or stabilize pollutants from water.

Different forms of phytoremediation can be applied depending on the mode of application of plants. Free-floating macrophyte systems use aquatic plants, such as Azolla and duckweed, which float on the surface of the water with roots hanging underneath (Figure 13.2b). However, they are not commonly used in wastewater treatment due to their susceptibility to washout and negative aesthetic appeal (Headley & Tanner, 2012). Floating treatment wetlands (FTWs) use buoyant mats to support emergent macrophytes, with their roots fully immersed in water (Figure 13.2c). This approach is more resilient to washout, limits the chance of parasitic growth, and the plants are less vulnerable to fluctuations in water depth.

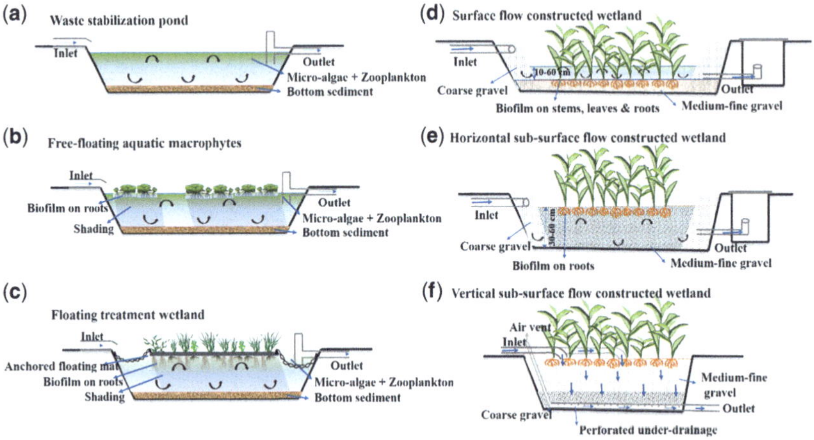

Figure 13.2 Schematics of different types of ENTS. (*Source*: modified from Headley and Tanner, 2012).

The organic matter removal in ENTS is primarily contributed by bacterial metabolism under aerobic conditions arising from photosynthetic aeration by micro and macro flora. A combination of several factors like denitrification, reduction by sulphate-reducing bacteria, microbial uptake, adsorption and assimilation by plant species have been attributed as the mechanisms for nutrient removal. Pathogenic microorganisms are removed due to UV irradiation and the release of anti-microbial compounds from plant roots.

13.3.1 Constructed wetland as an ENTS

CWs are among the most widely used ENTS for treating wastewater. CWs comprise a combination of substrate, vegetation, and microbes. They employ various physical and chemical processes in natural wetlands in a controlled setting. CWs are widely adopted for treating different wastewaters due to their simplicity and ease of operation. The application of CWs is governed by critical factors such as target pollutants, geographical location, availability of land, and financial gain.

13.3.2 Types and components of constructed wetlands

Constructed wetland systems come in two primary forms: surface flow and sub-surface flow (Figure 13.2d and e). Sub-surface flow CWs can be further classified as vertical, horizontal, and hybrid. The main components of CWs include basin structure, liner, substrate, plants, and inlet–outlet arrangements. A liner is used to prevent groundwater contamination. In horizontal flow-constructed wetlands (HFCW), the wastewater infiltrates slowly into the inlet zone and travels through the bed substrate horizontally until it reaches the outlet zone (Figure 13.2e). In contrast, in vertical flow-constructed wetlands (VFCW), wastewater enters from the top surface and gets collected at the bottom (Figure 13.2f). A hybrid wetland system combines vertical and horizontal flow systems (Ilyas & Van Hullebusch, 2020). CWs traditionally necessitate significant land area, making them costly to implement. Recently, modified designs, such as baffled flow CWs, artificial aerated CWs, step-feeding CWs, and green roof water recycling systems (GROW), have found their application in many field-scale scenarios.

13.3.3 Removal of organics, nutrients, and pathogens

The removal of organics in CWs can be achieved under aerobic and anaerobic conditions. Depending on the CW's design configurations, the biochemical activities of the microbial population can also vary. VFCW increases the redox potential and oxygen content of wastewater and also aids in the enhancement of chemical oxygen demand (COD) and ammonia removal. In contrast, a declining redox potential and dissolved oxygen (DO) trend were observed in HFCW, resulting in lower COD and ammonia removal. Nitrogen removal can occur through ammonification, nitrification, denitrification, plant uptake, volatilization, adsorption over materials, ion exchange, and incorporation into biomass. Ravichandran and Philip (2022a) recently documented the outstanding performance of natural zeolites and waste autoclaved aerated concrete (AAC) blocks for removing NH_3-N and TN due to their higher cation-exchange capacity. In the case of phosphate, sorption and deposition over the substrate materials, plant uptake, assimilation, and metabolism by microbes governs their removal in CWs. The phosphate removal is also associated with the level of Al, Ca, and Fe oxides in the filling materials (Karthik & Philip, 2021). The coliform removal in CWs is attributed to physical processes (sedimentation, filtration by roots), chemical processes (biocides secretion), biological processes (antibiosis, lytic, bacterial attack), and natural die-off (Zraunig *et al.*, 2019). In-situ pathogen inactivation processes may be enhanced by adding anti-microbial filter media and through various configurations of biofilters.

13.3.4 Removal of heavy metals

The removal of heavy metals in ENTS can occur through different pathways such as (1) adsorption of metals to the soils and sediments; (2) sedimentation; (3) exchange and complexation of cations and anions; (4) precipitation and co-precipitation as insoluble ions; (5) plant uptake, and (6) microbial metabolism. The efficiency of heavy metal removal can be improved by the utilization of appropriate filling materials, macrophyte species (poly-culture, hyperaccumulators), bioaugmentation of enriched microbial population, coupling with electrochemical systems, hybrid mode of CWs, the addition of chemicals (coagulants, chelating agents, external carbon source), and optimization of operating parameters (hydraulic loading, recirculation, water depth, tidal action) (Yu *et al.*, 2022).

13.4 FATE OF PPCPS IN ENTS AND THEIR REMOVAL MECHANISMS

13.4.1 Attenuation of PPCPs in CWs

The attenuation of PPCPs and their removal mechanisms observed in the constructed wetland vary with the characteristics of the compounds of interest (polar/ionizable and non-polar/non-ionizable). However, the pre-treatment systems (Imhoff tanks, septic tanks, and settling tanks) are usually responsible for removing 3–21% of PPCPs, probably due to their attachment to removed particulate matters (Ravichandran *et al.*, 2021).

In the case of polar hydrophilic compounds such as caffeine, acetaminophen, ciprofloxacin, methylparaben, atenolol, and so on, the average reduction was in the range of 80–99%. In contrast, the percentage removal of hydrophobic compounds such as triclosan, diclofenac, ibuprofen, triclocarban, gemfibrozil, and so on, was in the range of 76–91% in different CW configurations. The primary removal pathway for these pollutants is aerobic and anaerobic microbial degradation. Interestingly, in hybrid wetland systems, which comprise of aerobic and anaerobic conditions, a higher removal of PPCPs was documented for various ECs (Chen *et al.*, 2019). However, the neutral and hydrophilic pollutants can take up the plant roots via hydrogen bonding with water through the transpiration stream (Madikizela *et al.*, 2022). The hydrophobic nature of the contaminants influences their affinity for sorption over the substrate materials (Zraunig *et al.*, 2019). Due to the electrostatic interaction, the cationic PPCPs could adsorb over the biofilm of the filling medium and get degraded by the microbes in the planted wetland system.

The percentage removal of the non-polar pollutants such as erythromycin, diethyl phthalate, and bisphenol A can be in the range of 58–85%. These pollutants are moderately hydrophobic, and their removal in CWs is highly influenced by biodegradation and sorption over the filling materials. The stable structure of certain PPCPs and the presence of chlorine in them make them recalcitrant to degradation. However, some studies have reported that the higher redox potential could significantly promote the removal of recalcitrant compounds. Increased organic content caused a significant reduction of CBZ. Several studies have documented the seasonal effects on the performance of the CWs. Higher temperatures during the summer can enhance the microbial degradation rate of pollutants than during the winter (Ramprasad & Philip, 2018). However, the higher pollutant concentration in the influent and plant biomass may lead to an insignificant seasonal difference in PPCPs removal (Chen *et al.*, 2019). There is a significant correlation between PPCPs removal and the physicochemical parameters of wastewater.

13.4.2 The contributions of different removal mechanisms

A large fraction of PPCPs undergoes biodegradation in the wetland system (>85%), with sorption onto filling materials up to 12.5% and accumulation in plants up to 1% (Ravichandran *et al.*, 2021). However, the biodegradation of PPCPs involves several intermediated processes and organisms influenced by various environmental conditions. Although biodegradation and metabolization reduce the toxicity of parent compounds, the toxicity of the intermediates dominates in certain cases. The microbial community and its metabolic functions vary depending upon CW design (flow conditions, substrates, plant species, operating conditions, etc.). Microbial diversity could be more related to pollutant removal than microbial richness. Enhanced reduction of PPCPs was observed in bio-augmented CW (Ravichandran & Philip, 2022a). In the case of other factors (plant uptake and accumulation and the sorption over the substrates), the exposure concentration, log K_{ow}, water solubility, and molecular weight (decreasing order) are the influential factors (Ravichandran & Philip, 2022b).

13.5 FACTORS AFFECTING THE PERFORMANCE OF ENTS

The potential of ENTS to attenuate the micropollutants depends upon components such as substrates, vegetation, microbes, operating conditions (batch/continuous mode), redox potential (aerobic/anaerobic), and so on.

13.5.1 Flow configurations

Flow configurations in CWs can be horizontal, vertical, or a combination (hybrid). These configurations notably impact micropollutants removal since the flow pattern affects the micro-environmental conditions inside the bed. Relatively higher redox conditions and oxygen availability in the VFCW influence the removal of pollutants. However, an anoxic–anaerobic condition primarily exists in HFCW; therefore, the rate is generally slower than that for the aerobic process in VFCW. In the case of hybrid systems, the pollutants are exposed to the co-existing aerobic and anoxic–anaerobic conditions of the wetland bed, which results in higher removal (Ilyas & van Hullebusch, 2020). Furthermore, the presence and characteristics of plants and substrate materials significantly influence the micro-environmental conditions (pH, DO, redox potential) inside the wetland bed. The changes in these conditions can substantially affect microbial activity, abundance, species, and pollutant removal.

13.5.2 Substrate materials

Various materials (natural, agricultural or industrial waste materials, and artificial materials) are employed as CW supporting matrices. The characteristics of the substrate play essential roles in microbial development and pollutant removal (Ji *et al.*, 2022) (Figure 13.3). Thus, choosing appropriate substrates for CWs aids in effective pollutant removal. Most of the CWs employ gravel and sand as the filling materials. Though they perform satisfactorily for organics removal, researchers have been paying greater attention to identifying or developing cost-effective and efficient materials for removing nutrients and micropollutants. The selection of appropriate substrate material with high pollutant sorption capacity can enhance the ability of CWs to remove pollutants. Karthik and Philip (2021) screened the materials such as wood charcoal, LECA, natural zeolites and waste AAC blocks for the removal of three pharmaceuticals and nutrients under various operating conditions. The key characteristics identified for the suitable filling material are presented in Figure 13.3. Even though

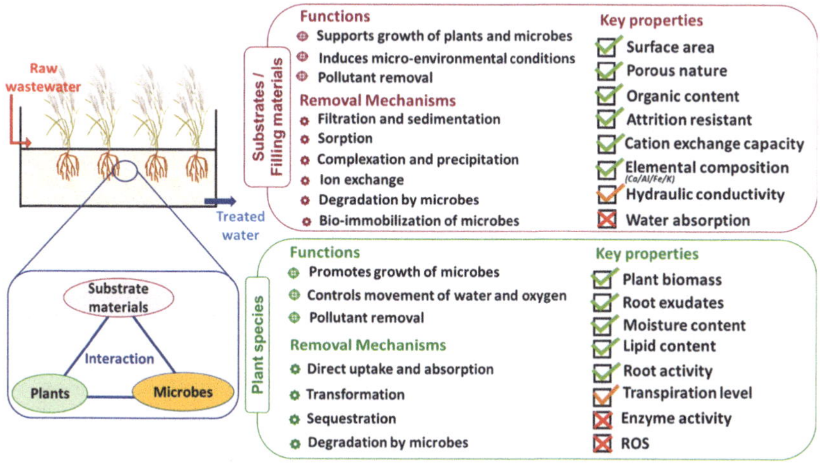

Figure 13.3 Functions, removal mechanisms and critical characteristics of filling materials and plant species of CWs.

the common materials have low sorption capacity for PPCPs, they would increase the functionality of microbial communities in CW. The highly porous nature of waste AAC blocks accommodated higher microbial biomass and resulted in enhanced removal of pharmaceuticals and nutrients (Ravichandran & Philip, 2022a). Additionally, such porous materials can be used as microbial carriers in bio-enhanced degradation systems. Also, using waste material as the substrate can be a cost-effective and environmentally friendly choice. It can be beneficial to use a combination of materials synergistically to prevent operational issues such as clogging and leaching of contaminants and improve the efficiency of micro-pollutant removal.

13.5.3 Plant species

Madikizela *et al.* (2022) extensively studied the efficacy of ENTS with different macrophytes for the removal of organics, nutrients, and toxic metals from wastewater. The widely used plant species in ENTS include *Phragmites*, *Canna*, *Typha*, *Scripus*, *Juncus*, and *Cyperus*. The specific criterion for selecting pertinent macrophytes may be established by understanding the relationship between the plant species and pollutant removal. In this regard, the plant features such as plant biomass, photosynthetic activity, evapotranspiration, root activity, and so on are essential for the removal of organics and nutrients. The bioaccumulation potential of the plants is important for the remediation of heavy metal-loaded waste streams. Ravichandran and Philip (2022b) identified plant biomass, transpiration rate, chlorophyll content, root exudates, and root activity as critical indicators for the selection of plants (Figure 13.3). Root exudates released by plants act as catalytic agents and assist in the removal of PPCPs. However, the secretion and characteristics of root exudates depend on the inherent physiology of plant species and the existing environmental conditions.

Once taken up by the plant roots, PPCPs can be degraded in the plant tissues through a series of biochemical pathways. The exposure of plants to PPCPs leads to the accumulation of reactive oxygen species (ROS) in plants, causing oxidative stress and growth defects by disrupting physiological and biochemical processes. It, in turn, diminishes the pollutant removal and degradation potential of plant species, and the variations in their physiological response can reduce the tolerance of plants to oxidative stress. However, certain plants can bear the over-production of ROS by upregulating antioxidant defence mechanisms (Ravichandran & Philip, 2022b). Also, the accumulation level and pattern of PPCPs in the plant tissues decline during the recuperation period, indicating the ability of the plant to recover from the stress. A recent study also found the influence of macrophytes on coliform removal through direct plant anti-microbial effects, the support of enhanced populations of antagonistic microbiota, and reduced populations of faecal microbe supporters/mutualists. Incorporating different macrophyte species also enriches the microbial diversity and results in the enhanced removal of PPCPs in CWs.

13.5.4 Operating conditions

Higher removal efficiencies can be achieved with longer HRT, associated with a longer hydraulic pathway and residence time, enabling better contact with the substrates and microbes (Ramprasad & Philip, 2018). Moreover, the degradation of PPCPs may be boosted by lowering the loading frequency in the planted microcosms. In addition, intensification of CWs through external or forced aeration, recirculation of treated or partially treated effluents, partial saturation of bed, co-plantation of various plant species, and so on, are effective. All these modifications influence the redox conditions in the bed, enhancing the diversity of microbial populations, which in turn degrade complex pollutants.

13.6 CASE STUDIES FOR THE APPLICATION OF ENTS

13.6.1 Decentralized rural wastewater treatment using constructed wetlands

Rural and peri-urban regions of the global south witness serious environmental issues due to poor infrastructure, weak governance, and inadequate waste management practices. To address such problems, the constructed wetland was adopted as a decentralized wastewater treatment option in the Vichoor panchayat of Tiruvallur district (Tamil Nadu, India) (Ravichandran *et al.*, 2021). It serves a population of nearly 450, where the greywater and effluent of septic tanks from each household were collected and transported to the CW system through a small-bore system. The treatment system includes a septic tank (three-chambered, 44.5 m^3), a settling tank (three-staged, 3.8 m^3) and a vertical flow subsurface wetland (23.2 m^2) with *Canna indica* and *Phragmites australis* (Figure 13.4).

The fate of 14 micropollutants was monitored in the treatment units. Stimulant drugs (caffeine), anti-microbial (triclosan) and plasticizer (bisphenol A and diethyl phthalate) were detected frequently in the samples. The average influent and effluent PPCPs concentrations were 256.4 and 39 μg/L, respectively. The treated effluent quality met the discharge standards of the Central Pollution Control Board (COD – 37.4 mg/L $<$ 50 mg/L; TN – 5.63 mg/L $<$ 10 mg/L; TSS – 22.4 mg/L). The risk of residual PPCPs in the reclaimed water was assessed for discharging into the water bodies (ecological

Figure 13.4 (a) treatment scheme adopted in Vichoor, Tiruvallur, Tamil Nadu, India, (b) photograph of VFCW planted with *Canna indica* and *Phragmites australis*.

risk) and reusing for gardening activities (human health risk). The ecological risk was observed as low-to-mild level whereas the human health risk (carcinogenic and non-carcinogenic risk) was at a safer level of exposure.

13.6.2 In-situ remediation of polluted lake using floating treatment wetland

Many cities in India have implemented FTW systems to treat wastewater. The FTW in Neknampur lake, located in Hyderabad, is an example of a successful application. The Neknampur lake, located in Telangana state, India, covers an area of 25 acres and is polluted by a variety of sources, including domestic wastewater from the settlements on the banks of the lake, untreated sewage from nearby townships and polluted runoff. In 2018, a Floating Treatment Wetland project was implemented to remediate the lake. The buoyant platform for supporting the plants was made of bamboo, Styrofoam, and gunny bags, and nearly 3500 saplings of macrophytes such as vetiver, cattails, bulrush, Canna, and Citronella were planted on the floating mats. The design, development, and implementation of the FTW were led by Dhruvansh, an NGO, the Hyderabad Metropolitan Development Authority (HMDA) and the district authorities. The FTW had significant success in reducing the water pollution of the lake. The outlet quality was DO – 8.3 mg/L, nitrate – 15 mg/L, phosphate – 2–6 mg/L, BOD – 3.8 mg/L and COD – 56 mg/L (TSPCB, 2019).

13.7 SUMMARY

CWs can be a cost-effective and sustainable alternative to traditional STPs. However, proper design and operation are key factors. The incorporation of pre-treatment systems can improve performance of CWs significantly with lesser area requirements and operational issues. In addition, through the proper selection of substrate materials and plants, CW could effectively eliminate the PPCPs and curtail the environmental risk. During extreme weather events, when CWs are vulnerable to flooding, floating wetlands may be a better option as they are self-adaptive to water depth and less prone to plant stress. FTWs may also be suitable for in-stream applications in areas with limited land availability. Post-treatment options such as microalgae and microbial fuel cells can be employed to improve the treated water quality, along with options like green roofs and living walls that provide economic and social benefits. Thus, there is immense scope for the application of ENTS as a decentralized treatment option in major metropolitan settlements, and the treated water can be effectively and safely used for secondary purposes.

REFERENCES

Chen J., Deng W. J., Liu Y. S., Hu L. X., He L. Y., Zhao J. L., Wang T. T. and Ying G. G. (2019). Fate and removal of antibiotics and antibiotic resistance genes in hybrid constructed wetlands. *Environmental Pollution*, **249**, 894–903, https://doi.org/10.1016/j.envpol.2019.03.111

CPCB (2021). National Inventory of Sewage Treatment Plants Central Pollution Control Board. Parivesh Bhawan, East Arjun Nagar, Delhi.

Cross K., Tondera K., Rizzo A., Andrews L., Pucher B., Istenič D., Karres N. and Mcdonald R. (2021). Nature-Based Solutions for Wastewater Treatment. London, UK: IWA Publishing, https://doi.org/10.2166/9781789062267

Headley T. R. and Tanner C. C. (2012). Constructed wetlands with floating emergent macrophytes: an innovative storm water treatment technology. *Critical Reviews in Environmental Science and Technology*, **42**(21), 2261–2310, https://doi.org/10.1080/10643389.2011.574108

Ilyas H. and Van Hullebusch E. D. (2020). A review on the occurrence, fate and removal of steroidal hormones during treatment with different types of constructed wetlands. *Journal of Environmental Chemical Engineering*, **8**, 103793, https://doi.org/10.1016/j.jece.2020.103793

Ji Z., Tang W. and Pei Y. (2022). Constructed wetland substrates: a review on development, function mechanisms, and application in contaminants removal. *Chemosphere*, **286**, 131564, https://doi.org/10.1016/j.chemosphere.2021.131564

Karthik R. M. and Philip L. (2021). Sorption of pharmaceutical compounds and nutrients by various porous low cost adsorbents. *Journal of Environmental Chemical Engineering*, **9**, 104916, https://doi.org/10.1016/j.jece.2020.104916

Madikizela L. M., Botha T. L., Kamika I. and Msagati T. A. M. (2022). Uptake, occurrence, and effects of nonsteroidal anti-inflammatory drugs and analgesics in plants and edible crops. *Journal of Agricultural and Food Chemistry*, **70**, 34–45, https://doi.org/10.1021/acs.jafc.1c06499

Ramprasad C. and Philip L. (2018). Greywater treatment using horizontal, vertical and hybrid flow constructed wetlands. *Current Science*, **114**, 155–165, https://doi.org/10.18520/cs/v114/i01/155-165

Ravichandran M. K. and Philip L. (2022a). Assessment of the contribution of various constructed wetland components for the removal of pharmaceutically active compounds. *Journal of Environmental Chemical Engineering*, **10**, 107835, https://doi.org/10.1016/j.jece.2022.107835

Ravichandran M. K. and Philip L. (2022b). Fate of carbamazepine and its effect on physiological characteristics of wetland plant species in the hydroponic system. *Science of the Total Environment*, **846**, 157337, https://doi.org/10.1016/j.scitotenv.2022.157337

Ravichandran M. K., Yoganathan S. and Philip L. (2021). Removal and risk assessment of pharmaceuticals and personal care products in a decentralized greywater treatment system serving an Indian rural community. *Journal of Environmental Chemical Engineering*, **9**, 106832, https://doi.org/10.1016/j.jece.2021.106832

TSPCB (2019). Furnishing information under RTI Act 2005. Accessed 28 August 2023. https://yourti.in/document/hu58j3iw/

Wilkinson J. L., Boxall A. B. A., Kolpin D. W., Leung K. M. Y., Lai R. W. S., Wong D., Ntchantcho R., Pizarro J., Mart J., Echeverr S., Garric J., Chaumot A., Gibba P., Kunchulia I., Seidensticker S., Lyberatos G., Morales-salda J. M. and Kang H. (2022). Pharmaceutical pollution of the world's rivers. *Proceedings of the National Academy of Sciences*, **119**(8), e2113947119, https://doi.org/10.1073/pnas.2113947119

Yu G., Wang G., Chi T., Du C., Wang J., Li P. and … Chen H. (2022). Enhanced removal of heavy metals and metalloids by constructed wetlands: a review of approaches and mechanisms. *Science of the Total Environment*, **821**, 153516, https://doi.org/10.1016/j.scitotenv.2022.153516

Zraunig A., Estelrich M., Gattringer H., Kisser J., Langergraber G., Radtke M., Rodriguez-Roda I. and Buttiglieri G. (2019). Long term decentralized greywater treatment for water reuse purposes in a tourist facility by vertical ecosystem. *Ecological Engineering*, **138**, 138–147, https://doi.org/10.1016/j.ecoleng.2019.07.003

doi: 10.2166/9781789063714_0155

Chapter 14

Carbon-based filters for water and wastewater treatment

G. Vajitha[1], K. P. Madan[1], C. Tewari[2], N. G. Sahoo[2] and Shihabudheen M. Maliyekkal[1]*

[1]Department of Civil and Environmental Engineering, IIT Tirupati, Tirupati, India
[2]Department of Chemistry, Kumaun University, Nainital, India
**Corresponding author: shihab@iittp.ac.in*

ABSTRACT

Carbon is a unique and exciting material with extensive use in various fields of science and engineering. Purification of water and wastewater is one of the earliest applications of carbon, which has now become an integral part of most household- and community-based water treatment units. Abundant availability, low cost and toxicity, high chemical inertness and thermal stability, and high surface area and pore volume make carbon and its derivatives ideal for water and wastewater treatment. Carbon has been used to remove excess chlorine, disinfection by-products and their precursors, taste and odour compounds, synthetic organic chemicals, heavy metals, and salts from water. Many bulk and nanoscale carbon modifications have been developed, expecting improved properties and enhanced water-purifying ability. This chapter reviews the evolution and application of carbon-based materials in water and wastewater treatment. The synthesis, removal mechanism, and factors affecting purification by carbon filters are discussed. The challenges and future prospects of carbon-based filtration technology are discussed briefly.

Keywords: activated carbon, adsorption, nanoscale carbon, treatment, water and wastewater

14.1 INTRODUCTION

The use of carbon filters dates back several hundred years and is considered one of the oldest and most effective ways of water purification. In its current form, activated carbon (AC) has a short history. Nevertheless, several records indicate that carbon usage in its earliest form, called wood char (charcoal), dates back to ancient times. Sanskrit writings dating to 200 C.E. describe the usage of charcoal-based filters for water purification. They recommend disinfecting water by filtering it through coal after storing it in copper vessels and exposing it to sunlight to remove impurities (Cecen & Aktas, 2011). The mechanism underlying charcoal's properties was not recognized until much later, in the eighteenth century, during Scheele's (1773) gas treatment using carbon. A decade later, Lipscombe prepared a carbon material to purify potable water, paving the way for the commercial applications of AC in water and wastewater treatment. AC was first produced at

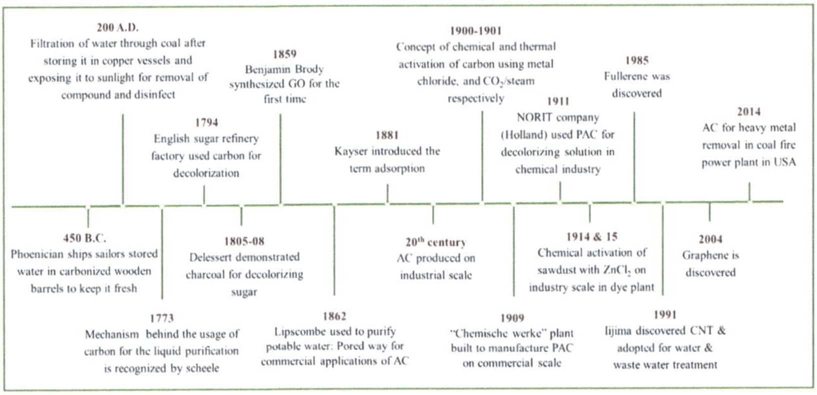

Figure 14.1 Evolution of carbon-based filters for water and wastewater treatment (information obtained from Cecen and Aktas, 2011, and represented pictorially).

the beginning of the twentieth century on an industrial scale. In 1901, Von Ostreyko, a Swedish chemist, obtained patents for developing AC on a commercial scale through chemical and thermal (or physical) activation, with metal chlorides and carbon dioxide and steam, respectively (Ostreyko, 1901). Nanoscale carbon was first used for water treatment at the end of the twentieth century. Now different derivatives of carbon in bulk and nanoforms are used in water and wastewater treatment. The major milestones in the evolution of carbon-based materials for water and wastewater treatment are shown in Figure 14.1.

Currently, carbon and its derivatives are employed as adsorbents, photocatalysts, dechlorinating agents, and membranes for removing diverse contaminants in water. This is primarily due to their unique physical and chemical properties, such as high specific surface area (SSA), good thermal stability, high porosity, presence of functional groups, low cost, and toxicity. The most extensively used carbon materials for water and wastewater treatment are AC, graphene (G), graphene oxide (GO), single-walled carbon nanotubes (SWCNT), and multi-walled carbon nanotubes MWCNT. This chapter reviews the diverse applications of these carbon materials in water and wastewater purification. The synthesis, properties, influencing factors, challenges, and opportunities of using carbon-based water purification units are discussed.

14.2 ACTIVATED CARBON

AC is a microporous form of carbon with a high surface area, well-developed pore structure, excellent physical and chemical stability, a high degree of surface reactivity, high pore volume, and adsorption capacity. These advantages of AC aid in the removal of pollutants such as pharmaceuticals, metallic and non-metallic pollutants, dyes, and taste and odour. Generally, two significant steps are involved in the synthesis of AC. The first step is carbonization and the second is activation. Carbonization is done through a pyrolysis process with a temperature ranging from 400 to 600°C in an inert atmosphere to produce biochar. This step removes undesirable by-products and volatile low molecular weight fractions from carbon precursors. Activation (physical, chemical, or physiochemical) is carried out after the carbonization to achieve high surface area, required pore size and volume, and functional groups and surface charge. The activation phase is required to enhance the pore volume and pore diameter by eliminating the disorganized carbon. Currently, AC accounts for 80% of carbon materials, and the global AC market is predicted to reach USD 8.9 billion by 2026 at a compound annual growth rate (CAGR) of 9.3%. Though the growth potential is high, the scarcity of raw materials like coconut shells can hamper growth and leads to production price hike (Fortune Business Insights, 2022). Therefore, a more sustainable and reliable carbon precursor is needed to sustain the growth and meet the demand.

14.3 NANOSCALE CARBON

Carbon nanomaterials (CNMs) are gaining interest in water and wastewater treatment due to their properties, such as the high surface-area-to-volume ratio, high thermal and chemical stability, ease of surface functionalization, and catalytic potential. The various CNMs employed for the treatment of wastewater include fullerenes (0D), carbon nanotubes (CNTs) (1D), and G (2D). The classification and significance of CNMs are presented in Table 14.1. Different synthesis methods are available for producing fullerenes and CNTs. These include the thermal evaporation of graphite in an inert atmosphere, the laser ablation of graphite, the microwave method, and the soot obtained after the combustion/pyrolysis of aromatic hydrocarbons. The most common fullerene produced is C_{60}, which contains 12 pentagons and 20 hexagons arrangement of 60 carbon atoms. In contrast, the structure

Table 14.1 Classification and significance of CNMs.

S. No	Classification of CNMs		Properties	Reference
1	0-D	Carbon quantum dots (CQD)	sp^2 or sp^3 hybridized carbon Diameter < 10 nm Strong electrochemical, electronic and quantum size effects, easy surface functionalization, and adjustable composition	Madima *et al.* (2020)
		Nanodiamond	sp^3 hybridized carbon Diameter < 100 nm High SSA > 300 m^2/g Tunable surface structure and hydrophilic	
		Fullerene, C_{60}	sp^2 hybridized carbon Diameter: 0.7 nm Surface–volume ratio: 3:77 Non-toxic and bio-compatible	
2	1-D	CNTs	sp^2 hybridized carbon Diameter: 0.8–4 nm High surface area of SWCNTs: 2–1587 m^2/g; MWCNTs: 22–1670 m^2/g Excellent aspect ratio: 2.6–3.3 Outstanding thermal conductivity: 2980–6000 W/m K	Birch *et al.* (2013)
3	2-D	G	sp^2 hybridized carbon High thermal conductivity: 3080–5150 W/m K Electrical conductivity: 0.53×10^6 to 0.84×10^6 S/cm	Ghosh *et al.* (2007)
		GO	sp^2 and sp^3 hybridized carbon Hydrophilic and highly dispersible	
		RGO	sp^2 and sp^3 hybridization Insoluble, hydrophobic, and mechanically stable	
4	3-D	Graphite	sp^2 hybridized carbon High thermal conductivity: < 2195 W/m K Electrical conductivity: < 2×10^2 S/cm Insoluble in water and organic solvents	Alofi and Srivastava (2013)

D: dimensional; SWCNTs: single-walled CNT; MWCNT: multi-walled CNT.

of CNTs is cylindrical with perfectly connected hexagonal atoms. G is a more recent addition to the family of carbon allotropes. Since its discovery in 2004, this wonder material has gained much attention among the research community due to its wide possible applications. Currently, many methods of G synthesis, including top-down and bottom-up approaches, have been developed, expecting distinct physical and chemical properties. These methods include but are not limited to scotch tape-based micro-cleaving of highly oriented pyrolytic graphite, liquid exfoliation of graphite crystal, epitaxial growth on silicon carbide, chemical reduction, and chemical vapour deposition (CVD) (Mohandoss *et al.*, 2017). However, the main challenges in using CNMs in water and wastewater treatment are lack of quality control, high cost of production, difficulty in solid–liquid separation, and poor yield. To reduce cost and improve sustainability, the current focus is on developing CNMs from waste-derived carbon materials as a precursor.

14.4 APPLICATION OF CARBON IN WATER AND WASTEWATER TREATMENT

CNMs can be used for various purposes in water and wastewater treatment, as shown in Figure 14.2.

14.4.1 Adsorption

Adsorption is an advanced tertiary treatment technique widely used to remove recalcitrant compounds, heavy metals, emerging contaminants, and other pollutants from water and wastewater. Adsorption is a popular treatment technology and is preferred over other technologies due to its cost-effectiveness, ease of use, reuse potential, and versatility, especially in removing trace concentrations of pollutants (Naga Jyothi *et al.*, 2021). A general schematic diagram of the adsorption process and possible mechanisms of removing pollutants are shown in Figure 14.3.

14.4.1.1 Removal of synthetic organic chemicals

Synthetic organic chemicals (SOCs) are manufactured carbon-based chemical substances that include pesticides, dyes, per- and poly-fluoroalkyl substances, pharmaceutically active compounds, and several endocrine-disrupting substances. The increasing societal and industrial dependence on these toxic chemicals poses health threats to humans due to their widespread occurrence across global aquifers. Therefore, to remove SOCs, carbon-based materials, including biological activated carbon (BAC), granular activated carbon (GAC), CNTs, and G and its derivatives are in use. Among these, CNTs and G have shown more promising results. In the case of AC, the presence of abundant micropores limits SOCs' uptake due to the quick saturation of large molecules. Unlike ACs, CNTs show promising uptake of SOCs due to their pore-filling mechanism, but the aggregation and natural organic matter (NOM) can hinder its performance.

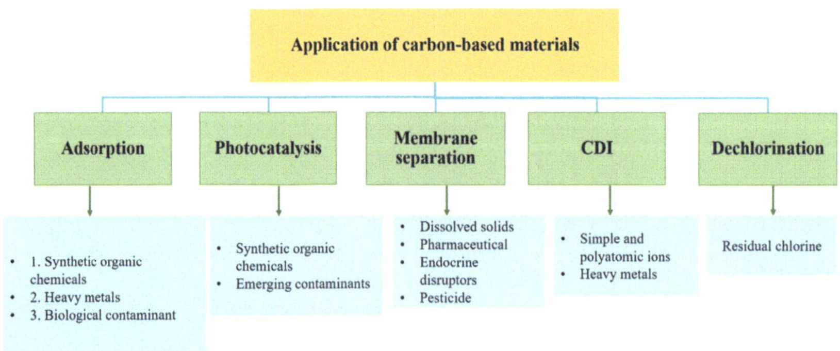

Figure 14.2 Application of CNM in water and wastewater treatment.

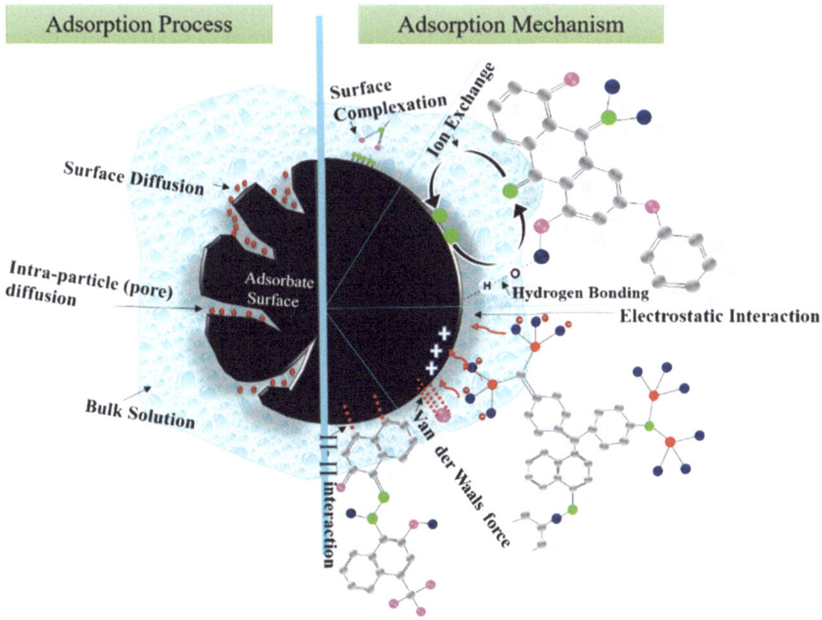

Figure 14.3 Schematic mechanism of a generic dye's adsorption from the bulk liquid by AC. The figure is adapted from Azam *et al.* with the permission from Elsevier. (*Source*: Azam *et al.*, 2022).

A study by Maliyekkal *et al.* (2013) demonstrated the removal of various pesticides, such as chlorpyrifos, endosulfan, and malathion, using reduced graphene oxide (RGO). They reported pesticide removal capacities more than the self-weight of the adsorbent. Following this study, Koushik *et al.* (2016) studied the removal of 1,1,1,3,3,3-hexafluoro-2-propanol using an RGO-silver (Ag) nanocomposite. The incorporation of Ag nanoparticles on the RGO facilitated the degradation of the adsorbed SOCs, which was unique to RGO–Ag nanocomposites. No such degradation effect was observed with AC–Ag nanocomposites.

14.4.1.2 Heavy metal removal

Heavy metals like Cu, Fe, Mn, Co, Zn, and Ni are essential to humans' growth, metabolism, and the development of different organs. However, heavy metals like chromium, lead, arsenic, mercury, and cadmium are toxic to humans, animals, and the environment, even at trace concentrations. The food chain and water are the ways for entry of these heavy metals and their removal from water is warranted. Sreeprasad *et al.* (2011) demonstrated the removal of Hg(II) from water using RGO–metal/metal oxide composites (RGO–MnO_2 and RGO–Ag). Both materials exhibited excellent adsorption capacity in real water under field conditions. Lee *et al.* (2016) investigated the removal of heavy metals from industrial plating wastewater using carbon foam on a pilot scale. They reported a higher adsorption capacity for Cr (73.64 g/kg) than Cu (14.86 g/kg) and Ni (7.74 g/kg) during 350 h of operation. The higher removal was due to the surface precipitation of Cr in the adsorption bed.

14.4.1.3 Radionuclides

The widespread presence of radionuclides, such as uranium and thorium, in groundwater is an emerging concern in the drinking water sector. Although the primary source of uranium pollution is natural, anthropogenic sources such as fertilizer manufacturing, mining and processing mineral ores, the use of uranium in medicine and research, and so on, are responsible for accelerating the entry of uranium into groundwater (Gandhi *et al.*, 2022). However, the rate of dissolution depends on the geological conditions of the region. It is also found that the chemical toxicity of uranium is significantly higher than its radionuclide toxicity, and long-term exposure to it can cause various cancerous and non-cancerous diseases. Studies show that carbon-based materials are promising due to their higher acid–base stability (compared with common inorganic adsorbents) and high stability and thermal resistance (compared with organic exchange resins). AC and functionalized AC were employed in radionuclide removal by adsorption. The presence of functional groups enhanced adsorption capacity (Lu *et al.*, 2019). Surface-modified CNTs and GO were also studied for the removal of uranium in water. However, problems such as aggregation, nanotoxicity, the yield of synthesis and production cost must be addressed to realize its practical utility.

14.4.2 Photocatalysis

Photocatalysis is a simple and green technology for the degradation of organic pollutants in water. The process involves light absorption, charge separation, electron–hole transfer to the catalyst surface, and charge utilization for a redox reaction. Generally, semiconducting materials such as TiO_2, ZnO, CdS, and ZrO_2 are used as photocatalysts due to their distinct low band gap and ability to generate charge carriers. Upon excitation of these photocatalysts with light energy, electron–hole pairs are created, which promote redox reactions with the surrounding liquid medium to generate hydroxyl radicals, superoxide, and peroxyl radicals (Kusuma *et al.*, 2021). However, metal oxides are limited due to the need for a high-energy light source, low chemical stability, high cost, and poor quantum efficiency. Hence, several strategies have been investigated to synthesize new materials or modify existing materials responsive to visible light. The three-step method proposed for degrading organic dye molecules (methylene blue) using G hybrid composite photocatalysts is depicted schematically in Figure 14.4.

One such approach is integrating a photocatalytic system with CNMs that enhance charge mobility, increase the adsorption site, reduce energy band gap, and suppress the recombination rate of electron–hole pairs. CNMs also act as a substrate for fabricating novel photocatalytic materials, which help in the easy separation of the catalyst from the solution and also improve its reusability (Harafan *et al.*, 2022). Apart from CNMs, CNTs, G, and graphitic carbon nitride are used in photocatalytic

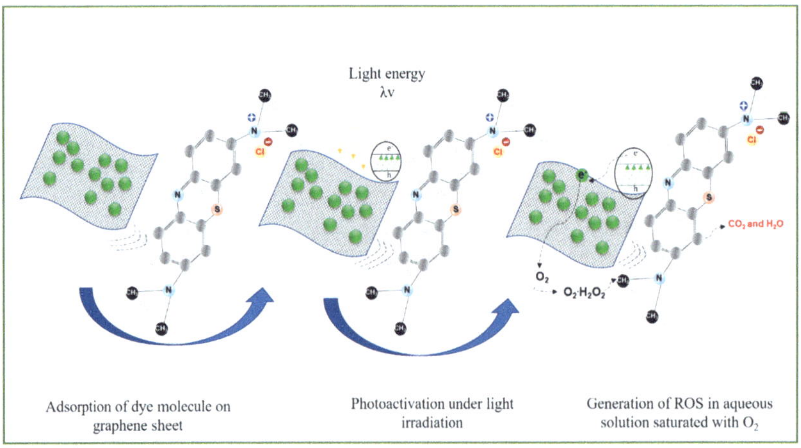

Figure 14.4 Three-step method proposed for degrading organic dye molecules (methylene blue) by G hybrid composite photocatalysts. (*Source*: Modified and redrawn from Perreault *et al.*, 2015).

degradation due to superfast electron transfer, high adsorption capacity, enhanced charge separation, charge transport properties, and high thermal conductivity (≈5000 W/m k) (Kumar *et al.*, 2017). Luna-Sanguino *et al.* (2020) investigated the degradation efficiency of pesticides such as methomyl, pyrimethanil, isoproturon, and alachlor using the TiO_2–RGO nanocomposite and studies showed 100% degradation efficiency within 30 min. This can be attributed to the increased surface area of RGO, which also acts as an electron mediator and acceptor inhibiting the recombination of electron–hole pairs, thereby improving the degradation rate.

Integrating CNMs and semiconducting materials enhances photocatalytic performance. However, surface and bulk blockage of photocatalysts and interference from various ions in the water need further attention. Another crucial factor is the production of CNMs on a large scale with high purity, few defects, and a controllable geometry. The deviation in purity can alter the band energy gap of CNM-semiconductor composites (CSC) in every batch, which can alter their performance in visible light. The optimization of each step, while synthesizing the CSC composite, solves the problem to some extent. Another challenge is to overcome the Schottky barrier, which is the energy difference between the valence band of the semiconductor and the Fermi energy level of G. This barrier thickness and G material thickness influence electron transportation from the Fermi level of G to the conduction band of the semiconductor. Therefore, a suitable method which can produce uniform G with only a few layers can help in solving the problem (Tang *et al.*, 2018).

14.4.3 Dechlorination

In the chlorination process, chlorine gas or chlorine-containing compounds are added to water to enable disinfection, following which it is necessary to remove the excess chlorine (>0.2 mg/L) as it is toxic and imparts taste and odour to the water. Although several dechlorinating agents are available in the market, a packed bed of GAC has been the most cost-effective and common in large treatment plants and point-of-use household water systems. Initially, chlorine is transported from the water to the pores of the AC by diffusion and physical adsorption. Then, the catalytic reduction (chemical reaction) transfers an electron from the AC surface to the residual disinfectant, thereby converting the chlorine to non-oxidative chloride ions. Kristiana *et al.* (2011) used powdered activated carbon (PAC) in addition to coagulation to reduce the formation of disinfection by-products (DBPs) in a water treatment plant in Western Australia. The addition of PAC before disinfection removed the NOM by about 70%, reducing the DBPs' formation by 80–95%. Many factors, including carbon particle size, solution pH, wastewater characteristics, type of free chlorine species present in water, and temperature, can affect the efficiency of the dechlorination. Therefore, optimizing the operating parameter is the key to maximizing dechlorination efficiency.

14.4.4 Desalination

Desalination techniques like reverse osmosis (RO), electrodialysis (ED), multistage flash (MSF), multi-effect distillation (MED), and capacitive deionization (CDI) have been employed extensively on a commercial basis for the production of freshwater worldwide. Membrane technologies have been employed in water and wastewater treatment due to their flexibility, no phase change, and high desalination efficacy. Currently, these membranes are made of inorganic or organic material. Even though inorganic membranes (ceramics/metals) have good mechanical, structural and thermal strength, their use is limited due to their constrained permeability and susceptibility to microbial attack. Organic (polymeric) membranes are frequently used because of their extrusive flexibility, chemical stability, good physicochemical properties, and better film-forming properties. However, the major problem that needs to be addressed is the high reject stream (50–60%), water permeability, chlorine resistance, and antifouling effectiveness. To overcome these problems, researchers modify membranes by incorporating CNMs that improve porosity and hydrophilicity, and attach various functional groups on the membrane surface. G and its derivatives were recently explored in membrane technology. It improves the membrane's hydrophilicity property and helps form a tightly bound hydration layer, preventing the foulants from displacing the water molecules and attaching to the membrane surface. Hence, it subsequently improves the fouling resistance and anti-adhesive property against the bacteria. Still, there is a challenge to decrease cost of producing CNMs and graphenic materials, as well as the viability and synthesis protocol reproducibility. It is also reported that single sheet of monolayer G was about $250 which is expensive than RO membrane's price range of \$20–40/m^2. Therefore, at present, researchers are trying to balance the cost of graphenic materials and improved membrane performance by incorporating these graphenic materials. The optimized conditions give better water flux at a decreased desalination cost.

To overcome some of the above-mentioned problems, CDI is found to be a promising sustainable technology to treat brackish water at lower operating costs and works based on the principle of electro-sorption. The schematic design of CDI using G electrodes is shown in Figure 14.5. It is proven that the performance of CDI depends significantly on the morphology and structure of the electrode material. So, a larger specific surface area and porosity, high electrical conductivity and capacitance, and better electrochemical stability are required for the electrode material. Due to this aspect, carbon-based materials such as AC, carbon aerogel (CA), carbon cloth, CNTs, G and its composites, and carbon fibres are used as electrode materials. Since the early 1970s, AC has been widely employed for CDI due to its extensive properties discussed in previous sections. To further improve AC's performance, it is surface modified to improve wettability and capacitance compared to its pristine counterpart. For example, nitric acid-modified AC enhances the oxygen-containing functional group, which reduces charge resistance and increases capacitance. In contrast, KOH-modified AC, having more hydroxyl groups, improves the wettability of the electrode (Oladunni *et al.*, 2018). In the later stage, CNMs are preferred over AC due to their high surface area and good electrical conductivity. As mentioned earlier, the selection of electrode material is most important in CDI.

Exceptionally suitable electrode materials for CDI must have the appropriate combination of high conductivity, good wettability, and adequate pore structure for the best desalination performance. When selecting a possible contender, the electrode's stability over a broad pH, voltage range, scalability and affordability, must be taken into account. Due to their accessibility, affordability, and stability, porous carbon compounds are used as electrode materials in CDI. Current and upcoming research looks at combining porous carbon with other components like G, conductive polymers, and metal oxides. To improve and optimize the performance of the electrodes, a thorough understanding of the electrochemistry at the electrode surface is required. For instance, materials made of G have outstanding mechanical flexibility and electrical conductivity, but their surface wettability is poor, which restricts ion mobility during the desalination process. To increase wettability, different surface enhancement approaches have been investigated, including metal oxide addition, nitrogen doping, and surface functionalization. These engineered/modified materials are a boon for Global South

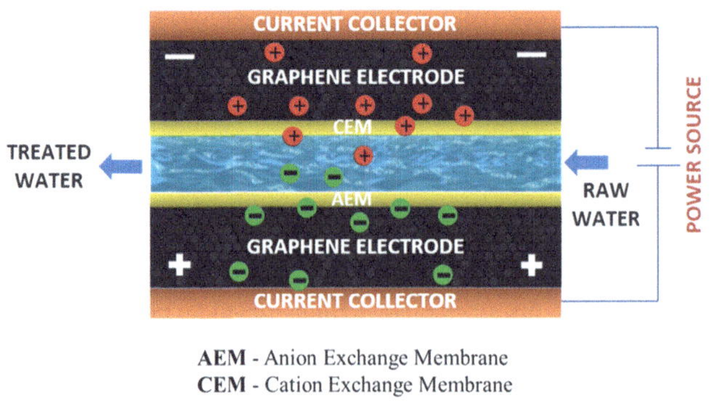

Figure 14.5 Schematic representation of membrane capacitive deionization using a graphenic electrode.

Table 14.2 Comparison of carbon-based materials used in various water and wastewater treatments.

S. No	Materials	Pollutant	Treatment Technology	Adsorption Capacity (mg/g)/Removal Efficiency (%)	Cost of Material ($/kg)	Difficulty in Operation and Maintenance	Operator Skill Requirement	References
1.	Biochar from groundnut shell	Basic red 09 dye	Adsorption	46.3 mg/g	141.1	Medium	Low	Praveen *et al.* (2021)
2.	GAC-TiO_2	TVS	Photocatalysis	43–63%	55–62	High	High	Asha *et al.* (2015)
3.	UV-GAC	TVS	Photocatalysis	100%	13.79	High	High	
4.	Graphenic material	Fluoride, As, Pb	CDI	>90%	0.4–0.54 ($/m^3)	Low	Low	Islam *et al.* (2022)
Field Scale/Point of Use Systems								
5.	Raffia palm shell AC- aluminium oxide composite	Fluoride	Adsorption	14 mg/g (55.85%)	–	Medium	Low	Iwar *et al.* (2022)
6.	Coconut shell	Pharmaceuticals	Adsorption	98.2%	-	Medium	Low	Li *et al.* (2018)

GAC: granular AC; TVS: total volatile solids; CDI: capacitive deionization.

countries as they often do not have a good economy or proper water resource management, and lack power connectivity and cooperation by people to adapt to advanced technology. Therefore, developing a material at an affordable cost and combining it with state-of-the-art technology like CDI is an immediate solution. The major reason is that CDI consumes very less energy compared to RO and 70% of the energy spent can be recovered. The system can be integrated with solar panels, which enables the deployment of units in remote locations. Other major advantages are low maintenance, the retention of essential minerals in treated water, and less water wastage (10–20%).

A comparison of carbon-based materials used in various water and wastewater treatment systems is presented in Table 14.2.

14.5 CONCLUSION, CHALLENGES, AND THE WAY FORWARD

The use of carbon-based materials in water and wastewater treatment is well established. The large surface area, high chemical, electrical, and thermal stability, high electrical surface charge, and good electrical conductivity make AC and its nanoscale counterparts an exciting component in various water purification technologies. Presently, the application of carbon and its derivatives includes removing natural organic compounds, SOCs, taste and odour-causing compounds, and residual chlorine. Carbon materials are also used in the photocatalysis and desalination of brackish water. Currently, carbon is the most used electrode material in capacitive deionization, an emerging non-membrane desalination technique. As a result, carbon-based materials, including AC, CNTs, carbon fibres, and G and its derivatives, are in high demand. The global market size of AC and CNT is anticipated to be USD 8.9 billion and 32.8 billion by 2030 at a compound annual growth rate (CAGR) of 9.3% (2021–2026) and 22.4% (2022–2030), respectively.

Among the various carbon-based materials, AC takes the front seat regarding practical applications. Though nanoscale carbon materials have many advantages over conventional AC, several challenges must be addressed before their viable application in the field. These challenges include bulk material production, cost, reusability, and nanotoxicity. Developing nanoscale material on a large scale without affecting quality, quantity, reliability, and cost is the need of the hour to realize their use in everyday mass applications like water and wastewater treatment. The effective re-formation of the active surface is a challenge with carbon-based filters, including AC, on a large scale, as regeneration by cleaning agents can cause permanent physical or chemical changes. In addition, their pH sensitivity and tendency to foul due to their high surface functionality can reduce the shelf life and efficiency of carbon-based filters.

On the contrary, the toxicity of nanoscale carbon materials can affect humans and the environment due to their nanoparticle size and reactivity. According to reports, nanocarbon can enter the bloodstream through various pathways, generating reactive oxygen species and causing inflammation and impairment of proteins, DNA, and cell membranes. According to various research findings, the presence of nanocarbons in soil can retard plant growth considerably. Therefore, the spent, pollutant-laden nanoscale carbon needs a careful disposal strategy. Besides, there needs to be more knowledge on the fate and transport of nanoscale carbon in the environment, which requires further research. In short, carbon-based materials are an integral part of water and wastewater treatment and have a large growth potential. However, developing affordable, reliable, and sustainable carbon materials is the need of the hour to sustain their use in environmental remediation.

REFERENCES

Alofi A. and Srivastava G. P. (2013). Thermal conductivity of graphene and graphite. *Physical Review B*, **87**(11), 115421, https://doi.org/10.1103/PhysRevB.87.115421

Asha R. C., Vishnuganth M. A., Remya N., Selvaraju N. and Kumar M. (2015). Livestock wastewater treatment in batch and continuous photocatalytic systems: performance and economic analyses. *Water, Air, & Soil Pollution*, **226**(5), 132, https://doi.org/10.1007/s11270-015-2396-4

Azam K., Shezad N., Shafiq I., Akhter P., Akhtar F., Jamil F., Shafique S., Park Y.-K. and Hussain M. (2022). A review on activated carbon modifications for the treatment of wastewater containing anionic dyes. *Chemosphere*, **306**, 135566, https://doi.org/10.1016/j.chemosphere.2022.135566

Birch M. E., Ruda-Eberenz T. A., Chai M., Andrews R. and & Hatfield R. L. (2013). Properties that influence the specific surface areas of carbon nanotubes and nanofibers. *The Annals of Occupational Hygiene*, **57**(9), 1148–1166, https://doi.org/10.1093/annhyg/met042

Cecen F. and Aktas Ö (2011). Activated Carbon for Water and Wastewater Treatment: Integration of Adsorption and Biological Treatment. Wiley–VCH, Germany, https://doi.org/10.1002/9783527639441

Fortune Business Insights (2022). Activated carbon market research report. Accessed 2 May 2023. https://www.fortunebusinessinsights.com/activated-carbon-market-102175

Gandhi T. P., Sampath P. V. and Maliyekkal S. M. (2022). A critical review of uranium contamination in groundwater: treatment and sludge disposal. *Science of the Total Environment*, **825**, 153947, https://doi.org/10.1016/j.scitotenv.2022.153947

Ghosh P., Afre R. A., Soga T. and Jimbo T. (2007). A simple method of producing single-walled carbon nanotubes from a natural precursor: eucalyptus oil. *Materials Letters*, **17**(61), 3768–3770, https://doi.org/10.1016/j.matlet.2006.12.030

Harafan A., Gafoor S. A., Kusuma T. D. and Maliyekkal S. M. (2022). Graphene modified photocatalysts for the abatement of emerging contaminants in water. In: New Trends in Emerging Environmental Contaminants, S. P. Singh, A. K. Agarwal, T. Gupta and S. M. Maliyekkal (eds), Springer, Singapore, pp. 371–406, https://doi.org/10.1007/978-981-16-8367-1_16

Islam M. R., Gupta S. S., Jana S. K. and Pradeep T. (2022). Industrial utilization of capacitive deionization technology for the removal of fluoride and toxic metal ions ($As^{3+/5+}$ and Pb^{2+}). *Global Challenges*, **6**(4), 2100129, https://doi.org/10.1002/gch2.202100129

Iwar R. T., Ogedengbe K. and Ugwudike B. O. (2022). Groundwater fluoride removal by novel activated carbon/aluminium oxide composite derived from raffia palm shells: optimization of batch operations and field-scale point of use system evaluation. *Results in Engineering*, **14**, 100407, https://doi.org/10.1016/j.rineng.2022.100407

Koushik D., Sen Gupta S., Maliyekkal S. M. and Pradeep T. (2016). Rapid dehalogenation of pesticides and organics at the interface of reduced graphene oxide–silver nanocomposite. *Journal of Hazardous Materials*, **308**, 192–198, https://doi.org/10.1016/j.jhazmat.2016.01.004

Kristiana I., Joll C. and Heitz A. (2011). Powdered activated carbon coupled with enhanced coagulation for natural organic matter removal and disinfection by-product control: application in a Western Australian water treatment plant. *Chemosphere*, **83**(5), 661–667, https://doi.org/10.1016/j.chemosphere.2011.02.017

Kumar S., Kumar A., Bahuguna A., Sharma V. and Krishnan V. (2017). Two-dimensional carbon-based nanocomposites for photocatalytic energy generation and environmental remediation applications. *Beilstein Journal of Nanotechnology*, **8**(1), 1571–1600, https://doi.org/10.3762/bjnano.8.159

Kusuma T. D., Naga Jyothi M. S. V., Rao C. P. and Maliyekkal S. M. (2021). Advanced oxidation processes: a promising route for abatement of emerging contaminants in water. In: Nanomaterials and Nanocomposites for Environmental Remediation, S. P. Singh, K. Rathinam, T. Gupta and A. K. Agarwal (eds), Springer, Singapore, pp. 275–305.

Lee C.-G., Song M.-K., Ryu J.-C., Park C., Choi J.-W. and Lee S.-H. (2016). Application of carbon foam for heavy metal removal from industrial plating wastewater and toxicity evaluation of the adsorbent. *Chemosphere*, **153**, 1–9, https://doi.org/10.1016/j.chemosphere.2016.03.034

Li J., Zhou Q. and Campos L. C. (2018). The application of GAC sandwich slow sand filtration to remove pharmaceutical and personal care products. *Science of the Total Environment*, **635**, 1182–1190, https://doi.org/10.1016/j.scitotenv.2018.04.198

Lu S., Sun Y. and Chen C. (2019). Chapter 4 – adsorption of radionuclides on carbon-based nanomaterials. In: Interface Science and Technology, C. Chen (ed.), Elsevier, Brazil, Vol. **29**, pp. 141–215, https://doi.org/10.1016/B978-0-08-102727-1.00004-2

Luna-Sanguino G., Ruíz-Delgado A., Tolosana-Moranchel A., Pascual L., Malato S., Bahamonde A. and Faraldos M. (2020). Solar photocatalytic degradation of pesticides over TiO_2-rGO nanocomposites at pilot plant scale. *Science of the Total Environment*, **737**, 140286, https://doi.org/10.1016/j.scitotenv.2020.140286

Madima N., Mishra S. B., Inamuddin I. and Mishra A. K. (2020). Carbon-based nanomaterials for remediation of organic and inorganic pollutants from wastewater. A review. *Environmental Chemistry Letters*, **18**(4), 1169–1191, https://doi.org/10.1007/s10311-020-01001-0

Maliyekkal S. M., Sreeprasad T. S., Krishnan D., Kouser S., Mishra A. K., Waghmare U. V. and Pradeep T. (2013). Graphene: a reusable substrate for unprecedented adsorption of pesticides. *Small*, **9**(2), 273–283, https://doi.org/10.1002/smll.201201125

Mohandoss M., Sen Gupta S., Nelleri A., Pradeep T. and Maliyekkal S. M. (2017). Solar mediated reduction of graphene oxide. *RSC Advances*, **7**(2), 957–963, https://doi.org/10.1039/C6RA24696F

Naga Jyothi M. S. V., Gayathri S., Pushparaj Gandhi T. and Maliyekkal S. M. (2021). Dissolved arsenic in groundwater bodies: a short review of remediation technologies. In: Pollution Control Technologies, S. P. Singh, K. Rathinam, T. Gupta and A. K. Agarwal (eds), Springer, Singapore, pp. 75–111, https://doi.org/10.1007/978-981-16-0858-2_5

Oladunni J., Zain J. H., Hai A., Banat F., Bharath G. and Alhseinat E. (2018). A comprehensive review on recently developed carbon based nanocomposites for capacitive deionization: from theory to practice. *Separation and Purification Technology*, **207**, 291–320, https://doi.org/10.1016/j.seppur.2018.06.046

Ostreyko (1901). German Patent 136 792 (1901).

Perreault F., Faria A. F. d. and Elimelech M. (2015). Environmental applications of graphene-based nanomaterials. *Chemical Society Reviews*, **44**(16), 5861–5896, https://doi.org/10.1039/C5CS00021A

Praveen S., Gokulan R., Pushpa T. B. and Jegan J. (2021). Techno-economic feasibility of biochar as biosorbent for basic dye sequestration. *Journal of the Indian Chemical Society*, **98**(8), 100107, https://doi.org/10.1016/j.jics.2021.100107

Sreeprasad T. S., Maliyekkal S. M., Lisha K. P. and Pradeep T. (2011). Reduced graphene oxide–metal/metal oxide composites: facile synthesis and application in water purification. *Journal of Hazardous Materials*, **186**(1), 921–931, https://doi.org/10.1016/j.jhazmat.2010.11.100

Tang B., Chen H., Peng H., Wang Z. and Huang W. (2018). Graphene modified TiO_2 composite photocatalysts: mechanism, progress and perspective. *Nanomaterials*, **8**, 2. Article 2.

doi: 10.2166/9781789063714_0167

Chapter 15

Nutrient recovery from wastewater for circular economy

V. Nagabalaji[1,2], M. Madhumidha[1], V. Mozhiarasi[3], Indrasis Das[1,2], U. Sathya[1], P.C. Sabumon[4] and S.V. Srinivasan[1,2]*

[1]Department of Environmental Engineering, CSIR – Central Leather Research Institute (CSIR-CLRI), Chennai, India
[2]Academy of Scientific and Innovative Research (AcSIR), Ghaziabad, Uttar Pradesh, India
[3]CLRI Regional Centre, CSIR-Central Leather Research Institute (CSIR-CLRI), Jalandhar, India
[4]School of Civil Engineering, Vellore Institute of Technology (VIT), Chennai, India
**Corresponding author: srinivasansv@yahoo.com*

ABSTRACT

In recent years, the focus of the wastewater management sector has shifted significantly from conventional treatment methods to resource and nutrient recovery techniques to promote a circular economy. The recovery of nutrients such as nitrogen and phosphorus from wastewater marks a sustainable approach to wastewater management and supports ecological and economic sustainability. This chapter includes a comprehensive overview of existing conventional technologies that are used to recover nutrients from different nutrient-rich wastewater generated from domestic, industrial, and agricultural sources as well as from anaerobic digestate. In addition, various advanced methods, such as chemical processes and biological technologies that are used to recover nutrients are also discussed. Furthermore, a few more unique applications of novel futuristic technologies that are in the budding stage or ready for piloting or commercialization are also included. Finally, future perspectives in terms of possible research directions and breakthroughs of a more economic and efficient alternative approach with minimal carbon footprints are also explored.

Keywords: nutrient recovery, phosphorus recovery, nitrogen recovery, wastewater, circular economy

15.1 INTRODUCTION

The advancement of science has facilitated economic and demographic expansion, which unfortunately leads to increased resource exploitation as well as environmental pollution. In this situation, it is necessary to embrace a paradigm shift in which the waste stream is viewed not only as means of eliminating contaminants to comply with environmental regulations but also as extractable resources. Circular economy (CE) denotes an economic model that focuses on the 3R's which are reducing, reusing materials, and recovery of 'waste' to manufacture new products. CE model can allow the recovery of resources such as nutrients, energy, water, and other materials from wastewater streams (Robles *et al.*, 2020). Nutrient (N and P) loading of freshwater systems is the major cause of eutrophication, methaemoglobinemia (blue baby syndrome) in infants and many other problems. Algal blooms could be caused by phosphorus at concentrations as low as 0.03 mg/L. Likewise, nitrogen levels higher than 60 mg N/L could cause severe human health effects (Rodhe, 1969). On the contrary, synthetic N and P are currently used in agriculture, and their production is carried out on an extensive scale globally. The conversion of atmospheric nitrogen into ammonia and nitrate using the Haber process consumes a substantial amount of energy resulting in global warming (Razon, 2014). Similarly, turning phosphorus-rich rocks into fertilizers is energy intensive (Cordell *et al.*, 2009; Craggs *et al.*, 1996). It is estimated that around 15% of phosphate fertilizer demand in the world can be fulfilled by recovering phosphates from wastewater treatment plants (WWTP) alone (Williams *et al.*, 2015).

Therefore, nutrient recovery is gaining increased attention due to both economic and environmental gains, such as improved treatment, reduced carbon footprints and the development of new natural capital in a safe manner. It also improves water quality, reduces sludge and unwanted precipitates, improves operation and performance at waste treatment facilities, and improves food security and social equity (Bradford-Hartke *et al.*, 2015). This chapter discusses the conventional technologies devoted to nitrogen (N) and phosphorus (P) recovery from wastewater and the extent of its contribution to the transition towards a CE-based development model. This chapter discusses various techniques, such as chemical, physical, and biological processes, as well as the challenges associated with N and P recovery. Other novel futuristic technologies that are in the early stages of development or commercialization are also highlighted. Finally, the future prospects of innovations and potential research directions are also investigated.

15.2 CONVENTIONAL PROCESS FOR N AND P RECOVERY

15.2.1 Chemical precipitation

Chemical precipitation of nutrients was investigated and it was found that mixing intensity, pH, and coagulant dosage have an effect on the nutrient recovery efficiency (Bratby, 2006). Various coagulants, synthetic organic polymers, and pre-hydrolysed metal salts, such as poly-aluminium chloride and poly iron chloride, are mostly used for the precipitation of phosphorus (Mehta *et al.*, 2015; Tchobanoglous *et al.*, 2003). The optimum pH depends upon the type of the coagulant. The addition of iron coagulant and calcium or magnesium salts was reported with 91.9% and 85% phosphorus removal (Barua *et al.*, 2019; Wang *et al.*, 2006). Similarly, nitrogen precipitation is observed as struvite along with phosphorus and magnesium as a nutrient source (Perera *et al.*, 2019).

Struvite ($MgNH_4PO_4{\cdot}6H_2O$) is a crystalline mineral composed of equimolar concentrations of magnesium (Mg), ammonium (NH_4), and phosphate (PO_4) and it is an efficient nutrient source for plant growth. Struvite precipitation from aerobically digested municipal wastewater was assessed and about 90% of total phosphate is recovered as struvite (Hallas *et al.*, 2019; Kataki *et al.*, 2016). Studies in the Netherlands reported that the main barriers to the implementation are the variations in the characteristics of struvite, the waste status of struvite, in addition to the high investment cost with uncertainty on return on investment for full-scale onsite plants (De Boer *et al.*, 2018). Also, the value of a single product, struvite, is too low to compete against the relatively low cost of mined P.

15.2.2 Air stripping

Air stripping is a desorption process in which wastewater and air are intensively brought in contact with each other, resulting in the release of the volatile compounds found in wastewater, such as ammonia, into the atmosphere. There are three types of stripping processes, namely, thermal alkali, hyper thermal alkali, and vacuum thermal alkali stripping, that are categorized based on the boiling temperature and applied pressure. Ammonium hydroxide is produced when water combines with ammonia. At pH of around 11, ammonium hydroxide ions are converted into ammonia gas based on two-film theory

$$NH_4^+ + OH^- + air \Rightarrow H_2O + NH_3\,(g) \tag{15.1}$$

In vacuum thermal alkali stripping, it can be coupled to a gas absorption column resulting in the production of ammonium sulphate crystals ($(NH_4)_2SO_4$) (Han *et al.*, 2022). Through thermal ammonia stripping, high ammonia concentrations can be reduced to levels below 100 mg/L in a single pass. Hyper thermal alkali (temperature >80°C) stripping can be coupled with struvite precipitation or adsorption for improving the recovery of ammonia.

15.2.3 Adsorption

Successful phosphate adsorption has been reported using modified activated carbon (AC) due to its intrinsic positive charge, high surface area, low cost, availability, and surface porosity. Natural porous adsorbents such as diatomite, clays, zeolite, and biochar are considered suitable for phosphate sorption due to their large specific surface area and low costs. Likewise, red mud, metal oxide/ hydroxide, bentonite, calcite, kaolinite and zeolite and zirconium sorbents are used for P recovery. *Ficus carica, Moringa oleifera*, and saw dust are some of the natural phosphorus adsorbents (Subha *et al.*, 2015). Modified zeolite and clinoptilolite can be used for N recovery (Almanassra *et al.*, 2021).

15.2.4 Ion exchange

Ion exchange (IEX) processes hold promise for recovering nutrients from municipal wastewater. Nutrients are recovered in the form of $(NH_4)_2SO_4$ and hydroxyapatite $Ca_5(PO_4)_3(OH)$ from spent regenerants, allowing regenerant reuse. According to Huang *et al.* (2020), compared to traditional WWTPs based on biological nitrogen removal, the whole life costs (WLC) of IEX combined with traditional activated sludge processes and IEX combined with anaerobic membrane were 17% and 27% lower over a 40-year period, respectively. In addition, 98 tonnes of $(NH_4)_2SO_4$ and 3.4 tonnes of $Ca_3(PO_4)_2$ could be recovered each year. The benefits of lower costs, reduction in greenhouse gas emissions, and nutrient recovery are also aligned with the CE module (Huang *et al.*, 2020).

15.2.5 Biological methods of nutrient recovery

Biological processes are utilized for the recovery of nutrients from wastewater due to its cost feasibility and environmental sustainability (Mantzavinos & Kalogerakis, 2005). Microalgae have been found to be a more effective biological process in the removal of nutrients such as nitrogen and phosphorous from a wide range of effluents (Christenson & Sims, 2011; Whitton *et al.*, 2015; Zhou *et al.*, 2012). Also, it generates a significant amount of biomass rich in proteins and lipids, whereas other biological treatments produce additional waste sludge (Chisti, 2013; Mantzavinos & Kalogerakis, 2005). It also minimizes the emission of CO_2, N_2 and greenhouse gases along with the production of biomass of up to 500 kT/year. In addition, microalgae-based treatments require less than half of the energy spent (<0.2 kWh/m^3) during the conventional treatments for nutrient removal (Nagarajan *et al.*, 2020). Microalgae also have the potential to be used as bio-fertilizer.

The two main types of microalgal cultivation are open and closed systems. Open systems or raceway ponds are preferred because they allow an adequate amount of sunlight to penetrate the effluent, are commercially available, and are inexpensive (Oswald & Golueke, 1960). However, its atmospheric exposure results in a temperature rise which leads to water loss (Cai *et al.*, 2013). Closed systems contain flat panel reactors, tubular or column photobioreactors and bag systems, mainly designed for improved light availability and gas exchange, reduced water loss by evaporation, and minimized contamination and biomass production. But comparatively, its maintenance is labour-intensive and expensive (Borowitzka, 1999).

The major challenges faced are extended hydraulic retention time (HRT) and the harvesting of grown microalgae (Matamoros *et al.*, 2015). Longer generation time requires more HRT and the same can be overcome by co-culturing with bacteria, increasing the surface area, and settling velocity (Chevalier & de la Noüe, 1988; Manganaro *et al.*, 2015). Due to its smaller size and negative charge, harvesting can be done by using less energy-intensive processes like flocculation and sedimentation (Cromar & Fallowfield, 1997; Udaiyappan *et al.*, 2017). In addition, turbidity and effluent with high suspended solids will reduce the light penetration which in turn reduces the photosynthetic activity and microalgal productivity (Larsdotter, 2006; Zeng *et al.*, 2022). This can be fixed by using flocculants and coagulants to pretreat effluents (Sher *et al.*, 2013). Adding proper mixing and turbulence to the effluents also aids in light penetration (Kumar *et al.*, 2015).

15.2.6 Electrochemical system

An electrochemical setup of magnesium anode and titanium cathode was used for electrochemical nutrient recovery from nutrient-rich wastewater at alkaline pH of 8.8. Leaching of Mg^{2+} ions from the anode and successful synthesis of struvite was reported along with the reaction mechanism (Cai *et al.*, 2022). It was reported that over 90% phosphate removal with higher purity is possible in this technology at elevated pH levels (Bagastyo *et al.*, 2022). For efficient precipitation of struvite at neutral pH in electrochemical method, surface-bound amelogenin peptide was used and a promising recovery of pure struvite was observed (Wu *et al.*, 2022a, 2022b). Struvite precipitation was also evaluated in acidic wastewater in an electrochemical method using magnesium–aluminium alloy (AZ31) anode and a 4.5-fold increase in struvite recovery was reported (Kékedy-Nagy *et al.*, 2020).

Advantages and disadvantages of various N and P recovery processes are compared in Table 15.1.

15.3 HYBRID TECHNOLOGY

15.3.1 Bio-electrochemical method

The integration of one or more nutrient recovery technologies has gained popularity, owing to increased efficiency and economic feasibility. In this method, wastewater treatment, energy in terms of electricity or biofuel recovery, and nutrient recovery are possible (Potter, 1911). A microbial fuel cell (MFC) is a two-chambered setup where wastewater treatment is possible in an anaerobic anodic chamber and the cathodic chamber is aerobic and oxygen acts as a terminal electron acceptor (Logan *et al.*, 2006). In MFC, specific reactor orientation can be designed for anodic carbonaceous material oxidation and cathodic precipitation of the nutrient as struvite (Monetti *et al.*, 2019; Neethu *et al.*, 2018; Ye *et al.*, 2019). The membrane systems, which include forward osmosis, membrane distillation, and electrodialysis process, are favourable for combining with MFCs while applying MFCs to recover nutrients from diluted wastewater (Mahmoud *et al.*, 2022). Other methods such as capacitive deionization, metal organic frameworks, and microbial desalination cells are also employed in nutrient recovery (Nancharaiah *et al.*, 2016).

15.3.2 Hybrid membrane–microbial fuel cells

These systems include bioelectro-Fenton-MFC, microbial desalination cell, MFC-electrosorption cell, microbial solar cell, microbial reverse-electrodialysis cell, plant-MFC, and constructed wetland-MFC. Although MFC-hybrid systems are more promising than standalone MFCs due to their capacity for reducing major obstacles, such as low power densities, high reactor construction and operating costs, further research is needed to overcome significant hurdles for practical deployment (Zhang *et al.*, 2019).

15.3.3 Photobioreactor–membrane filtration

Membrane photobioreactors (MPBRs) not only produce highly concentrated biomass but also achieve considerable reduction of nutrient level, mainly due to the ability of MPBRs to enable

Table 15.1 Comparison of N and P recovery process.

Process	Nutrient Recovery	Advantages	Disadvantages
IEX	P	• Can achieve high-quality effluent 100× concentration.	• 100% desorption of nutrients cannot be achieved during media regeneration. • Leaching of other metal cations affects the purity of subsequent struvite precipitation. • Biofouling of resin. • Continuous resin requirement due to reduction of regeneration capacity over time. • Limited selectivity.
Struvite precipitation	P	• Minimum leaching. • Can be crystallized with lower impurities.	• Requires chemical addition and pH adjustment. • Cost of P removal increases with chemical requirement.
Chemical precipitation	P	• Can meet low P discharge limits. • Easy operation. • Can handle shock loadings.	• Primarily used more for P removal than recovery. • Requires chemical dosing. • Excess sludge generation and handling costs. • Heavy metal contamination.
Biological	P	• Economical • PAOs can accumulate P up to 20% of their dry biomass weight.	• Sensitive to influent water characteristics and operational parameters. • Operationally complex.
Algae harvesting	P	• Less land requirement • Can achieve P content up to 3.4% of dry biomass weight.	• Requires post-processing. • Pathogens and micropollutants are a concern. • Lack of information on potential effects of allelochemicals and/or cyanotoxins.
Electrochemical	P	• Economical. • Continuous cation dosing. • Can use pH shifting for direct P recovery.	• Co-precipitation of $CaCO_3$. • Energy requirement Is inversely proportional to wastewater conductivity.
IEX /Adsorption	N	• Can meet very low N discharge limits. • Operationally simple. • Can withstand shock loadings.	• Mainly used more for N removal than recovery. • Limited adsorption/ desorption capacity and resin life. • Adsorption capacity impaired by competing ions. • Requires a lot of chemicals.
Electrochemical	N	• N can be concentrated to reduce chemical and operational costs in post-recovery steps.	• Reaction time limited by anode pH. • N recovery requires stripping and absorption into an acid solution.
Bioelectrochemical	N	• Can generate energy. • Concentrate N in cathode compartment for post-recovery. • Eliminate chemical pH adjustment.	• Sensitive to pH, influent toxicity, and substrate loading. • N recovery requires stripping and absorption into an acid solution.
Struvite precipitation	N	• Lower evaporative losses.	• High operational costs due to chemical requirement.
Stripping and absorption	N	• Easy installation and simple operation. • Energy requirement is economical.	• Require pH adjustment. • Typically used for high strength wastewater. • Energy requirement for stripping is inversely proportional to influent NH_3 concentration.

Source: Modified from Perera *et al.* (2019).

independent control of HRT and solids retention time (SRT). Despite their advantage over conventional algal systems, the relative complexity derived from the operation and maintenance of the additional membrane separation process of MPBR systems is a key challenge to implementation (Luo *et al.*, 2018).

15.4 NUTRIENT RECOVERY FROM DIFFERENT WASTEWATERS

15.4.1 Urine

Although urine constitutes less than 1% volume of total wastewater, it adds up to 50%–80% of the nutrient load in WWTPs. However, urine could be a potential cheap substitute for chemical fertilizers. Direct application of urine as fertilizer has been discouraged due to various hygienic pitfalls, thus driving researchers to explore alternative nutrient recovery methods. Urine source separation with P recovery as struvite had lower life cycle environmental impacts than P removal during wastewater treatment combined with synthetic fertilizer production, even when accounting for the new infrastructure required for urine source separation (Wu *et al.*, 2022a, 2022b).

15.4.2 Anaerobic digestate

There is a focus on nutrient recovery technologies from anaerobic digestate (AnD) (Table 15.2). AnD is normally rich in NPK and also contains stabilized carbon, magnesium, calcium, zinc, manganese, sulphur, heavy metals, and so on (Table 15.3). The resultant digestate is widely applied as manure onto farmlands.

Substrates for anaerobic digestion process include food waste; agro residues; fish farm sludge; catering waste; waste activated sludge and vegetable waste; cattle manure; and cow dung and pig slurry (Kuusik *et al.*, 2017; Sheets *et al.*, 2015; Tampio *et al.*, 2016; Walsh *et al.*, 2018; Xia & Murphy, 2016).

15.4.3 Domestic wastewater

Domestic wastewater consists of blackwater (water used to flush toilet and human faeces), greywater (water from showers, laundries, dishwashers, and tubs), and yellow water (human urine). Lately, research has focused on recovering nutrients from domestic wastewater, due to the abundant presence of N and P in it. Various technologies adopted for nutrient recovery from domestic wastewater are summarized in Table 15.4.

Table 15.2 Digestate characteristics.

S. No.	Characteristics	Unit	Range
	pH	–	7.1–8.9
	Moisture	%	65.8–98.6
	Organic matter	% of total solids	12.3–81.5
	Carbon content	g/kg TS	300–452
	Nitrogen content	g/kg	3.5–157
	Ammoniacal nitrogen	g/kg	1.5–108
	Phosphorus content	g/kg	0.06–66
	Potassium content	g/kg	0.02–100
	Calcium	g/kg	3–10
	Sulphur	g/kg	6–10
	Magnesium	g/kg	10–52

Source: Modified from Selvaraj and Velvizhi (2021).

Table 15.3 Nutrient recovery technologies from anaerobic digestate.

Process	Performance Outcome	References
Ammonia stripping	93.3%–99.9% of ammonia recovery	Ukwuani and Tao (2016)
Ammonia stripping and struvite crystallization	Ammonia – 80%–90% and phosphorus – 90%–94% recovery	Hidalgo *et al.* (2016)
Membrane filtration	4.7–6.1 kg phosphorus/ton; 6–9.7 kg total nitrogen/ton	Gienau *et al.* (2018)
Membrane system	71.6% ammonia recovery	Rivera *et al.* (2021)
Electrodialysis + struvite precipitation	95.8%–100%,ammonia and 86.1%–94.4% phosphate	Wang *et al.* (2015)

Table 15.4 Nutrient recovery technologies from domestic wastewater.

Wastewater	Recovery Method	N Recovery (%)	P Recovery (%)	References
Domestic wastewater	Struvite precipitation	–	87	Mehta *et al.* (2015)
Municipal wastewater	Adsorption with natural adsorbents	100	93	Bhattacharya *et al.* (2018)
Semiconductor wastewater	Struvite precipitation	98	70	Kim *et al.* (2009)
Cola beverage	Struvite precipitation	–	97	Foletto *et al.* (2013)

Table 15.5 Nutrient recovery technologies from agricultural wastewater.

Wastewater	Process	Performance Outcome	References
Swine wastewater	Biofocculant prepared from anaerobic sludge	54.5% ammonia recovery	Guo *et al.* (2018)
Swine wastewater	Modified zeolite + biofloculant	85.8% ammonia recovery	Guo *et al.* (2018)
Aquaculture wastewater	Using a thermally treated gastropod shell	99% P recovery	Oladoja *et al.* (2015)
Digested swine wastewater	Ammonia stripping with struvite precipitation	88.03% Ammonia and 96.07% total phosphorus recovery	Cao *et al.* (2019)

15.4.4 Agriculture

Among farm-based wastes, nutrients have been recovered successfully from dairy manure, poultry manure, swine manure, cattle urine, aquaculture wastewater, abattoir wastewater, and so on (Kataki *et al.*, 2016). Various technologies adopted for nutrient recovery from agriculture wastewater is summarized in Table 15.5.

15.5 ECONOMIC VIABILITY OF THE NUTRIENT RECOVERY PROCESS

In 1000 m^3 of domestic wastewater, chemical precipitation yields 33.8 kg struvite, while micro-algae results in 299.1 kg (dry powder). Energy consumption was lowest for the fuel cells at 216.2 kWh/1000 m^3, while microalgae used the highest energy at 943.3 kWh/1000 m^3. Cost-saving analysis showed that microalgae as a nutrient recovery choice was the most economic than the rest (Gowd *et al.*, 2022). Only a few industries, such as Ostara, Colsen Water & Environment and Algalwheel have implemented pilot systems on nutrient recovery across the world. In a few cases, nutrient recovery can reduce the maintenance costs by generating revenue (Egle *et al.*, 2016; Van der Hoek *et al.*, 2016). In Europe, the cost of phosphate recovery was economical (€2–3 ($\approx$ US\$2.22–3.33)/kg·P as compared to phosphate removal (Ashley *et al.*, 2009). Additionally, effluent quality has a significant impact (De Vrieze *et al.*, 2019; Etter *et al.*, 2011; Sánchez, 2020; Ye *et al.*, 2020). However, in many cases, the commercial viability of the product remains a challenge.

On a large scale, biological phosphorus removal combined with chemical and electrochemical struvite precipitation and chemical precipitation alone can be recommended for centralized plants. Whereas for onsite and packaged treatment plants, electrochemical and chemical precipitation, and IEX may be adapted. For nitrogen recovery, struvite precipitation and acid absorption following separation by gas stripping or gas permeable membrane have an edge over other methods. Electrochemical precipitation of struvite and calcium phosphate is preferred to minimize the chemical addition (Morales *et al.*, 2013; Perera *et al.*, 2019; Wei *et al.*, 2018).

15.6 CHALLENGES AND FUTURE PERSPECTIVES

The need for nutrient recovery is urgent due to the expensive process of ammonium production and the quick depletion of phosphate-based rocks (Sartorius *et al.*, 2012; Ye *et al.*, 2018). But the main problems are contamination, high operational costs, low value of end products, and end-user perception of recovered nutrients made from waste (Verstraete *et al.*, 2016). To overcome technology bottlenecks, successful business models are needed to transform economic returns into commercial successes, and policy and education strategies must be implemented to make nutrient recovery socially acceptable (Ye *et al.*, 2020).

15.7 SUMMARY

In wastewater treatment, nutrient recovery is crucial for reducing pollution and environmental harm and for improving sustainability. The selection of nutrient recovery technology plays an important role. Struvite and microalgae are the most frequently recovered products that are highly recommended as an alternative to fertilizers. But compared to commercially available fertilizers, the recovered nutrient-rich product is still expensive. In future, even if the economic returns are insufficient, considering the sustainable development goals (SDGs) and its environmental benefits and discharge norms stipulated by regulators, nutrient recovery from wastewater streams will become mandatory.

REFERENCES

Almanassra I. W., Kochkodan V., Mckay G., Atieh M. A. and Al-Ansari T. (2021). Review of phosphate removal from water by carbonaceous sorbents. *Journal of Environmental Management*, **287**, 112245, https://doi.org/10.1016/j.jenvman.2021.112245

Ashley K., Mavinic D. and Koch F. (2009). International Conference on Nutrient Recovery From Wastewater Streams (Vancouver, 2009). IWA Publishing, London, 2009. https://doi.org/10.2166/9781780401805

Bagastyo A. Y., Anggrainy A. D., Khoiruddin K., Ursada R., Warmadewanthi I. D. A. A. and Wenten I. G. (2022). Electrochemically-driven struvite recovery: prospect and challenges for the application of magnesium sacrificial anode. *Separation and Purification Technology*, **288**(February), 120653, https://doi.org/10.1016/j.seppur.2022.120653

Barua S., Zakaria B. S., Chung T., Hai F. I., Haile T., Al-Mamun A. and Dhar B. R. (2019). Microbial electrolysis followed by chemical precipitation for effective nutrients recovery from digested sludge centrate in WWTPs. *Chemical Engineering Journal*, **361**(December 2018), 256–265, https://doi.org/10.1016/j.cej.2018.12.067

Bhattacharya A., Behari Jan B., Kumar Mand S., Bhakta J., Ghosh D. and Lahiri S. (2018). Nutrient deportation efficiency of selected natural and synthetic adsorbents from municipal wastewater for environmental and economic benefits. *Research Journal of Environmental Sciences*, **12**(5), 234–246, https://doi.org/10.3923/rjes.2018.234.246

Borowitzka M. A. (1999). Commercial production of microalgae: ponds, tanks, tubes and fermenters. *Journal of Biotechnology*, **70**(1–3), 313–321, https://doi.org/10.1016/S0168-1656(99)00083-8

Bradford-Hartke Z., Lane J., Lant P. and Leslie G. (2015). Environmental benefits and burdens of phosphorus recovery from municipal wastewater. *Environmental Science & Technology*, **49**(14), 8611–8622, https://doi.org/10.1021/es505102v

Bratby J. (2006). Coagulation and Flocculation in Water and Wastewater Treatment, 2nd edn (IWA Publishing, London), https://doi.org/10.2166/9781780407500

Cai T., Park S. Y. and Li Y. (2013). Nutrient recovery from wastewater streams by microalgae: status and prospects. *Renewable and Sustainable Energy Reviews*, **19**, 360–369, https://doi.org/10.1016/j.rser.2012.11.030

Cai Y., Han Z., Lin X., Du J., Lei Z., Ye Z. and Zhu J. (2022). Mechanisms of releasing magnesium ions from a magnesium anode in an electrolysis reactor with struvite precipitation. *Journal of Environmental Chemical Engineering*, **10**(1), 106661, https://doi.org/10.1016/j.jece.2021.106661

Cao L., Wang J., Xiang S., Huang Z., Ruan R. and Liu Y. (2019). Nutrient removal from digested swine wastewater by combining ammonia stripping with struvite precipitation. *Environmental Science and Pollution Research*, ***26***(7), 6725–6734, https://doi.org/10.1007/s11356-019-04153-x

Chevalier P. and de la Noüe J. (1988). Behaviour of algae and bacteria co-immobilized in carrageenan, in a fluidized bed. *Enzyme and Microbial Technology*, **10**(1), 19–23, https://doi.org/10.1016/0141-0229(88)90093-2

Chisti Y. (2013). Raceways-based production of algal crude oil. *Green*, **3**(3–4), 195–216.

Christenson L. and Sims R. (2011). Production and harvesting of microalgae for wastewater treatment, biofuels, and bioproducts. *Biotechnology Advances*, **29**(6), 686–702, https://doi.org/10.1016/j.biotechadv.2011.05.015

Cordell D., Drangert J.-O. and White S. (2009). The story of phosphorus: global food security and food for thought. *Global Environmental Change*, **19**(2), 292–305, https://doi.org/10.1016/j.gloenvcha.2008.10.009

Craggs R. J., Adey W. H., Jenson K. R., John M. S. S., Green F. B. and Oswald W. J. (1996). Phosphorus removal from wastewater using an algal turf scrubber. *Water Science and Technology*, **33**(7), 191–198, https://doi.org/10.2166/wst.1996.0138

Cromar N. J. and Fallowfield H. J. (1997). Effect of nutrient loading and retention time on performance of high rate algal ponds. *Journal of Applied Phycology*, **9**(4), 301–309, https://doi.org/10.1023/A:1007917610508

De Boer M. A., Romeo-Hall A. G., Rooimans T. M. and Slootweg J. C. (2018). An assessment of the drivers and barriers for the deployment of urban phosphorus recovery technologies: a case study of the Netherlands. *Sustainability*, **10**(6), 1790, https://doi.org/10.3390/su10061790

De Vrieze J., Colica G., Pintucci C., Sarli J., Pedizzi C., Willeghems G., Bral A., Varga S., Prat D. and Peng L. (2019). Resource recovery from pig manure via an integrated approach: a technical and economic assessment for full-scale applications. *Bioresource Technology*, **272**, 582–593, https://doi.org/10.1016/j.biortech.2018.10.024

Egle L., Rechberger H., Krampe J. and Zessner M. (2016). Phosphorus recovery from municipal wastewater: an integrated comparative technological, environmental and economic assessment of P recovery technologies. *Science of the Total Environment*, **571**, 522–542, https://doi.org/10.1016/j.scitotenv.2016.07.019

Etter B., Tilley E., Khadka R. and Udert K. M. (2011). Low-cost struvite production using source-separated urine in Nepal. *Water Research*, **45**(2), 852–862, https://doi.org/10.1016/j.watres.2010.10.007

Foletto E. L., Dos Santos W. R. B., Mazutti M. A., Jahn S. L. and Gündel A. (2013). Production of struvite from beverage waste as phosphorus source. *Materials Research*, **16**(1), 242–245, https://doi.org/10.1590/S1516-14392012005000152

Gienau T., Brüß U., Kraume M. and Rosenberger S. (2018). Nutrient recovery from anaerobic sludge by membrane filtration: pilot tests at a 2.5 MWe biogas plant. *International Journal of Recycling of Organic Waste in Agriculture*, **7**(4), 325–334, https://doi.org/10.1007/s40093-018-0218-6

Gowd S. C., Ramakrishna S. and Rajendran K. (2022). Wastewater in India: an untapped and under-tapped resource for nutrient recovery towards attaining a sustainable circular economy. *Chemosphere*, **291**, 132753, https://doi.org/10.1016/j.chemosphere.2021.132753

Guo J., Du J., Chen P., Huang X. and Chen Q. (2018). Enhanced efficiency of swine wastewater treatment by the composite of modified zeolite and a bioflocculant enriched from biological sludge. *Environmental Technology*, **39**(23), 3096–3103, https://doi.org/10.1080/09593330.2017.1375017

Hallas J. F., Mackowiak C. L., Wilkie A. C. and Harris W. G. (2019). Struvite phosphorus recovery from aerobically digested municipal wastewater. *Sustainability*, **11**(2), 1–. https://doi.org/10.3390/su11020376

Han Y., Agyeman F., Green H. and Tao W. (2022). Stable, high-rate anaerobic digestion through vacuum stripping of digestate. *Bioresource Technology*, **343**, 126133, https://doi.org/10.1016/j.biortech.2021.126133

Hidalgo D., Corona F., Martín-Marroquín J. M., del Álamo J. and Aguado A. (2016). Resource recovery from anaerobic digestate: struvite crystallisation versus ammonia stripping. *Desalination and Water Treatment*, **57**(6), 2626–2632, https://doi.org/10.1080/19443994.2014.1001794

Huang X., Guida S., Jefferson B. and Soares A. (2020). Economic evaluation of ion-exchange processes for nutrient removal and recovery from municipal wastewater. *NPJ Clean Water*, **3**(1), 7. https://doi.org/10.1038/s41545-020-0054-x

Kataki S., West H., Clarke M. and Baruah D. C. (2016). Phosphorus recovery as struvite from farm, municipal and industrial waste: feedstock suitability, methods and pre-treatments. *Waste Management*, **49**, 437–454, https://doi.org/10.1016/j.wasman.2016.01.003

Kékedy-Nagy L., Teymouri A., Herring A. M. and Greenlee L. F. (2020). Electrochemical removal and recovery of phosphorus as struvite in an acidic environment using pure magnesium vs. the AZ31 magnesium alloy as the anode. *Chemical Engineering Journal*, **380**(May 2019), 122480, https://doi.org/10.1016/j.cej.2019.122480

Kim D., Kim J., Ryu H.-D. and Lee S.-I. (2009). Effect of mixing on spontaneous struvite precipitation from semiconductor wastewater. *Bioresource Technology*, **100**(1), 74–78, https://doi.org/10.1016/j.biortech.2008.05.024

Kumar K., Mishra S. K., Shrivastav A., Park M. S. and Yang J.-W. (2015). Recent trends in the mass cultivation of algae in raceway ponds. *Renewable and Sustainable Energy Reviews*, **51**, 875–885, https://doi.org/10.1016/j.rser.2015.06.033

Kuusik A., Pachel K., Kuusik A. and Loigu E. (2017). Possible agricultural use of digestate. *Proceedings of the Estonian Academy of Sciences*, **66**(1), 64–74, https://doi.org/10.3176/proc.2017.1.10

Larsdotter K. (2006). Wastewater treatment with microalgae – a literature review. *Vatten*, **62**(1), 31.

Logan B. E., Hamelers B., Rozendal R., Schröder U., Keller J., Freguia S., Aelterman P., Verstraete W. and Rabaey K. (2006). Microbial fuel cells: methodology and technology. *Environmental Science and Technology*, **40**(17), 5181–5192, https://doi.org/10.1021/es0605016

Luo Y., Le-Clech P. and Henderson R. K. (2018). Assessment of membrane photobioreactor (MPBR) performance parameters and operating conditions. *Water Research*, **138**, 169–180, https://doi.org/10.1016/j.watres.2018.03.050

Mahmoud R. H., Wang Z. and He Z. (2022). Production of algal biomass on electrochemically recovered nutrients from anaerobic digestion centrate. *Algal Research*, **67**(May), 102846, https://doi.org/10.1016/j.algal.2022.102846

Manganaro J. L., Lawal A. and Goodall B. (2015). Techno-economics of microalgae production and conversion to refinery-ready oil with co-product credits. *Biofuels, Bioproducts and Biorefining*, **9**(6), 760–777, https://doi.org/10.1002/bbb.1610

Mantzavinos D. and Kalogerakis N. (2005). Treatment of olive mill effluents: part I. Organic matter degradation by chemical and biological processes – an overview. *Environment International*, **31**(2), 289–295, https://doi.org/10.1016/j.envint.2004.10.005

Matamoros V., Gutiérrez R., Ferrer I., García J. and Bayona J. M. (2015). Capability of microalgae-based wastewater treatment systems to remove emerging organic contaminants: a pilot-scale study. *Journal of Hazardous Materials*, **288**, 34–42, https://doi.org/10.1016/j.jhazmat.2015.02.002

Mehta C. M., Khunjar W. O., Nguyen V., Tait S. and Batstone D. J. (2015). Technologies to recover nutrients from waste streams: a critical review. *Critical Reviews in Environmental Science and Technology*, **45**(4), 385–427, https://doi.org/10.1080/10643389.2013.866621

Monetti J., Ledezma P., Virdis B. and Freguia S. (2019). Nutrient recovery by bio-electroconcentration is limited by wastewater conductivity. *ACS Omega*, **4**(1), 2152–2159, https://doi.org/10.1021/acsomega.8b02737

Morales N., Boehler M. A., Buettner S., Liebi C. and Siegrist H. (2013). Recovery of N and P from urine by struvite precipitation followed by combined stripping with digester sludge liquid at full scale. *Water*, **5**(3), 1262–1278, https://doi.org/10.3390/w5031262

Nagarajan D., Lee D.-J., Chen C.-Y. and Chang J.-S. (2020). Resource recovery from wastewaters using microalgae-based approaches: a circular bioeconomy perspective. *Bioresource Technology*, **302**, 122817, https://doi.org/10.1016/j.biortech.2020.122817

Nancharaiah Y. V., Venkata Mohan S. and Lens P. N. L. (2016). Recent advances in nutrient removal and recovery in biological and bioelectrochemical systems. *Bioresource Technology*, **215**, 173–185, https://doi.org/10.1016/j.biortech.2016.03.129

Neethu B., Bhowmick G. D. and Ghangrekar M. M. (2018). Enhancement of bioelectricity generation and algal productivity in microbial carbon-capture cell using low cost coconut shell as membrane separator. *Biochemical Engineering Journal*, **133**, 205–213, https://doi.org/10.1016/j.bej.2018.02.014

Oladoja N. A., Adelagun R. O. A., Ahmad A. L. and Ololade I. A. (2015). Phosphorus recovery from aquaculture wastewater using thermally treated gastropod shell. *Process Safety and Environmental Protection*, **98**, 296–308, https://doi.org/10.1016/j.psep.2015.09.006

Oswald W. J. and Golueke C. G. (1960). Biological transformation of solar energy. *Advances in Applied Microbiology*, **2**, 223–262, https://doi.org/10.1016/S0065-2164(08)70127-8

Perera M. K., Englehardt J. D. and Dvorak A. C. (2019). Technologies for recovering nutrients from wastewater: a critical review. *Environmental Engineering Science*, **36**(5), 511–529, https://doi.org/10.1089/ees.2018.0436

Potter M. C. (1911). Electrical effects accompanying the decomposition of organic compounds. *Proceedings of the Royal Society of London. Series B, Containing Papers of a Biological Character*, **84**, 260–276, https://doi.org/10.1098/rspb.1911.0073

Razon L. F. (2014). Life cycle analysis of an alternative to the Haber-Bosch process: non-renewable energy usage and global warming potential of liquid ammonia from cyanobacteria. *Environmental Progress & Sustainable Energy*, **33**(2), 618–624, https://doi.org/10.1002/ep.11817

Rivera F., Muñoz R., Prádanos P., Hernández A. and Palacio L. (2021). A systematic study of ammonia recovery from anaerobic digestate using membrane-based separation. *Membranes*, **12**(1), 19, https://doi.org/10.3390/membranes12010019

Robles Á., Aguado D., Barat R., Borrás L., Bouzas A., Giménez J. B., Martí N., Ribes J., Ruano M. V., Serralta J., Ferrer J. and Seco A. (2020). New frontiers from removal to recycling of nitrogen and phosphorus from wastewater in the circular economy. *Bioresource Technology*, **300**, 122673, https://doi.org/10.1016/j.biortech.2019.122673

Rodhe W. (1969). Crystallization of eutrophication concepts in Northern Europe. In: Eutrophication: Causes, Consequences, Correctives. National Academy of Sciences, Washington, DC, pp. 50–64.

Sánchez A. S. (2020). Technical and economic feasibility of phosphorus recovery from wastewater in São Paulo's metropolitan region. *Journal of Water Process Engineering*, **38**, 101537, https://doi.org/10.1016/j.jwpe.2020.101537

Sartorius C., von Horn J. and Tettenborn F. (2012). Phosphorus recovery from wastewater – expert survey on present use and future potential. *Water Environment Research*, **84**(4), 313–322, https://doi.org/10.2175/106143012X13347678384440

Selvaraj D. and Velvizhi G. (2021). Sustainable ecological engineering systems for the treatment of domestic wastewater using emerging, floating and submerged macrophytes. *Journal of Environmental Management*, **286**, 112253, https://doi.org/10.1016/j.jenvman.2021.112253

Sheets J. P., Yang L., Ge X., Wang Z. and Li Y. (2015). Beyond land application: emerging technologies for the treatment and reuse of anaerobically digested agricultural and food waste. *Waste Management*, **44**, 94–115, https://doi.org/10.1016/j.wasman.2015.07.037

Sher F., Malik A. and Liu H. (2013). Industrial polymer effluent treatment by chemical coagulation and flocculation. *Journal of Environmental Chemical Engineering*, **1**(4), 684–689, https://doi.org/10.1016/j.jece.2013.07.003

Subha E., Sasikala S. and Muthuraman G. (2015). Removal of phosphate from wastewater using natural adsorbents. *International Journal of ChemTech Research*, **7**, 3095–3099.

Tampio E., Salo T. and Rintala J. (2016). Agronomic characteristics of five different urban waste digestates. *Journal of Environmental Management*, **169**, 293–302, https://doi.org/10.1016/j.jenvman.2016.01.001

Tchobanoglous G., Burton F. L. and Stensel H. D. (2003). Meltcalf & Eddy, Inc.'s Wastewater Engineering: Treatment, Disposal, and Reuse, 4th Edition. McGraw-Hill, Inc., New York. 1819 pp.

Udaiyappan A. F. M., Hasan H. A., Takriff M. S. and Abdullah S. R. S. (2017). A review of the potentials, challenges and current status of microalgae biomass applications in industrial wastewater treatment. *Journal of Water Process Engineering*, **20**, 8–21, https://doi.org/10.1016/j.jwpe.2017.09.006

Ukwuani A. T. and Tao W. (2016). Developing a vacuum thermal stripping – acid absorption process for ammonia recovery from anaerobic digester effluent. *Water Research*, **106**, 108–115, https://doi.org/10.1016/j.watres.2016.09.054

Van der Hoek J. P., de Fooij H. and Struker A. (2016). Wastewater as a resource: strategies to recover resources from Amsterdam's wastewater. *Resources, Conservation and Recycling*, **113**, 53–64, https://doi.org/10.1016/j.resconrec.2016.05.012

Verstraete W., Clauwaert P. and Vlaeminck S. E. (2016). Used water and nutrients: recovery perspectives in a 'panta rhei' context. *Bioresource Technology*, **215**, 199–208, https://doi.org/10.1016/j.biortech.2016.04.094

Walsh J. J., Jones D. L., Chadwick D. R. and Williams A. P. (2018). Repeated application of anaerobic digestate, undigested cattle slurry and inorganic fertilizer N: impacts on pasture yield and quality. *Grass and Forage Science*, **73**(3), 758–763, https://doi.org/10.1111/gfs.12354

Wang X. J., Xia S. Q., Chen L., Zhao J. F., Renault N. J. and Chovelon J. M. (2006). Nutrients removal from municipal wastewater by chemical precipitation in a moving bed biofilm reactor. *Process Biochemistry*, **41**(4), 824–828, https://doi.org/10.1016/j.procbio.2005.10.015

Wang X., Zhang X., Wang Y., Du Y., Feng H. and Xu T. (2015). Simultaneous recovery of ammonium and phosphorus via the integration of electrodialysis with struvite reactor. *Journal of Membrane Science*, **490**, 65–71, https://doi.org/10.1016/j.memsci.2015.04.034

Wei S. P., van Rossum F., van de Pol G. J. and Winkler M.-K. H. (2018). Recovery of phosphorus and nitrogen from human urine by struvite precipitation, air stripping and acid scrubbing: a pilot study. *Chemosphere*, **212**, 1030–1037, https://doi.org/10.1016/j.chemosphere.2018.08.154

Whitton R., Ometto F., Pidou M., Jarvis P., Villa R. and Jefferson B. (2015). Microalgae for municipal wastewater nutrient remediation: mechanisms, reactors and outlook for tertiary treatment. *Environmental Technology Reviews*, **4**(1), 133–148, https://doi.org/10.1080/21622515.2015.1105308

Williams A. T., Zitomer D. H. and Mayer B. K. (2015). Ion exchange-precipitation for nutrient recovery from dilute wastewater. *Environmental Science: Water Research & Technology*, **1**(6), 832–838, https://doi.org/10.1039/C5EW00142K

Wu H., Foster X., Kazemian H. and Vaneeckhaute C. (2022a). N, P, K recovery from hydrolysed urine by Na-Chabazite adsorption integrated with ammonia stripping and (K-) struvite precipitation. *Science of the Total Environment*, **857**(July 2022), 159277, https://doi.org/10.1016/j.scitotenv.2022.159277

Wu I., Hostert J. D., Verma G., Kuo M. C., Renner J. N. and Herring A. M. (2022b). Electrochemical struvite precipitation enhanced by an amelogenin peptide for nutrient recovery. *ACS Sustainable Chemistry & Engineering*, **10**(43), 14322–14329, https://doi.org/10.1021/acssuschemeng.2c04691

Xia A. and Murphy J. D. (2016). Microalgal cultivation in treating liquid digestate from biogas systems. *Trends in Biotechnology*, **34**(4), 264–275, https://doi.org/10.1016/j.tibtech.2015.12.010

Ye Y., Ngo H. H., Guo W., Liu Y., Chang S. W., Nguyen D. D., Liang H. and Wang J. (2018). A critical review on ammonium recovery from wastewater for sustainable wastewater management. *Bioresource Technology*, **268**, 749–758, https://doi.org/10.1016/j.biortech.2018.07.111

Ye Y., Ngo H. H., Guo W., Liu Y., Chang S. W., Nguyen D. D., Ren J., Liu Y. and Zhang X. (2019). Feasibility study on a double chamber microbial fuel cell for nutrient recovery from municipal wastewater. *Chemical Engineering Journal*, **358**(October 2018), 236–242, https://doi.org/10.1016/j.cej.2018.09.215

Ye S., Gao L., Zhao J., An M., Wu H. and Li M. (2020). Simultaneous wastewater treatment and lipid production by Scenedesmus sp. HXY2. *Bioresource technology*, **302**, 122903, https://doi.org/10.1016/j.biortech.2020.122903

Zeng W., Ma S., Huang Y., Xia A., Zhu X., Zhu X. and Liao Q. (2022). Bifunctional lighting/supporting substrate for microalgal photosynthetic biofilm to bio-remove ammonia nitrogen from high turbidity wastewater. *Water Research*, **223**, 119041, https://doi.org/10.1016/j.watres.2022.119041

Zhang Y., Liu M., Zhou M., Yang H., Liang L. and Gu T. (2019). Microbial fuel cell hybrid systems for wastewater treatment and bioenergy production: synergistic effects, mechanisms and challenges. *Renewable and Sustainable Energy Reviews*, **103**, 13–29, https://doi.org/10.1016/j.rser.2018.12.027

Zhou W., Min M., Li Y., Hu B., Ma X., Cheng Y., Liu Y., Chen P. and Ruan R. (2012). A hetero-photoautotrophic two-stage cultivation process to improve wastewater nutrient removal and enhance algal lipid accumulation. *Bioresource Technology*, **110**, 448–455, https://doi.org/10.1016/j.biortech.2012.01.063

doi: 10.2166/9781789063714_0179

Chapter 16

Water pollution abatement using waste-derived materials: a sustainable approach

*Vaishali Choudhary[1], Sumit Kumar[1], Chetna Tewari[2], Nandagopal Sahoo[2] and Ligy Philip[1]**

[1]*Department of Civil Engineering, IIT Madras, Chennai, India*
[2]*Prof. Rajendra Singh Nanoscience and Nanotechnology Centre, Department of Chemistry, Kumaun University, Nainital, India*
**Corresponding author: ligy@iitm.ac.in*

ABSTRACT

The rapidly growing population has resulted in increased water demand and generation of high volume of solid wastes. Both excess solid waste and poor water reuse efficiency primarily result from improper management. Accordingly, to reduce the associated environmental burdens, the utilization of waste-derived material for water treatment can be considered as a sustainable approach. The current chapter aims to provide a holistic approach to solid waste management by generating value-added materials and their potential application for water pollution abatement. Different classes of waste, including agriculture, industrial, and electronic, and their possible activation methods are discussed. Also, the potential applications of such waste-derived products in different water treatment techniques, such as adsorption, catalysis, and electrochemical application, are detailed. Overall, the possibilities of utilizing waste to derive value-added products that can be employed for pollution abatement of contaminated water and achieve circular economic concepts are reviewed.

Keywords: water treatment, solid waste, activation, adsorbents, catalysts

16.1 INTRODUCTION

Globally, environmental sustainability and conservation are at stake due to rapid urbanization and industrialization. The whole world is confronted with the major challenges of a clean and safe water supply for an increasing population. In recent years, several pollutants, such as dyes, heavy metals, pharmaceuticals, pesticides, detergents, phenols, and oils, have been detected in the waterways. The exposure and accumulation of these contaminants pose serious damage to living bodies and jeopardize natural resources. Water and wastewater technologies, including ion exchange, adsorption, electrochemical oxidation/reduction, membrane separation, catalysis, and reverse osmosis, have been tested to abate the pollutant load (Crini & Lichtfouse, 2019). These technologies require materials such as adsorbents, catalysts, or membranes to facilitate pollutant removal or degradation. In this regard, a series of metal oxides, nanomaterials, transition metals, double-layered hydroxides, carbonaceous materials, polymers, and their composites has been developed.

Out of the aforementioned materials, the carbonaceous and metal oxide-based materials appear as attractive options due to their diverse physicochemical properties, abundance as renewable resources, and low cost. Most importantly, these materials can be derived from various classes of waste generated in the environment. For instance, a total of 998 million tonnes of agricultural waste is produced annually, which is highly rich in carbon content. This makes agricultural waste an excellent source for the synthesis of carbon-based materials. In addition to carbon richness, agricultural biomass has abundant functional groups, such as amino, acetamido, carbonyl, alcoholic, phenolic, and sulphhydryl groups. This helps the biomass-derived materials to act as suitable adsorbents for removing different pollutants via complexation, chelation, chemisorption, and ion exchange (Al-Rumaihi *et al.*, 2022). In a similar line, food waste, ash, sewage sludge, distilleries, and bauxite residues (red mud) can aid in the removal of water contaminants. Moreover, the improper management and handling of such waste have led to disposal in landfills, dumpsites, and water bodies, causing threats to living organisms and increasing environmental burdens. Therefore, upcycling such waste into value-added products for water management seems feasible and essential. The current book chapter attempts to give a comprehensive strategy for producing value-added materials from waste with potential applications in water pollution abatement. Modification and activation strategies of different types of wastes are examined. Also, possible uses of such waste-derived materials in various water treatment techniques, such as adsorption, catalysis, and electrochemical applications, are discussed. Overall, the potential for using waste to restore water resources is communicated.

16.2 STRATEGIES FOR MODIFICATION OF WASTE INTO FUNCTIONAL MATERIALS

Solid waste can rarely be utilized directly for water and wastewater treatment due to the leaching of phenolic groups over time, poor regeneration, and stability. Thus, effective strategies are required to turn waste into useful functional materials prior to its use in water and wastewater treatment processes.

16.2.1 Co-precipitation

Co-precipitation is the standard and most straightforward modification technique, whereby metal-based nanomaterials are formed via nucleation, growth, coarsening, and/or agglomeration of metal salt in an alkaline solution. The co-precipitation process primarily extracts metal from electronic waste and loads it on waste-derived carbon materials. For instance, magnetic reed and agriculture biochar can be developed by chemical co-precipitation of iron oxides on the surface of the biomass and subsequent pyrolysis method. Santamaría *et al.* (2022) developed nanomaterials like double-layered hydroxides from the metal extraction of the slag. The primary constraints in using co-precipitation as a modification process are the high probability of metal leaching, poor stability of the developed materials, and non-uniformity in materials (adsorbent or catalyst) structural properties.

16.2.2 Hydrothermal synthesis

Hydrothermal synthesis is a promising chemical process for converting waste to value-added materials. The chemical reactions are carried out in the aqueous phase (water) at 180–280°C (Yang & Park, 2019). Hydrochar and metal-based functional are the two primary ways of valorization of sewage sludge, food and distillery, and industrial waste. The advantage of the process over thermal modification is that the waste with high moisture content does not require a separate drying procedure. Aside from the nature of waste, the hydrothermal reaction's temperature, solid-to-liquid ratio, pH, and reaction duration have a significant impact on the structural and chemical properties of the material. Hydrothermal synthesis can also be integrated with thermal conversion to prepare hierarchical activated porous carbon microspheres and porous oxides like γ-Al_2O_3 microspheres from red mud. Moreover, it has been claimed that the material generated from hydrothermal synthesis has the potential to replace traditional catalysts and electrode materials. However, with limited information on the techno-economics of the process, the scale of the process or integration with other technologies, such as anaerobic digestion, cannot be confirmed.

16.2.3 Thermal conversion

Pyrolysis and calcination of wastes are the two frequently used thermal modification techniques. Pyrolysis is the thermochemical breakdown process that results in the production of carbon-enriched biochar, hydrocarbons (bio-oils), and volatile gases. Unlike pyrolysis, calcination involves heating solid wastes at a high temperature (900°C) in the presence of oxygen. It is typically used to create metal oxide-based valuable products from waste, such as photocatalysts (Al-Rumaihi *et al.*, 2022). In both cases, the physicochemical characteristics of initial waste and reaction temperatures significantly influence the properties and yields of the developed value-added product. For instance, increasing the pyrolysis temperature (>500°C) reported the destruction of the oxygen-containing group, and the production of aromatic thin-layered graphitic structure has been reported. Likewise, the biomass produced from agricultural waste has a low C:N ratio compared to biochar produced from animal waste. Though the biochar produced at high temperatures can be used as a persulphate activator via graphite electron donor-transfer complex, the pore blockage and low yield can adversely affect the probable use of the biochars as an adsorbent. The calcination temperatures also have severe effects on the nanostructure and crystallinity of materials developed from waste. For instance, higher temperatures increase crystallinity and severe particle aggregation. Overall, there is a need to examine the operating parameters for pyrolysis/calcinations and waste characteristics individually to develop biochar with suitable surface chemistry, conductivity, and porosity. In addition, special care should be taken in producing biochar at low temperatures (300°C) due to the formation of harmful environmentally persistent free radicals, which may negatively affect the ecosystems (Bi *et al.*, 2022).

16.2.4 Sol–gel technique

The sol–gel method is commonly used for the valorization of inorganic-rich waste into colloidal suspension of inorganic particles, which transformed into the gelatinous network. The gel formed can be converted into oxide materials by high-temperature calcination. Waste like spent Ni–Cd batteries can be used to develop mixed metal oxide. Similarly, metal-organic frameworks like MIL-53 (Al) have been synthesized using waste aluminium and polyethylene terephthalate bottles (Panda *et al.*, 2020). Although the concept of sol–gel transformation seems appealing, the structures often generated have lower Brunauer–Emmett–Teller (BET) surface areas, lower crystallinity, and chaotic morphologies. Thus, reasonable experimental design and application are highly recommended.

16.3 ACTIVATION METHOD OF WASTE-DERIVED MATERIALS

The modification of solid waste into functional material seems a feasible way to develop engineered materials for water treatment applications. However, poor surface area, inadequate number of functional groups, or high hydrophobicity limits their application in water/wastewater treatment. Moreover, poor stability of metal oxide and leaching is a common problem. It should be noted that often waste materials are carbonized using pyrolysis or calcinations to produce carbon-based adsorbents and nanocomposites. Accordingly, the current section discusses various pre- and post-modification strategies (Figure 16.1) and their suitability for carbon-made materials.

16.3.1 Physical activation of waste-derived carbon materials

16.3.1.1 Ball milling

Ball milling is a chemical-free and economical modification strategy whereby mixing two or more solids or solid–liquid is carried under mechanical force. During the process, the dynamic energy of

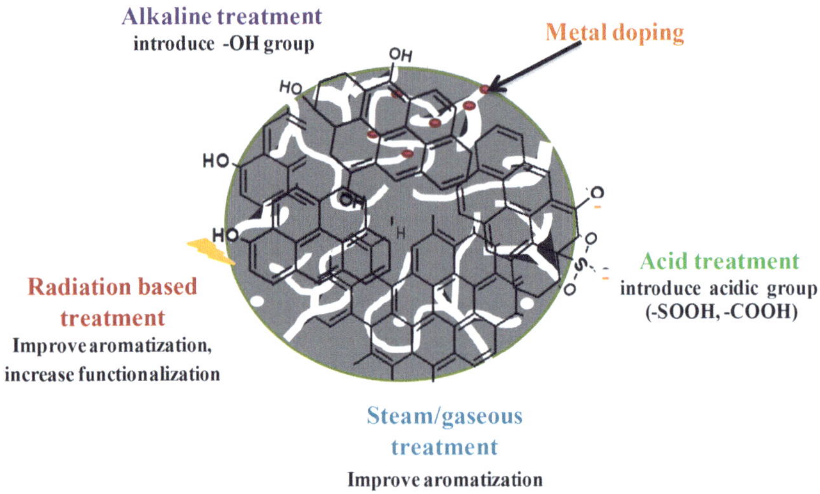

Figure 16.1 Modification method of carbon-based materials.

the moving ball induces chemical and structural changes in the feedstock via grinding and breaking (Amusat *et al.*, 2021). Ball-milling plays a major role in breaking the material down to the nanoscale without damaging the crystal structure. The use of ball-milling for nanoparticle production has been studied widely in recent years because of its low cost, large-scale applicability, and low energy for production. However, the specific relationships between the feedstock, temperature, and other properties with changes in particle size still need to be explored. Moreover, the dispersibility of the nano biochar is adversely affected, restricting its application in water treatment.

16.3.1.2 Gaseous activation

Gas activation using steam, nitrogen, and carbon dioxide has been used to enhance the porosity and specific surface area of the materials. The process also allows the removal of impurities such as ash, organic matter, and products of incomplete combustion from the surface (Sajjadi *et al.*, 2019a). During the steam activation process, the carbonized waste material (like biochar) is subjected to incomplete gasification, the continuous steam supply allows the conversion of CO into CO_2, and the released H_2 weakly combines with carbon to produce a weakly bound carbon–hydrogen complex. Accordingly, this disintegration increases the aromaticity, formation of crystalline carbon, and the opening of embedded porous structures. The steam-activated biochar has the potential to act as a slow-release fertilizer and CO_2 capture. However, the reduction in the carboxylic and phenolic groups on the surface adversely affects the polarity and metal removal potential. Thus, gaseous activation processes are favourable in pre-treatment and improving the porosity of carbon material rather than functionalization.

16.3.1.3 Radiation-based activation

The activation of carbon materials using microwave, ultrasound, and plasma-based radiation is gaining attention. The microwave-based activation forms micropores and increases surface area and polarity ((O + H)/C content). It should be noted that microwave irradiation is not feasible for the pyrolysis of pure biomass; thus, microwave absorbers (e.g., metallic oxides or charcoal) should always be impregnated into biomass structure (Sajjadi *et al.*, 2019a). Unlike microwave heating, sonication (ultrasound treatment) aids in leaching metal salt pores and increasing the internal surface area.

Additionally, carboxylation, hydrogenation, water splitting, and exfoliation reaction at the surface allow the formation of a graphitic (sp^2) layer and increase the swelling properties of the biochar. Thermal plasma is another promising method for activation or an alternative to pyrolysis. The plasma-based activation and modification strategies provide higher energy density, faster reaction rate, and lower formation of heavy tarry compounds. Moreover, pre-treatment, like chemical washing and prolonged drying, is not required for plasma-based systems. However, the major problem with the process is in introducing oxygen-containing functional groups into a sample surface (Sajjadi *et al.*, 2019a). Overall, the energy throughput of sonication-based treatment is the least (even lower than the traditional methods).

16.3.2 Chemical activation of waste-derived carbon materials

Chemical modification methods like acidic/alkaline treatment, metal salt doping, or modification with oxidizing reagents are the most commonly employed and easy-to-use techniques. The wet chemistry-based methods can improve the functionalization of waste-derived materials (Table 16.1).

16.3.2.1 Acid and alkali treatment

The acid and alkali modification of waste-derived materials like biochar and carbon nanocomposite mainly alters the oxygen-containing functional groups. Strong acids like H_3PO_4 and H_2SO_4 alter the elemental composition and surface area of the materials, but treatment using high concentrations (2% sulphuric acid) can sometimes cause a reduction in the surface area. While weak acids such as oxalic acid and citric acid can introduce the carboxyl group on the surface through esterification. Accordingly, optimization of the nature of acid and concentration to increase functional groups and specific surface area is highly essential (Wang & Wang, 2019). Similar to acid modification, alkaline modification involves the usage of strong bases like sodium hydroxide (NaOH) and potassium hydroxide (KOH). On treatment with the alkali, the aromaticity of the carbon composite is increased due to the formation of a positive charge, and hydroxyl functional groups are easily incorporated (Sajjadi *et al.*, 2019b). It should be noted that the effect of acid/alkaline modification relies on the feedstock and preparation methods. Hence, it is too difficult to obtain a consistent result.

16.3.2.2 Oxidizing agents

Chemical oxidants like hydrogen peroxide, potassium permanganate, and zinc chloride have been used to increase the number of lactones, carboxylic, and hydroxyl groups and the decomposition of the aromatic carbons (Wang & Wang, 2019). It should be noted that the decrease in aromaticity of the structure can often weaken the π–π interaction and adversely affect the sorption of hydrophobic compounds. Thus, the properties of targeted pollutants should be considered before using the modification methods.

16.3.2.3 Metal doping

The doping of metal oxides/nanoparticles such as iron oxide, nZVI, and metal chloride salts onto feedstock/biochar has been studied owing to the benefit of increased surface area, adsorption capacity, catalytic and magnetic properties, and ease of separation (Figure 16.1). Additionally, using substrates like rice straw or char prevents the agglomeration of nanomaterials. It should be noted that most of the waste-derived material had lower pK_a (~2). Accordingly, the high negative charge on the surface makes the adsorption of anionic pollutants like phosphate and dyes difficult. Thus, modification with metal/metal oxides adjusts the surface charge and allows the effective removal of negatively charged contaminants (Wang *et al.*, 2019). The magnetic modification method provides a solution for adsorbent recovery. However, significant increases in the adsorption capacity are not always noted. Overall, the incorporation of metal/metal-oxide/nanomaterials depends on the use of

Table 16.1 Summary of physiochemical properties of activated waste-derived carbon material.

Feedstock	Carbonization Technique	Activation Type	Properties	Remarks
Peanut hull	Pyrolysis	Acid activation	S_{BET} = 952.6 m^2/g Pore vol. = 0.88 cm^3/g	• The biochar found rich in oxygenated functional groups • The adsorption capacity was reported at 149.25 mg/g
Rice husk	Microwave pyrolysis		S_{BET} = 1165 m^2/g Pore vol. = 0.78 cm^3/g	• Acid treatment increases S_{BET} from 752 to 1165 m^2/g • Significant adsorption capacity was reported at 441.52 mg/g
Cinnamomum camphora	Pyrolysis	Alkali activation	S_{BET} = 75.92 m^2/gPore Vol. = 0.13 cm^3/g	• Alkali treatment jointly with ultrasound increases S_{BET} from 29.86 to 75.92 m^2/g, and the pore vol. increased from 0.056 to 0.128 cm^3/g • Significant surge in adsorption capacity was reported from 39.06 to 98.33 mg/g
Sugarcane waste	Pyrolysis		S_{BET} = 291.8 m^2/g Pore vol. = 0.24 cm^3/g	• Alkali treatment has improved oxygenated functionalities and microporosity • Acid treatment reduced the ash content.
Peach stone	Hydrothermal	Gaseous activation	S_{BET} = 1247 m^2/g Pore vol. = 0.12 cm^3/g	• Microporosity was almost absent • Very low functional groups attached • Very high surface area was achieved
Eucalyptus globulus	Pyrolysis		S_{BET} = 801.2 m^2/g Pore vol. = 0.08 cm^3/g	• Richness in functional groups is maintained after pyrolysis • The biochar possesses more micropore structures
Bamboo	Solvothermal	Irradiation activation	S_{BET} = 88.52 m^2/g Pore vol. = 0.18 cm^3/g	• 2D and 3D graphene was synthesized • 100% removal of pollutants was achieved • This material showed bioimaging application
Willow	Steam pyrolysis		S_{BET} = 443.2 m^2/g Pore vol. = 0.242cm^3/g	• Microwave activation showed high potential in toxicity reduction • Ash content was reduced, and the carbon content increased drastically
Microalgae	Hydrothermal	Doping	S_{BET} = 62.2 m^2/g Pore vol. = 0.1 cm^3/g	• TiO_2 is doped over microalgae biochar • 99.2% removal of pollutants was achieved • Reusability up to the 8th cycle noted to be 80%
Jute fibres	Pyrolysis		S_{BET} = 62.2 m^2/g Pore vol. = 0.1 cm^3/g	• ZnO was dispersed over jute biochar • 99% of degradation achieved • Reusability till three cycles for 78.6% degradation

chemicals for modification, activation, and function. Aside from the fact that it is a costly method, the high chances of re-introduction of toxic waste from the chemicals used are a cause for concern.

16.4 APPLICATION OF WASTE-DERIVED MATERIALS AS ADSORBENT

The adsorption technique is widely employed to remove the trace/low concentration of contaminants from water. This section will discuss the different forms of waste-derived absorbents and their suitability for water and wastewater remediation.

16.4.1 Biochar and hydrochar

The popularity of biochar arises due to some key advantages, such as it restricts greenhouse gases during biomass decomposition and the richness of physical and chemical properties such as high surface area, various functionality, and so on. Compared to biochar, the abundance of hydrogen and oxygen makes hydrochar more suitable for adsorption. The diverse physicochemical properties of biochar and hydrochar mediate a broader range of reactions over its surfaces and facilitate higher sorption capabilities than activated carbon. A meta-analysis conducted by many research groups to compare the economic and ecological implications of biochar/hydrochar and activated carbon showed that biochar is more efficient in pollutant removal, with the additional advantage of lower greenhouse gas emissions. For instance, 97 MJ/kg of energy is required for activated carbon synthesis, while just 6.1 MJ/kg is needed for biochar. In addition, Choudhary and Philip (2022) conducted a sustainability assessment of biochar and granular activated carbon and found that the total environmental impact of biochar was 20 times less than commercial carbon. This brings one of the key motivations to switch existing adsorbents with low-cost waste-derived carbonaceous materials like biochar and hydrochar. A significant amount of studies are reported on the uses of such biochars for removing organic (polyaromatic hydrocarbons, carbofuran, polychlorinated dibenzo-*p*-dioxins, perfluoroctane, pentachlorophenol) and inorganic (Cd^{2+}, Zn^{+}, As, Cr, Pb^{2+}) pollutants from water and wastewater. All earlier studies showed the removal efficiency of 30–100% at an adsorbent dose of 0.03–10 g/L, which promotes the extensive use of such carbonaceous materials in the remediation of water and wastewater (Amalina *et al.*, 2022). Therefore, it can be affirmed that biochar/hydrochar can be a sustainable, low-cost alternative to commercial activated carbon.

16.4.2 Nanomaterials

With advancements in nanoscience and nanotechnology, a large class of nanomaterials has been developed using waste materials. Broad applications of waste-derived nanomaterials have been reported due to their high thermal, mechanical stability, optical, and electrical properties. The critical aspects of nanomaterials are a very high surface-to-volume ratio, high porosity, and small size, making these materials superior adsorbents. Particularly engineered carbon-based nanomaterials have attracted attention as sensors, energy storage devices, and air pollution control due to their tunable and multifunctional nature. The application of carbon-based nanomaterials in wastewater remediation has also shown high potential. The nanomaterials like nanotubes, nanofibres, nanoparticles, quantum dots, and nanosheets have been synthesized from rice husk, wheat straw, sugarcane bagasse, and so on., and showed very high adsorption potential for water and wastewater abatements (Jun *et al.*, 2018). These functionalized nanomaterials are highly prone to the adsorption of organic contaminants via electrostatic interactions, hydrogen bonding, π–π interactions like electron donor–accepter mechanism, and Van der Waals interactions (Figure 16.2). These nanomaterials have proven to be more effective adsorbents than traditional materials such as zeolites, clays, and polymers. Also, few studies reported the development of two-dimensional (2D) and three-dimensional (3D) graphene nanomaterials using bamboo biomass and their promising applicability in water and wastewater remediation (Mubarik *et al.*, 2021). Therefore, it would be a great opportunity for upcycling waste and its applications in water and wastewater purification.

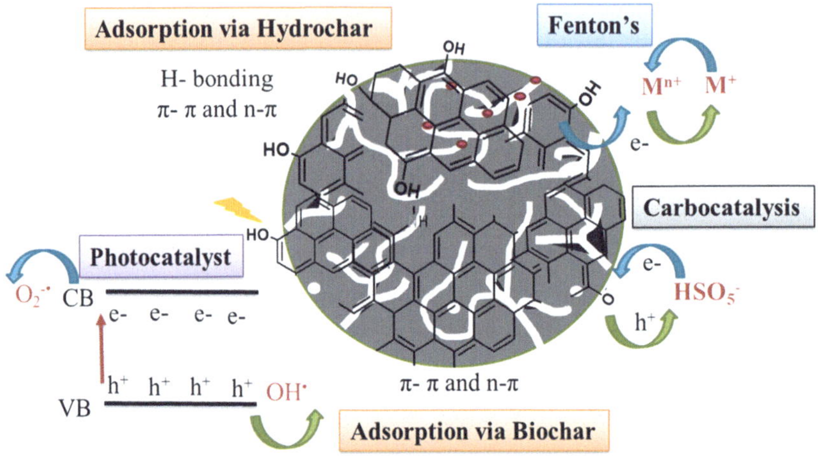

Figure 16.2 Schematics of sorption and degradation mechanism using waste-derived carbon materials and composite.

16.5 APPLICATION OF WASTE-DERIVED MATERIALS AS CATALYSTS

Low-cost waste-derived carbonaceous materials have recently been tested for catalytic application in water treatment, fuel cells, and soil abatements. However, an understanding of the practical uses in respective fields is still unclear. In this section, we will discuss the use of waste-derived materials as a potential catalyst in water treatment applications.

16.5.1 Fenton-like catalysts

The Fenton processes like conventional Fenton, electro-Fenton, and photo-Fenton are among the oldest advanced oxidation processes where Fe^{2+} is used as a catalyst to activate the oxidant (H_2O_2). However, this process is identified with key demerits like non-recycling of catalyst, large production of sludge, and lower operational pH (<3). In recent years, the 'Fenton-like' catalysts have been getting attention to overcome the limitations of the classical Fenton's systems. This utilizes reusable and recyclable catalysts and different oxidants like H_2O_2 and $S_2O_8^{2-}$. The porous carbon materials and composites are a preferred option as they can enrich the pollutants and disperse the reactive species, resulting in better efficiencies. Additionally, such porous structures facilitate mass transfer owing to shorter diffusion pathways. Despite lower catalytic efficiencies reported for pristine carbon materials towards H_2O_2 activation, the doping/modification of carbon materials with metal catalysts can enhance the crystallinity and performance of the composite catalysts. Moreover, carbon-based composite provides radical and non-radical pathways for degradation, which alleviates the performance in real wastewater. It should be noted that carbon composite developed from high-temperature carbonization shows improved oxygen reduction reaction and $S_2O_8^{2-}$ activation due to the richness in carbon defects. While mild temperature carbonization results in the distribution of oxygen functionalities and promotes the generation of hydroxyl radicals. Thus, the generation of diverse reactive radical species enhances the degradation and mineralization of a wide range of organic pollutants from water and wastewater (Pan & Qian, 2022).

16.5.2 Photocatalysts

Photocatalysis is proven to be a very effective in-situ water treatment technology in the last few decades. The process efficiency is significantly dependent on the photocatalyst used. Generally, semiconducting metals are recommended as photocatalysts; however, quick recombination of electron–hole pairs and poor response towards visible lights adversely affects the performance. Moreover, due to scarce semiconducting metals, such catalysts put an extra burden on the treatment technology. So, considerable efforts have been made to develop metal oxides and non-metal-based photocatalysts. Such developed catalysts are reported to reduce band gap and improve photo absorbance. However, photo corrosion, thermal instability, and dopant leaching are the major drawbacks that still need to be tackled. Various non-metallic and composite catalysts are being studied these days. In this regard, modified carbonaceous materials such as TiO_2-coconut shell, TiO_2-reed straw, TiO_2-corn cob, and TiO_2-bamboo are focused on due to their structure tunability, electronic conductivity, and good absorptive nature. Moreover, the surface functionality of biochar has shown active participation in reactions and promoted catalytic efficiency. For instance, the larger surface area can facilitate better absorption, due to which the pollutants could be prone to attack by short-lived radical species. Carbonaceous material also helps in suppressing the recombination of electron–hole pairs and narrowing the band gap to improve photo-absorption by forming a bridge and acts as a photosensitizer, charge carrier, and electron transport (Li *et al.*, 2022).

16.5.3 Carbonaceous catalysts

The term carbocatalyst describes the use of carbonaceous materials in catalysis as a catalyst. Few carbon forms, primarily graphene, carbon nanotubes, and fullerenes, have been focused in the field of catalysis. However, in recent years, other carbonaceous materials have emerged as better alternatives. Biochar, which is widely available and inexpensive, has shown chemical and physical properties similar to traditional carbon catalysts. Furthermore, differing electronic structures of edges or defects can be created by post treatment of biochar/carbonaceous materials. The important criterion for defining the use of biochar as carbocatalyst is the presence of an sp^2 network. This rigid carbon framework provides high chemical stability and assists in free electron generation. The carbonaceous interface allows an effective transfer of reagents for oxidation and offers sufficient space around functional groups to act as landing sites (adsorption space). The inertness of the carbon allows a spill-over mechanism for the reactive species (active radicals) without getting attacked by them (Schlögl, 2021). In addition, the anisotropic nature of carbonaceous materials makes them easily regenerable in case of deactivation or contamination, which is absent in the case of metal-based catalysts.

16.5.4 Electrochemical catalyst

In the past few years, electrochemistry-based water and wastewater treatment technologies have attracted huge attention due to their lower chemical consumption, high efficiency, ease in handling, and low emissions. Metal-based materials are extensively used as electrocatalysts, and all such catalysts have shown excellent effectiveness. But the constant exploitation of natural resources has limited its large-scale application due to the high cost of catalysts, by-product formation, metal leaching, and so on. Therefore, green and low-cost materials are needed as electrodes. Electrochemical catalysts have shown tremendous potential due to their multitude applications, such as energy storage, electro-sorption, and electrocatalytic oxidation/reduction (Alkhadra *et al.*, 2022). Agricultural wastes like straw, lotus leaves, hemp stem, Platanus fruit, peels, and rice husk are among the common low-cost green materials used for electrode development. For instance, carbonized cornstalk and modified biochar with Fe and Zn resulted in effective nitrobenzene degradation in less time than the pure metal electrode. Electrocatalytic performance is also demonstrated to be highly dependent on electronic properties, surface area, and nano-structure of electrode materials. Accordingly, dopants such as Pd, Co, Pt, S, and N on biochar have been studied. The presence of dopant and carbon substrate enhances the adsorption of pollutants and allows effective reduction/oxidation by providing electron transport channels and enhancing the radical species generation (Macchi *et al.*, 2022). Overall, electrochemical technology offers the advantage of modular design and effective removal of contaminants; however more efforts in the direction of developing and optimising the performance of waste-derived electrode and electrocatalyst are needed.

16.6 SUMMARY

This chapter systematically presented the overview of waste-derived water and wastewater treatment materials. The conversion of common wastes to valuable products such as metal oxide, double-layered hydroxides, nanomaterials, biochar, and hydrochar provides an option for the treatment of wastes and contributes towards environmental sustainability. Different activation methods can be selected to modulate the physiochemical properties of waste-derived carbon material (specifically biochar). In terms of performance, waste-derived adsorbents and catalysts show a wide prospect. However, most studies have been conducted in a laboratory scale. To facilitate the application of these adsorbents and catalysts in the real field: (1) optimization of activation/ modification methods, (2) study in wastewater and mixed systems, (3) investigation of the stability of produced materials, and (4) reuses and post-treatment of spent materials are needed.

REFERENCES

Alkhadra M. A., Su X., Suss M. E., Tian H., Guyes E. N., Shocron A. N., Conforti K. M., de Souza J. P., Kim N. and Tedesco M. (2022). Electrochemical methods for water purification, ion separations, and energy conversion. *Chemical Reviews*, **122**(16), 13547–13635, https://doi.org/10.1021/acs.chemrev.1c00396

Al-Rumaihi A., Shahbaz M., Mckay G., Mackey H. and Al-Ansari T. (2022). A review of pyrolysis technologies and feedstock: a blending approach for plastic and biomass towards optimum biochar yield. *Renewable and Sustainable Energy Reviews*, **167**, 112715, https://doi.org/10.1016/j.rser.2022.112715

Amalina F., Abd Razak A. S., Krishnan S., Sulaiman H., Zularisam A. W. and Nasrullah M. (2022). Biochar production techniques utilizing biomass waste-derived materials and environmental applications – a review. *Journal of Hazardous Materials Advances*, **7**, 100134, https://doi.org/10.1016/j.hazadv.2022.100134

Amusat S. O., Kebede T. G., Dube S. and Nindi M. M. (2021). Ball-milling synthesis of biochar and biochar-based nanocomposites and prospects for removal of emerging contaminants: a review. *Journal of Water Process Engineering*, **41**, 101993, https://doi.org/10.1016/j.jwpe.2021.101993

Bi D., Huang F., Jiang M., He Z. and Lin X. (2022). Effect of pyrolysis conditions on environmentally persistent free radicals (EPFRs) in biochar from co-pyrolysis of urea and cellulose. *Science of the Total Environment*, **805**, 150339, https://doi.org/10.1016/j.scitotenv.2021.150339

Choudhary V. and Philip L. (2022). Sustainability assessment of acid-modified biochar as adsorbent for the removal of pharmaceuticals and personal care products from secondary treated wastewater. *Journal of Environmental Chemical Engineering*, **10**(3), 107592, https://doi.org/10.1016/j.jece.2022.107592

Crini G. and Lichtfouse E. (2019). Advantages and disadvantages of techniques used for wastewater treatment. *Environmental Chemistry Letters*, **17**(1), 145–155, https://doi.org/10.1007/s10311-018-0785-9

Jun L. Y., Mubarak N. M., Yee M. J., Yon L. S., Bing C. H., Khalid M. and Abdullah E. C. (2018). An overview of functionalised carbon nanomaterial for organic pollutant removal. *Journal of Industrial and Engineering Chemistry*, **67**, 175–186, https://doi.org/10.1016/j.jiec.2018.06.028

Li Z., Li K., Du P., Mehmandoust M., Karimi F. and Erk N. (2022). Carbon-based photocatalysts for hydrogen production: a review. *Chemosphere*, **308**, 135998, https://doi.org/10.1016/j.chemosphere.2022.135998

Macchi S., Denmark I., Le T., Forson M., Bashiru M., Jalihal A. and Siraj N. (2022). Recent advancements in the synthesis and application of carbon-based catalysts in the ORR. *Electrochem*, **3**(1), 1–27.

Mubarik S., Qureshi N., Sattar Z., Shaheen A., Kalsoom A., Imran M. and Hanif F. (2021). Synthetic approach to rice waste-derived carbon-based nanomaterials and their applications. *Nanomanufacturing*, **1**(3), 109–159, https://doi.org/10.3390/nanomanufacturing1030010

Pan Z. L. and Qian X. F. (2022). Porous carbons for use in electro-Fenton and Fenton-like reactions. *Xinxing Tan Cailiao/New Carbon Materials*, **37**(1), 180–195, https://doi.org/10.1016/S1872-5805(22)60578-X

Panda D., Patra S., Awasthi M. K. and Singh S. K. (2020). Lab cooked MOF for CO_2 capture: a sustainable solution to waste management. *Journal of Chemical Education*, **97**(4), 1101–1108, https://doi.org/10.1021/acs.jchemed.9b00337

Sajjadi B., Chen W.-Y. and Egiebor N. O. (2019a). A comprehensive review on physical activation of biochar for energy and environmental applications. *Reviews in Chemical Engineering*, **35**(6), 735–776, https://doi.org/10.1515/revce-2017-0113

Sajjadi B., Zubatiuk T., Leszczynska D., Leszczynski J. and Chen W. Y. (2019b). Chemical activation of biochar for energy and environmental applications: a comprehensive review. *Reviews in Chemical Engineering*, **35**(7), 777–815, https://doi.org/10.1515/revce-2018-0003

Santamaría L., Korili S.A. and Gil A. (2022). Layered double hydroxides from slags: Closing the loop. *Journal of Environmental Chemical Engineering*, **10**(1), 106948.

Schlögl R. (2021). Quo vadis carbocatalysis? *Journal of Energy Chemistry*, **61**, 19–27.

Wang J. and Wang S. (2019). Preparation, modification and environmental application of biochar: a review. *Journal of Cleaner Production*, **227**, 1002–1022, https://doi.org/10.1016/j.jclepro.2019.04.282

Wang L., Wang Y., Ma F., Tankpa V., Bai S., Guo X. and Wang X. (2019). Mechanisms and reutilization of modified biochar used for removal of heavy metals from wastewater: a review. *Science of the Total Environment*, **668**, 1298–1309, https://doi.org/10.1016/j.scitotenv.2019.03.011

Yang G. and Park S.-J. (2019). Conventional and microwave hydrothermal synthesis and application of functional materials: a review. *Materials*, **12**(7), 1177, https://doi.org/10.3390/ma12071177

doi: 10.2166/9781789063714_0191

Chapter 17

Evaluating sustainability for water and wastewater treatment technologies

Sritama Mukherjee[1], *Tanmayaa Nayak*[2] *and T. Pradeep*[2*]

[1]*Department of Fibre and Polymer Technology, KTH Royal Institute of Technology, Stockholm, Sweden*
[2]*DST Unit of Nanoscience (DST-UNS) and Thematic Unit of Excellence, Department of Chemistry, Indian Institute of Technology Madras, Chennai, India*
**Corresponding author: pradeep@iitm.ac.in*

ABSTRACT

High-performance water and wastewater treatment have become feasible with the development of innovative and advanced technologies that may have the potential to have an impact on health and the environment. At the same time, they can be helpful in protecting the environment and meet specific social needs in a more sustainable way. New purification techniques have the potential to lead to the sustainable development of clean water, understand global environmental challenges and seek ways to improve them. Water treatment processes become more environmentally friendly by reducing energy consumption, removing toxic and essential substances and making products sustainable during the synthesis phase. This chapter discusses water treatment in the context of India, the necessity and methods of assessing sustainability, as well as examples of technological assessments for sustainability.

Keywords: wastewater treatment, sustainability metrics, life-cycle assessment, risk assessment, social impact, green chemistry

17.1 THE WATER AND WASTEWATER TREATMENT SCENARIO IN INDIA

Owing to India's growing population, intensive agriculture, climate change, water pollution, and depleting natural water supplies, water management in India is challenging. Being the world's most populous country, 35% of the Indian population is concentrated in metropolitan regions, while 65% resides in rural areas. According to the 2018 report of NITI (National Institution for Transforming India, Government of India) Aayog, consumption of water could double by 2030, posing serious water scarcity to millions of people and a 6% decline in the country's GDP. Therefore, it is essential to understand and effectively manage our water resources. A sustainable future essentially requires treatment, recycling, and reuse of water. Wastewater treatment becomes crucial because of rapid urbanization and industrialization in the country. This has significant negative impacts on public health and the environment. NITI Aayog's report on urban wastewater scenario in India (2022) estimated that for 2020–2021, the wastewater generation in rural areas is predicted to be around 39,604 million litres per day (MLD), whereas it is estimated to be 72,368 MLD in urban areas; out of which 28% of the wastewater is actually treated and the remaining 72% is discharged into aquifers, rivers, and lakes. Inadequate infrastructure, underutilization of treatment capacity, inefficient operation and maintenance and lack of water quality monitoring are some of the primary issues related to the sewage treatment plants.

The Government of India has implemented several policies and programmes to address the issue of water stress in the country. Some of the key policies and programmes include: (1) National River Conservation Plan (NRCP), which aims to improve water quality in the country's rivers by reducing pollution from municipal and industrial sources; (2) Atal Bhujal Yojana, which aims to improve groundwater management by strengthening community-based water management institutions and promoting sustainable groundwater use; (3) National Mission for Clean Ganga (NMCG), which aims to rejuvenate the river Ganga by promoting conservation and management of the river and its tributaries, and (4) National Water Policy, which aims to ensure the conservation and development of water resources through sustainable water management practices. Government and private organizations emphasize on implementing various wastewater treatment methods over the past 10 years. Wetlands, aquifer systems, membrane technology, and biological filters, which are becoming more and more popular alternatives to the traditional activated sludge process (Jadeja *et al.*, 2022). Other initiatives include construction of sewage treatment plants, the promotion of better waste management practices, and the protection of water resources.

At many places, domestic or treated wastewater is used as a source of irrigation, which contains various types of nutrients such as phosphorus, nitrogen, potassium, and sulphur improving the soil fertility that can be easily accumulated by the plants. In that process, it is essential to make sure that the resultant output from the treatment systems conforms to environmental standards particularly with regard to the quality of the treated effluents that are discharged into nature. Centralized sewage and wastewater treatment systems are often too expensive and sophisticated to serve only a section of big cities, while the smaller towns and low-income areas get overlooked. Decentralized wastewater treatment system (DEWATS) are simplified systems for community-based sanitation centres having capacity of 1–1000 m^3/day for treating organic wastewater that is both cost effective and efficient for developing nations. It uses aerobic treatment systems and constructed wetlands. Some working examples of DEWATS in India are: 500 KLD capacity soil biotechnology (SBT)-based DEWATS under New Delhi Municipal Corporation; Phytorid-based DWWTs at Bangalore of capacity 250 KLD; 307 KLD DEWATS in Puducherry; 10 KLD system in Gohri, Uttarakhand, a 12 KLD system serving people in Leh, Ladakh, and many more such treatment plants in bigger and smaller cities (Varma *et al.*, 2022).

Apart from wastewater management, India also suffers from groundwater contamination cases, such as higher fluoride and arsenic concentration in groundwater which exceed the permissible limits. The temporal and spatial distribution of these ions depend on the sources (natural and anthropogenic), process and interactions with environment, and the nature of recharge. AMRIT (Arsenic and Metal Removal by Indian Technology) which is instrumental in supplying arsenic-free water for over 1.2 million people every day, has new nanostructured materials with large adsorption capacities for arsenite and arsenate ions in water in field conditions (Mukherjee *et al.*, 2022). Similarly, filters composed of ferrihydrite nanoparticle (FeOOH)-impregnated hybrid polymeric (polyaniline/nanocellulose) composite to remove fluoride from drinking water using nanochemistry have the potential to be scaled up. This technology is able to reduce fluoride levels below the safety standards prescribed for drinking water (Mukherjee *et al.*, 2020).

17.2 TECHNOLOGICAL ADVANCEMENTS IN WATER TREATMENT

In the last decade, the solar-powered treatment of wastewater has gained vital importance in several parts of the world. In many countries, the application of solar energy for sewage water treatment is demonstrated for high efficiency anaerobic digestion of sludge absorption, advanced oxidation processes for decontamination of industrial effluents, solar photo-Fenton with anaerobic activated sludge in sequencing batch reactor for degrading non-biodegradable pollutants, new photo-catalysts, and photo-Fenton and photo reactor configurations. The solar desalination has the dominated contribution of 57% in desalination market world. Various direct solar desalinations are solar still, solar humidification–dehumidification (HDH) and solar chimney and solar-powered membrane desalination (MD). Solar photo-catalysis using semiconductor nanoparticles and metal complexes for destroying water contaminants and disinfection inactivation of waterborne microbial species has been recently developed (Sansaniwal, 2022).

Cellulose nanomaterials have inherent fibrous nature and remarkable mechanical properties. Due to its low cost, biocompatibility and sustainable source, it has huge potential as a component in water filtration membranes, adsorbents, and scaffolds for removal of inorganic and organic pollutants. Various clay (kaolinite, montmorillonite, bentonite, etc.), biopolymers like chitosan, cellulose, alginates; other polymers such as polyvinyl alcohol, polyvinylpyrrolidone, zeolites, and so on have been used to prepare (nano) composites with graphene oxides, CNTs, alumina, iron oxides, and other metal oxides for removal of water contaminants, most commonly heavy metals. Activated carbon derived from wide range of materials has become one of the cheap and versatile porous alternatives for pollutant scavengers. Some common materials that are used to make activated carbon are bituminous coal, bones, pine sawdust, coconut shells, lignite, petroleum-based residues, peanut, sugarcane bagasse, wastewater treatment sludge, and wood (Jaspal & Malviya, 2020).

Biocatalytic membranes combine the enzyme-based chemical reactions and membrane separation into one single unit, by the method of immobilizing enzymes on the membrane surfaces that results in higher selectivity and process efficiency. Its dual function of catalysis and separation has lower energy consumption leading to smaller footprint, and reduced equipment cost. Oxidoreductase enzymes such as peroxidases and laccases are most commonly used for the catalytic degradation of hazardous organic molecules, while hydrolases including a-amylase, trypsin, lipase, and so on are reported for fouling mitigation. These enzymes hydrolyse various organic compounds and bio-foulants, including starch, proteins, amino acids, lipids, and pectins (Barbhuiya *et al.*, 2022).

Electrospinning method has emerged in recent times to produce new generation membranes of nanofibres that have higher porosity than the conventional membranes used for nano or ultrafiltration. They can perform adsorption and filtration depending upon the nanofibre structure and chemical functionality, for example removal of heavy metals using cellulose acetate nanofibre ion-exchange membranes. Cellulose nanomaterial–polymer blends are also popularly electrospun into fibres, where high nanocellulose concentration gives rise to highly viscous polymer blend that results in larger diameter of fibres with narrower pore-size distributions (Carpenter *et al.*, 2015).

Attempts have been made to resolve some of the problems faced by conventional membranes such as polarized concentration, pressure drop, low mass transfer and high fouling tendencies using three-dimensional (3D)-printed membranes. These patterned and customized membranes can be used for desalination or other water/wastewater treatment-related processes. However, cost, time taken, and ease of production of membranes with 3D printing need to be evaluated before scale up (Yusuf *et al.*, 2020).

Forward osmosis technology naturally transports water through a semi-permeable membrane because of osmotic pressure difference between the draw solution (DS) and feed solution (FS) is the driving force. The increased longevity of the used membrane is due to its ability to prevent fouling and the system requires lower pressure as compared to reverse osmosis. Capacitive deionization (CDI) uses functionalized electrodes and an electric field to separate and recover heavy metal ions and salt ions from wastewater. Selectivity and efficiency of the electrodes are enhanced by adding selective ion-exchange resins on the surface of electrodes for the removal of target ions, for example. CDI is also recently being developed to address the organic contaminants (Bolisetty *et al.*, 2019).

Recent developments in bio-mimetic water treatment membrane design include aquaporins that are highly selective water channel proteins integrated into plasma membranes of biological cells. They allow the cell to maintain its volume and internal osmotic pressure in line with the hydrostatic and osmotic pressure differences. Rejection of pollutants occur based on charge and size. The aquaporin channel's distinct architecture is its hourglass structure which features wide entrance vestibules and a narrow hydrophobic centre that allows the passage of water molecules and blocks all other compounds (Werber *et al.*, 2016). Another technology of bioremediation concerns microbial production of enzymes that can cause destruction and effective breakdown of hazardous water pollutants. Genetically engineered microbes have become a point of interest in environmental biotechnology to remediate hazardous chemicals, xenobiotics, and pesticides.

17.3 CONCEPT OF SUSTAINABILITY, NEED AND WAYS OF TECHNOLOGY EVALUATION

Advanced analysis, development of clean technologies, and environmental policies can mitigate the environmental burden directly and indirectly. Water treatment technologies can be an energy-intensive process that is associated with many economic, environmental, and social impacts, hence any technology, in order to be sustainable, must address all three components of the sustainability triple bottom line: environmental, economic, and societal.

To understand the fundamentals of these parameters, we must first consider some key concepts that make a material or process 'green'. The mnemonics PRODUCTIVELY and IMPROVEMENTS, which represent the essence of the 12 principles of green chemistry and green engineering have been reported by Poliakoff and co-workers (Tang *et al.*, 2005, 2008).

PRODUCTIVELY – (a) **P**revent wastes, (b) **R**enewable materials, (c) **O**mit derivatization steps, (d) **D**egradable chemical products inputs, (e) **U**se of safe synthetic methods, (f), **C**atalytic reagents, (g) **T**emperature, pressure ambient, (h) **I**n-process monitoring, (i) **V**ery few auxiliary substances, (j) **E** factor, maximize feed in product, (k) **L**ow toxicity of chemical products, and (l) **Y**es, it is safe.

IMPROVEMENTS – (a) **I**nherently non-hazardous and safe, (b) **M**inimize material diversity, (c) **P**revention instead of treatment, (d) **R**enewable materials and energy, (e) **O**utput-led design, (f) **V**ery simple, (g) **E**fficient use of mass, energy, space, and time, (h) **M**eet the need, (i) **E**asy to separate by design, (j) **N**etworks for exchange local mass and energy, (k) **T**est the life cycle of the design, and (l) **S**ustainability throughout product life cycle.

17.3.1 Sustainability metrics

Sustainability or green metrics is used as the standard of measurement, based on criteria and indicators, to compare or track performance of a given technology, thereby an attempt of quantifying the progress towards the broader goal of environmental sustainability. Some of the common parameters used to determine sustainability metrics of various technologies are summarized in Table 17.1.

17.3.2 Life-Cycle Assessment

Life-cycle assessment (LCA) is a significant methodology that is commonly used to assess the environmental impact of a technology or product from the extraction of raw materials to its end of life (Chee *et al.*, 2022). LCA was first implemented in the wastewater treatment industry in the 1990s (Corominas *et al.*, 2013). LCA is considered as an effective tool for identifying potential life-cycle impacts of nanotechnologies, and its application in this field has expanded during the past 10 years (Klöpffer *et al.*, 2007).

According to the existing International Standards Organization (ISO) 14040 and 14044 standards, an LCA has four major phases such as: (1) goal and scope, (2) life-cycle inventory, (3) life-cycle impact assessment (LCIA), and (4) interpretation. The goal and scope definition of LCA defines the objective of a study and provides a description of the product system in terms of the system boundaries and a functional unit. Since LCA studies for wastewater treatment were first performed, sludge treatment and disposal have been considered in the system boundaries because of their considerable contribution to the total impacts. In case of LCA of waste water treatment process, many factors

Table 17.1 Various metrics and their corresponding expressions used for sustainability assessment.

Metrics	Expressions
Material efficiency	
Environmental factor (E factor)	$\frac{\text{Mass of the total waste}}{\text{Mass of the final product}}$
Atom economy	$\frac{\text{MW(product)}}{\sum \text{MW(reagents)}}$
Carbon efficiency	$\frac{\text{Amount of carbon in product}}{\text{Total carbon present in reactants}} * 100$
E-factor based on molecular weight	$\frac{\text{MW of waste by-product}}{\text{MW of final product}}$
Effective mass yield	$\frac{\text{Mass of product}}{\text{Mass of hazardous reagent}} * 100$
Actual atom economy	Reaction yield* AE
Environmental quotient	E factor * Q Q = quotient If a material is considered to be harmless, Q is 1 and EQ = E, while for toxic materials like heavy metal salts, Q can have value in the range of 100–1000, depending on ease of recycling, toxicity, and other factors.
Mass intensity, or process mass intensity	$\frac{\text{Total mass used in a process or process step}}{\text{Mass product}}$
Material recovery parameter	Materials used in the work-up, reaction, and purification are considered by the material recovery parameter (MRP), including catalysts, reaction solvents, washings for extractions, additives, solvents, and drying agents used in purification methods (MRP in the range of 0–1).
Reaction mass efficiency	For a particular reaction A + B → C, $\frac{\text{Mass(C)}}{\text{Mass(A)} + \text{Mass(B)}}$
Energy efficiency	
Energy intensity	$\frac{\text{Total process energy}}{\text{Mass of the final product}}$
Waste treatment energy	$\frac{\text{Waste treatment requirements}}{\text{Mass of the final product}}$
Solvent recovery energy	$\frac{\text{Solvent recovery requirement}}{\text{Mass of the final product}}$

Source: Adapted from Mukherjee *et al.* (2022).

such as operation of the plant and environmental impact of construction, demolition phases can be included in system boundaries. The second stage, life-cycle inventory analysis, combines inputs and outputs for every process in the life cycle and adds them up for the entire system (Hellweg & Canals, 2014). The inventory phase data are obtained from actual plants, pilot or lab facilities, expert estimation, related literature, and LCA databases. In the third phase, LCIA, potential toxicological impacts on the ecosystem and human health can be derived and identified with the help of life-cycle inventory phase. Finally, the final phase connects the first three to provide a summary and conclusion of all findings. The interpretation should, in accordance with ISO 14040:2006, include the following: (a) identification of significant issues based on the outcomes of the LCI and LCIA phases of an LCA; (b) evaluation of the study considering sensitivity, completeness, and consistency checks; and (c) limitations, recommendations, and conclusions.

17.3.3 Risk Assessment

Risk assessment (RA) stands for the qualitative and quantitative assessment of the risk that a certain pollutant or mixture of contaminants poses to the environment and human health (Boersema & Reijnders, 2009). RA is used as a path to prevent unacceptably adverse effects that technology systems might have on the environment, human health, resources, economy, and the society. To make sure that the preferred alternative or the generated solution complies with current requirements, it is also used proactively in the development of solutions that represent a controllable risk level (Hauschild *et al.*, 2022). When environmental chemists first recognized the widespread presence of persistent chemicals in the environment and wildlife in the 1960s, the concept of environmental RA was developed (van Straalen *et al.*, 2022). RA comprises of four main phases such as: (a) identification of hazard; (b) effect assessment; (c) exposure assessment, and (d) characterization of risk. Hazard identification is the process of identifying the adverse effects of a certain chemical and evaluation of that particular effect. The pathways, emissions, degradation, and rates of movement of a contaminant are all assessed through exposure assessment in order to determine a predicted environmental concentration (PEC). Effect assessment is the estimation of the causality between a substance's dose or level of exposure and the occurrence and severity of an effect. A predicted no effect concentration (PNEC) can be obtained by extrapolating this causality to till date untested species (Guinée *et al.*, 2017). Risk characterization is an estimation of the PEC/PNEC ratio and also quantifying its uncertainties. The REACH program (registration, evaluation, authorization, and restriction of chemicals) rules state that the ratio of PEC/PNEC greater than 1 triggers action and typically forbids approval.

17.3.4 Societal Acceptance

One of the key requirements for promoting a sustainable technology for commercialization is societal acceptability (Kamali *et al.*, 2019a). It is considered as the overall public's opinion about the relevance of the method to daily life (Kamali *et al.*, 2019b). Therefore, in addition to the environmental, economic, and technical performance of a particular technology, social benefits should be taken into consideration to achieve sustainable growth of wastewater treatment industries. The effectiveness of technology can have a favourable impact on how people perceive it, which could lead to an improvement in the economy including possibilities of new job opportunities due to large-scale productions of nanomaterials and their transportation in an industrial effluent treatment plant. Additionally, it is essential to compare the impacts of several types of spent materials being released from the wastewater treatment plant to the environment.

17.3.5 Economical Assessment

This aspect makes sure that an economically feasible system can continue to create goods and services (Assefa & Frostell, 2007). It is based on the three important factors namely: (a) initial investments; (b) operation cost, and (c) maintenance costs. When it comes to determining the feasibility of a certain technology for an industry, as in the case of wastewater treatment plants, initial investments such as civil works, land cost, and equipment are of crucial significance. Specifications of techniques can be useful for the estimation of the initial investment needed. For example, depending on the type of nanomaterials being employed, the reactor size can be designed using the kinetics of the treatment process as an indicator. When treatments based on nanotechnology are to be implemented, various contributions from the production or acquisition of engineered nanomaterials (ENMs), from labour, and from energy (such as the power needed for ENM-based photocatalysis treatments) should all be considered when estimating operating costs. The implementation of ENMs for the industrial wastewater treatment plants has the potential to significantly lower the maintenance cost of plants as compared to membrane system in which biofilms and dissolved organic compounds arises due to biological activities, fouling by inorganic suspended solids are major contributors to maintenance cost.

17.4 SUSTAINABILITY EVALUATION FOR TREATMENT TECHNOLOGIES

Ibrahim *et al.* (2018) designed a methodology to integrate the different framework components of the environmental, technical, economic, and social factors for assessing sustainability of desalination technologies employed in different places. They determined a set of sub-factors related to the technical details of the desalination technology and assigned them under the most relevant aforementioned three parameters. Each sub-factor was evaluated and valuation was done based on calculation, expert opinion, and literature data (for details please refer to the article), which were finally summed up leading to an aggregated score. Realization of the sub-factors was an important step in the whole process. The relevant sets of sub-factors were as follows: for (a) environmental – extraction of seawater (m^3/day), discharged brine impacts (temperature and salinity increase) (°C, ppm), CO_2 emission (kg CO_2-eq/m^3), other environmental impacts (g/m^3), land use (m^2/m^3) product water/day; (b) techno-economic – technology reliability and robustness, quality of water produced (mg TDS/L), scaling and fouling, levelized cost of water production (\$/m^3), the sensitivity of levelized cost of water production (% increase/decrease), the internal rate of return (IRR) %; (c) social – level of aesthetic acceptability, level of noise, provision of employment (number of employees), technology safety, consumption of fossil fuel (kWh/m^3), where the sub-factors are calculated for the product water.

Mukherjee *et al.* (2019, 2020) calculated the sustainability metrics of ferrihydrite incorporated microcellulose (MCCFH) and nanocellulose/PANI nanocomposites (CNPFH) for the removal of arsenic and fluoride from water, respectively. Composites (Na–Zn–Se) for release of essential minerals to purified (often desalinated) water were also considered for similar evaluation by Ravindran *et al.* (Ravindran *et al.*, 2019). Another ternary oxide-based composite CAlFeC developed for defluoridation was also assessed for the same (Egor *et al.*, 2021). All the four materials were assessed by the parameters, including mass intensity, solvent intensity, reaction mass efficiency, energy consumption, E-factor, and CO_2 emission. The nanocomposites were products of water-based green synthesis, hence exhibited excellent numbers with respect to waste generation (E-factor) and energy consumption, while they still require improving their yield, thus the reaction mass efficiency. These studies on sustainability metrics are summarized in Table 17.2.

The careful selection of treatment technology is crucial for effective mitigation of water issues, considering various factors such as location, weather conditions, and types and quantity of influents (Bassi *et al.*, 2022). Eco-based treatment systems were performed by mimicking natural processes,

Table 17.2 Sustainability metrics for various nanomaterials used in applications for clean water.

Nanomaterials	Sustainability Metrics					
	Mass Intensity (kg/kg)	Water Intensity (kg/kg)	Reaction Mass Efficiency (%)	Energy Intensity (kW h/kg)	E factor (kg/kg)	CO_2 emission (mg/kg)
MCCFH	1.9	29.2	52	2.3	0.3	–
CNPFH	1.84	38.8	54	1.8	0.6	–
Na–Zn–Se	1.204	4.7	83	3.2	0.02	40
CAlFeC	5.68	36.67	17.6	2.0	0.12	–

Source: Mukherjee & Pradeep (2023).

such as using the physiological processes of aquatic plants to filter and adsorb wastewater pollutants. Bigger land requirement, rising land prices, and long operation tenures make these systems very expensive. On the contrary, electro-mechanical treatment systems are often compact in size but are energy intensive. Each water/wastewater treatment technology needs to be evaluated for its initial capital and operational costs to understand the actual cost over a system's lifespan, which concerns the economic viability of the investments in such systems, that is further connected with environmental sustainability concerns. The National Green Tribunal's (NGT) Government of India order (dated 30 April 2019) specified stringent effluent discharge standards on pH, BOD, TSS, COD, N/P total, FC for all the existing and upcoming WWT plants in the country. Economic criteria considered for WWT systems assessment should include: investment, energy requirements, net profit value, maintenance and operational costs, land requirements, transportation, sludge disposal and production, resource consumption and recovery, and cost–benefit analysis. Proper implementation of water treatment technologies in communities requires a thorough analysis of opportunities to benefit from the technology by assessing interest in establishing and maintaining such services. Improved relations between governments and the private sector can develop guidelines for the transition from the typical single-use materials of flow economy to a green circular economy, and to address regulatory conflicts.

REFERENCES

Assefa G. and Frostell B. (2007). Social sustainability and social acceptance in technology assessment: a case study of energy technologies. *Technology in Society*, **29**(1), 63–78, https://doi.org/10.1016/j.techsoc.2006.10.007

Barbhuiya N. H., Misra U. and Singh S. P. (2022). Biocatalytic membranes for combating the challenges of membrane fouling and micropollutants in water purification: a review. *Chemosphere*, **286**, 131757, https://doi.org/10.1016/j.chemosphere.2021.131757

Bassi N., Kumar S., Kumar M. D., Van Ermen S. and Campling P. (2022). Promoting wastewater treatment in India: critical questions of economic viability. *Water and Environment Journal*, **36**(4), 723–736, https://doi.org/10.1111/wej.12810

Boersema J. J. and Reijnders L. (2009). Principles of Environmental Sciences. Springer, Springer Dordrecht, Netherlands.

Bolisetty S., Peydayesh M. and Mezzenga R. (2019). Sustainable technologies for water purification from heavy metals: review and analysis. *Chemical Society Reviews*, **48**(2), 463–487, https://doi.org/10.1039/C8CS00493E

Carpenter A. W., de Lannoy C.-F. and Wiesner M. R. (2015). Cellulose nanomaterials in water treatment technologies. *Environmental Science & Technology*, **49**(9), 5277–5287, https://doi.org/10.1021/es506351r

Chee P. L., Toh W. L., Yew P. Y., Peng S. and Kai D. (2022). Chapter 1. Introduction of nanotechnology and sustainability. In: Nanoscience & Nanotechnology Series, Z. Li, J. Zheng and E. Ye (eds.), Royal Society of Chemistry, Cambridge, CB4 0WF, UK, pp. 1–32.

Corominas L., Foley J., Guest J. S., Hospido A., Larsen H. F., Morera S. and Shaw A. (2013). Life cycle assessment applied to wastewater treatment: state of the art. *Water Research*, **47**(15), 5480–5492, https://doi.org/10.1016/j.watres.2013.06.049

Egor M., Kumar A. A., Ahuja T., Mukherjee S., Chakraborty A., Sudhakar C., Srikrishnarka P., Bose S., Ravindran S. J. and Pradeep T. (2021). Cellulosic ternary nanocomposite for affordable and sustainable fluoride removal. *ACS Sustainable Chemistry and Engineering*, **9**(38), 12788–12799, https://doi.org/10.1021/acssuschemeng.1c03272

Guinée J. B., Heijungs R., Vijver M. G. and Peijnenburg W. J. G. M. (2017). Setting the stage for debating the roles of risk assessment and life-cycle assessment of engineered nanomaterials. *Nature Nanotechnology*, **12**(8), 727–733, https://doi.org/10.1038/nnano.2017.135

Hauschild M. Z., McKone T. E., Arnbjerg-Nielsen K., Hald T., Nielsen B. F., Mabit S. E. and Fantke P. (2022). Risk and sustainability: trade-offs and synergies for robust decision making. *Environmental Sciences Europe*, **34**(1), 11, https://doi.org/10.1186/s12302-021-00587-8

Hellweg S. and Canals L. M. (2014). Emerging approaches, challenges and opportunities in life cycle assessment. *Science*, **344**(6188), 1109–1113, https://doi.org/10.1126/science.1248361

Ibrahim Y., Arafat H. A., Mezher T. and AlMarzooqi F. (2018). An integrated framework for sustainability assessment of seawater desalination. *Desalination*, **447**, 1–17, https://doi.org/10.1016/j.desal.2018.08.019

Jadeja N. B., Banerji T., Kapley A. and Kumar R. (2022). Water pollution in India – current scenario. *Water Security*, **16**, 100119, https://doi.org/10.1016/j.wasec.2022.100119

Jaspal D. and Malviya A. (2020). Composites for wastewater purification: a review. *Chemosphere*, **246**, 125788, https://doi.org/10.1016/j.chemosphere.2019.125788

Kamali M., Persson K. M., Costa M. E. and Capela I. (2019a). Sustainability criteria for assessing nanotechnology applicability in industrial wastewater treatment: current status and future outlook. *Environment International*, **125**, 261–276, https://doi.org/10.1016/j.envint.2019.01.055

Kamali M., Costa M. E., Aminabhavi T. M. and Capela I. (2019b). Sustainability of treatment technologies for industrial biowastes effluents. *Chemical Engineering Journal*, **370**, 1511–1521, https://doi.org/10.1016/j.cej.2019.04.010

Klöpffer W., Curran M. A., Frankl P., Heijungs R., Köhler A. and Olsen S. I. (2007). Nanotechnology and Life Cycle Assessment. A Systems Approach to Nanotechnology and the Environment: Synthesis of Results Obtained at a Workshop Washington, DC, 2–3 October 2006. European Commission, DG Research, jointly with the Woodrow Wilson International Center for Scholars, Washington DC, U.S.A.

Mukherjee S. and Pradeep T. (2023). Nanomaterials-enabled technologies for clean water and their sustainability aspects. In: Industrial Applications of Nanoparticles: A Prospective Overview, 1st edn, M. Irene Litter and A. Ahmad (eds.), CRC Press, Tailor & Francis, Boca Raton, U.S.A, pp. 16–31.

Mukherjee S., Kumar A. A., Sudhakar C., Kumar R., Ahuja T., Mondal B., Srikrishnarka P., Philip L. and Pradeep T. (2019). Sustainable and affordable composites built using microstructures performing better than nanostructures for arsenic removal. *ACS Sustainable Chemistry & Engineering*, **7**(3), 3222–3233, https://doi.org/10.1021/acssuschemeng.8b05157

Mukherjee S., Ramireddy H., Baidya A., Amala A. K., Sudhakar C., Mondal B., Philip L. and Pradeep T. (2020). Nanocellulose-reinforced organo-inorganic nanocomposite for synergistic and affordable defluoridation of water and an evaluation of its sustainability metrics. *ACS Sustainable Chemistry & Engineering*, **8**(1), 139–147, https://doi.org/10.1021/acssuschemeng.9b04822

Mukherjee S., Shantha Kumar J., Nagar A. and Pradeep T. (2022). Concepts of sustainability in clean water technologies. In: ACS Symposium Series, P. J. Boul (ed.), American Chemical Society, **1412**, Washington DC, U.S.A, pp. 625–657.

Ravindran J. S., Mahendranath A., Srikrishnarka P., Anil Kumar A., Islam M. R., Mukherjee S., Philip L. and Pradeep T. (2019). Geologically inspired monoliths for sustainable release of essential minerals into drinking water. *ACS Sustainable Chemistry & Engineering*, **7**(13), 11735–11744, https://doi.org/10.1021/acssuschemeng.9b01902

Sansaniwal S. K. (2022). Advances and challenges in solar-powered wastewater treatment technologies for sustainable development: a comprehensive review. *International Journal of Ambient Energy*, **43**(1), 958–991, https://doi.org/10.1080/01430750.2019.1682038

Tang S. L. Y., Smith R. L. and Poliakoff M. (2005). Principles of green chemistry: productively. *Green Chemistry*, **7**(11), 761, https://doi.org/10.1039/b513020b

Tang S. Y., Bourne R. A., Smith R. L. and Poliakoff M. (2008). The 24 principles of green engineering and green chemistry: 'improvements productively'. *Green Chemistry*, **10**(3), 268, https://doi.org/10.1039/b719469m

van Straalen N. M., den Haan K. H., Hermens J. L. M., van Leeuwen K., van de Meent D., Parsons J. R., de Voogt P. and de Zwart D. (2022). Risk assessment acknowledging variability in both exposure and effect. *Environmental Science & Technology*, **56**(20), 14223–14224, https://doi.org/10.1021/acs.est.2c06088

Varma G. V., Jha S., Raju H. K., Kishore L. and Ranjith V. (2022). A review on decentralized wastewater treatment systems in India. *Chemosphere*, **300**, 134462, https://doi.org/10.1016/j.chemosphere.2022.134462

Werber J. R., Osuji C. O. and Elimelech M. (2016). Materials for next-generation desalination and water purification membranes. *Nature Reviews Materials*, **1**(5), 16018, https://doi.org/10.1038/natrevmats.2016.18

Yusuf A., Sodiq A., Giwa A., Eke J., Pikuda O., De Luca G., Di Salvo J. L. and Chakraborty S. (2020). A review of emerging trends in membrane science and technology for sustainable water treatment. *Journal of Cleaner Production*, **266**, 121867, https://doi.org/10.1016/j.jclepro.2020.121867

Section 4
Sensors for Water Quality Monitoring

INTRODUCTION

A fundamental requirement of any water supply system, both urban and rural, is to assure the supply of an adequate quantity of water that matches the required quality. This can be achieved only if water quality parameters are monitored regularly at various critical points of the water supply system. It is also essential to regularly monitor the quality of the influent and effluent water of any wastewater treatment plant to ensure its proper functioning, protect the receiving water bodies from contamination, and safeguard public health, in case the treated wastewater is recycled and reused. Large-scale and continuous monitoring is possible only by using online sensor-based equipment. Therefore, the four chapters of Section IV are devoted to the discussion of sensors for water quality monitoring.

One of the basic requirements for any sensor, especially for use in countries in the Global South, is that it should be inexpensive and, at the same time, provide reasonably accurate measurements. Presently, various sensing techniques are in use to monitor contaminants in raw water sources. However, the main challenges for large-scale field deployment are: a complex synthesis approach, long-term reliability, poor portability, intensive maintenance, and unavailability of skilled personnel. Therefore, the chapter on **Low-cost colourimetric sensors for water quality monitoring includes a focused discussion** on sensors for measuring concentrations of eutrophying ions, heavy metals, pathogens, and emerging contaminants. The chapter on **Conductivity sensors for water quality monitoring** reviews conductivity measurement, as the quality of water can be measured, to some extent, by the conductivity of water. The review includes conductivity sensors for flow-through applications as well as immersion-type sensors. Non-contact-type conductivity sensors based on inductive techniques and capacitively coupled techniques are included in the discussion. The advantages and limitations of such sensors are discussed.

In many field situations, it may not be necessary to determine water composition very precisely. Implementation of regular water quality monitoring at a large scale can be planned only to assess the requirement of in-depth lab-based analysis. This can be achieved by developing methods that are reagent free, independent of high-end instrumentation, and facilitative of multiplexed detection, based on the identification of hidden signals through a probabilistic/predictive approach. This issue is addressed in the chapter on **Multi-analyte assessment of water quality**. The discussion in this chapter includes data analysis using AI and ML-based grouping techniques. As opposed to the traditional analytical techniques, molecule or material-based identification, that is sensing of specific target species, has been found to be the cost-effective method. However, most of the presently available probes are specific to one target, which is a major drawback when analysing field samples that may contain several toxic species. In this context, simultaneous analysis of different toxic species with distinctly different signalling pathways offers a practical solution. The chapter on **Point-of-use single probe multi-analyte sensors** focuses on this aspect.

doi: 10.2166/9781789063714_0203

Chapter 18

Low-cost colorimetric sensor for water quality monitoring

*Vaishali Choudhary[1], Kowsalya Vellengiri[2] and Ligy Philip[1]**

[1]Department of Civil Engineering, Indian Institute of Technology Madras, Chennai, India
[2]Larsen and Toubro Limited, Chennai, India
**Corresponding author: ligy@iitm.ac.in*

ABSTRACT

With the increased concentrations of eutrophying ions, heavy metals, and organic contaminants in the water bodies, there is a need to develop tools for regular water quality monitoring. Various sensing techniques such as colorimetric, electrochemical, optical fluidic, and luminescent have been developed to monitor contaminants in water bodies. Many of the reported sensors enable high sensitivity (i.e., nM or ppb detection levels); however, the primary challenges in their field application include complex synthesis approach, reliability in long-term use, poor portability, intensive maintenance, and skilled labour requirement. This chapter reviews the research and development of various low-cost sensors to detect eutrophying anions (nitrate, nitrite, and phosphate), heavy metals (lead, chromium, cadmium, and mercury), pathogens, and emerging contaminants (ECs). The analytical performance, cost, and ease of deployment of sensors for detecting pollutants are critically evaluated. The feasibility of developed sensors for deployment on the Internet of Things platform is also identified. Lastly, a recommendation on the applicability of the developed sensors for field-scale application is presented.

Keywords: monitoring, contaminants, optical, electrochemical, sensors

18.1 INTRODUCTION

Water quality monitoring is essential to manage pollutant loading in the waterways and maintain quality checks of existing resources. Water quality sensors help in the identification and quantification of polluting substances, which in turn will help in strategic management, decision-making, and policy formulation. However, such interventions are missing in developing countries due to the limited availability of technologies for regular and continuous water quality monitoring and data collection. Traditionally, water quality analysis includes sample collection, preservation, transport, and analysis in laboratory environments using wet chemical routes, costly equipment, and qualified professionals (Silva *et al.*, 2022). Although these methods offer high accuracy in detection, the high cost, labour, and time-intensive procedures cause a delay in contaminant detection. Also, the chances of sample degradation during transport make them futile for regular water quality monitoring. In countries like India with large watersheds, conventional water monitoring techniques seem cumbersome. Thus, the development of low-cost, accessible, and easy-to-handle sensors for water quality analysis will provide a viable alternative for regular on-site monitoring and collecting real-time data.

With the progress of nanotechnology and the advance in engineering materials, various types of sensors, such as biosensors, colorimetric, electrochemical, optical fluidic, and luminescent, have been developed to monitor contaminants in water bodies. Figure 18.1 shows the sensing mechanism in electrochemical and optical sensors. A combination of the sensors with microcontrollers and cloud computing enables the development of smart water quality monitoring systems (Paepae *et al.*, 2021). Nonetheless, the feasibility of the sensor for water quality monitoring in the field is dependent on the (1) limit of detection (LOD), (2) selectivity, (3) cost, (4) robustness, (5) portability, (6) response time, (7) ease of sampling and automation, (8) reproducibility, and (9) user-friendliness. This chapter reviews various portable low-cost sensors available for the detection of eutrophying anions (nitrate, nitrite, and phosphate), heavy metals (lead, chromium, cadmium, and mercury), pathogens, and emerging contaminants. The analytical performance, cost, and ease of deployment of sensors for detecting pollutants are critically evaluated. The feasibility of developed sensors for deployment on the Internet of Things (IoT) platform is also identified. Lastly, a recommendation on the applicability of the developed sensors for field-scale application is presented.

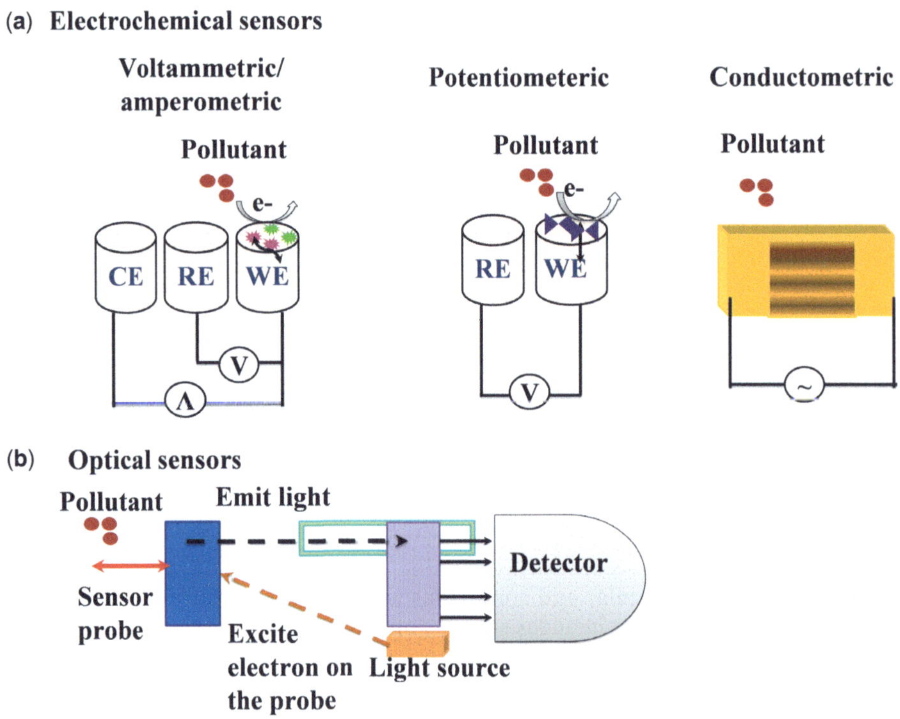

Figure 18.1 Pictorial representation of sensing mechanism in electrochemical and optical sensors. (*Source*: Modified from Huang *et al.*, 2022)

18.2 SENSORS FOR THE DETECTION OF EUTROPHYING ANIONS

Developing sensors for detecting nitrate, nitrite, and phosphate is essential, as their excess concentrations cause eutrophication and harmful algal blooms. The pros and cons of optical and electrochemical sensors are discussed in this section.

18.2.1 Electrochemical sensors for the detection of eutrophying ions

Commercially available ion-selective electrodes (ISE) have been utilized to measure nitrate, nitrite, and phosphate concentration by estimating the difference in the potential of the electrode. However, due to the poor selectivity of ISE in real wastewater, modifications using scaffolds, ionophores, and nanomaterials have been suggested (Mahmud *et al.*, 2020). However, the high cost, short durability, and stability of nanomaterials in real water are some major obstacles in the field application of these sensors. Recently, Nag *et al.* (2019) developed a portable IoT-based sensing system for online phosphate monitoring. The sensor uses graphite electrodes decorated on the flexible polydimethylsiloxane (PDMS)-based substrate. The detection is in the range of 0.03–30 mg/L with LOD 0.03 mg/L (Nag *et al.*, 2019). However, the primary challenge in measuring phosphate using electrochemical sensors is the non-electro-active nature of the molecule. Accordingly, additives like acidic molybdate or vanadate must be added externally for all measurements. The biofilm growth and electrode corrosion in real water are two primary challenges that need intervention. Also, much of the new advances remain at the laboratory scale, with limited to no option for online monitoring and field deployment. Thus, development of a sensor with extended shelf life and faster and more sensitive electron transduction is needed.

18.2.2 Optical sensors for the detection of eutrophying ions

Optical sensors, such as fluorescent, colorimetric, and luminescent sensors, are used to detect nitrate, nitrite, and phosphate. The first online colorimetric sensor for nitrate detection was developed using Griess assay as a colorimetric probe. The detection was based on the conversion of nitrate (NO_3^-) into nitrite (NO_2^-), followed by a reaction of NO_2^- with the Griess reagent to obtain colour. Though the Griess reagent enables naked-eye detection (endpoint: colourless to pink) of nitrite and nitrate, the extended response time, poor sensitivity, precaution in storage (at 4°C), and interference of transition metals like Fe^{3+} and Cu^{2+} often limit its applicability (Table 18.1). Accordingly, stable colour-inducing probes such as aza-BODIPY, pyrrole and diazo-quinone, and amine-functionalized graphene oxide nanoparticles have also been studied. Likewise, metal-organic frameworks (MOFs) and graphene quantum dots (GQDs) have been used for phosphate detection. Major constraints in nanoparticle-based sensors are: poor detection limits, the high cost of noble metals, and interference from natural organic matter. Also, principal challenges in applying the dye-based approach for colorimetric phosphate detection arise from the formation of insoluble complexes, poor stability, and interference in the multi-ion systems. The use of paper-based colorimetric sensors such as BF@W and W-BG allows the selective and sensitive nitrite (LOD: 0.005 mg/L) and phosphate (0.07 mg/L) detection (Choudhary & Philip, 2021; Vellingiri *et al.*, 2019). Moreover, the high molar extinction coefficient of the probe, simple synthesis, low cost, portability, and ease of use make BF@W and W-BG feasible alternatives for water quality monitoring. The advantage of naked eye colour change with ease of operation allows using the colorimetric sensor in remote, rural, and peri-urban areas without technical guidance. However, the selectivity of sensors and the development of portable and automated instruments are critical.

18.3 SENSORS FOR THE DETECTION OF HEAVY METALS

The presence of heavy metals such as mercury (Hg), cadmium (Cd), arsenic (As), chromium (Cr), and lead (Pb) in aqueous systems poses a serious concern to human health. Conventional techniques like

Table 18.1 Some of the standards and common colorimetric-based methods used for the detection of pollutants.

Pollutant	Method	LOD	Reaction Time Probe Stability	Remarks
Phosphate (PO_4^{3-} as P)	Vanadmolybdo-phosphoric acid	200 μg/L	>10 min >1 year	Most frequently used for automated sensors, poor sensitivity
	Stannous chloride	3 μg/L	10–12 min >6 month	Cannot be used in microfluidic system due to the high viscosity of reagents
	Malachite Green	10 μg/L	10 min >1 year	Used for biological analysis, dye precipitate in the solution
Nitrate (NO_3^- as N)	UV spectrophotometer	Concentration range (0–11 mg/L	<1 min	Expensive, interference due to organic matter
	Nitrate selective electrode	Concentration range (0.14–1400 mg/L	~1 min	Multiple time calibration required, prone to biofilm growth
Nitrite	Griess reagent		12 min >6 months at 4°C	High interference from metal Fe^{2+} and Cu^{2+}, poor stability, and high toxicity of reagents
Mercury (Hg^{2+})	Polyaniline nanofibres	5 nM	NA	Sensor can be used as strip; sensitivity in environmental samples is not checked
Cadmium (Cd^{2+})	Apt (4 × glutathione (GSH))-functionalized AuNPs	1.12 μg/L	NA	Complex and costly synthesis, High selectivity
E. coli, Salmonella enterica	Magnetic nanoparticle microarray-based assay	316 × 10^3 CFU/L	15 min	High selectivity; however, sensors are affected due to refractive indices and turbidity of wastewater
S. aureus	MIL–53	31 × 10^3 CFU/L	20 min	
Methyl parathion	Lanthanum-based gold nanoparticle (AuNPs-La)	0.1 nM	NA	Complex and costly synthesis, Tested in the laboratory, sensitivity in environmental samples is not checked
Ciprofloxacin	AuAg Nps	80 nM	NA	
Perfluoroctanoic acid (PFOA)	Thiol-terminated polystyrene gold nanoparticles	~241 μM	NA	

atomic absorption spectroscopy (AAS), flame or furnace spectroscopy, inductively coupled plasma mass spectrometry (ICP-MS), and energy-dispersive x-ray fluorescence (EDXRF) spectrometry remain the gold standard for their analyses. Though these techniques are selective and sensitive to the nanogram level, the systems suffer due to long response time, high cost, lack of field portability, and requirement of a skilled workforce (Mukherjee *et al.*, 2021). Therefore, the pros and cons of optical (fluorescence) and electrochemical methods for heavy metal detection are summarized in this section.

18.3.1 Electrochemical sensors for the detection of heavy metals

Various nanomaterial-based electrodes such as Fe_3O_4 coupled MWCNTs and bismuth nanoparticles have been used for the detection of Cd^{2+} (LOD: 0.05), Pb^{2+} (LOD: 0.08 nM), Cu^{2+} (LOD: 0.02 nM), and Hg^{2+} (LOD: 0.05 nM) (Mukherjee *et al.*, 2021). However, the selectivity of nanomaterials-based sensors is generally low. Accordingly, ion-imprinted polymers (IIP), aptamers, and DNA enzyme-based biosensors have been used as sensing probes. Recently, an electrochemical aptasensor has been developed for selective and sensitive detection (LOD: 312 pM) of Pb^{2+}, based on gold nanoparticles (AuNPs), complementary strands of aptamer (CS) and thionine. The ion-imprinted polymers (IIP) and nanoparticles modified IIPs have been used for the electrochemical detection of Pb^{2+}, Cd^{2+}, Hg^{2+}, and Cr(VI) (Mukherjee *et al.*, 2021). Though this technology has yet to be improved to attain LOD values equivalent to traditional detection methods, its benefits are: high output, low cost, and quick screening of samples. Recently, electrochemical detection methods using an electronic tongue (e-tongue) and electronic nose (e-nose) have been developed to enhance portability (Teodoro *et al.*, 2019). An automated electronic tongue allows the detection of a mixture of heavy metals like Pb^{2+}, Cd^{2+}, and Cu^{2+} in water. Major challenges for the field-scale applicability of the above-mentioned sensors are poor reproducibility and signal stability. Also, issues regarding uniform size distribution, homogeneity of nanostructures and multiplexing need to be resolved. The commercialized systems mainly lack the bio-recognition elements like aptamers making them less selective, and hampering multiple quantification.

18.3.2 Optical sensors for the detection of heavy metals

The application of sodium diphenylamine sulphonate (DPH) reagents is the most well-established colorimetric technique for the detection of Cr (VI). Unlike Cr (VI), the detection of other heavy metals like Cd^{2+} and Pb^{2+} is not possible by dye-based approaches. Mostly, aptamers (Apt) and surface plasmon resonance (SPR)-based nanoparticles such as gold nanoparticles (AuNp) are used for colorimetric detection (Figure 18.2). For instance, Cd^{2+} detection (LOD: 1.12 μg/L) was enabled using Apt (4 × glutathione (GSH))-functionalized AuNPs (Gan *et al.*, 2020). Likewise, Hg^{2+}-specific Apt and enzymatic sensors like GR-5 DNAzyme can enable the detection of Hg^{2+} and Pb^{2+}. However, the application of the above sensors in real water samples has not been examined.

Efforts have been made to develop fluorescent sensors for the detection of heavy metals using quantum dots (QDs), carbon dots (CDs), modified aptamers (Apt), and DNA enzymes. Biocompatible graphene QDs and graphene oxide can be used as fluorescent turn-on nanosensors for fast detection. Since carbon- and graphene-based QDs have weak fluorescence intensity, the modification of CD with gold (Cd_{Au}) and the use of DNAzyme aptamer such as haem can be adopted. Though these sensors provide very high sensitivity, the major challenge in the use of nanomaterials is the difficulty in separation. Accordingly, magnetic biosensors are recommended. Highly sensitive (ng/L) detection of Pb^{2+} and Cr^{3+} can also be attained by dual channel dual emission CDs. The primary advantages of the sensor are high selectivity and multiplexed detection of ions in a short time. Though the CDs and GQDs showed very high selectivity, biocompatibility, and detection in the ng/L range, the application of the sensor in the field is difficult due to its non-portability. Recently, a portable sensor based on loading of ZnSe quantum dots on the three-dimensional (3D) rotary paper-based microfluidic and a 3D origami ion-imprinted polymer microfluidic device has been developed for the detection of Pb^{2+}, Cd^{2+}, Cu^{2+}, and Hg^{2+} (Mukherjee *et al.*, 2021). The large-scale synthesis of QDs and Apt can be complex and costly. The sensitivity of fluorescent sensors depends on the proper dispersion, temporal stability, and ambient storability. Also, developed sensors are yet to be deployed on IoT-based systems. Therefore, such investigations are needed before the scale-up of fluorescent-based water quality monitoring tools.

18.4 SENSORS FOR THE DETECTION OF EMERGING CONTAMINANTS (ECs)

Due to their diverse chemistry, structure, properties, and biological activity, the detection of emerging contaminants (ECs) is conducted using advanced analytical methods like high-performance liquid chromatography (HPLC), mass spectrometry, gas chromatography–mass spectrometry (GC-MS), time

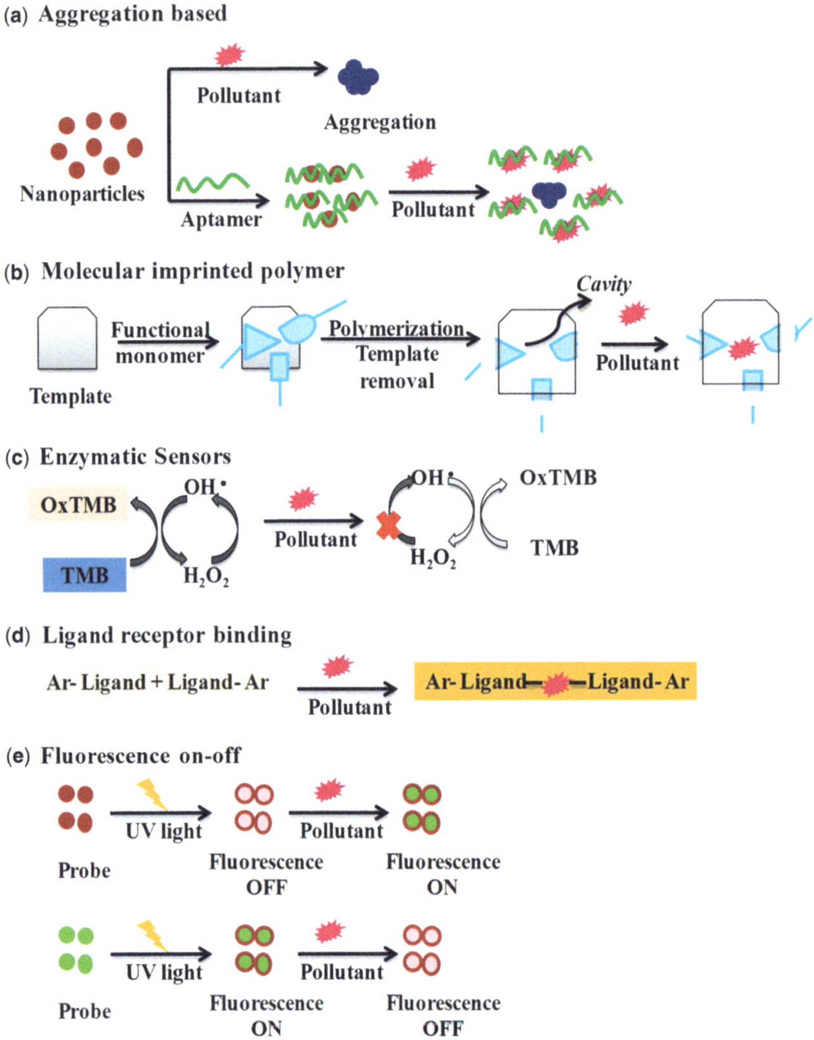

Figure 18.2 Summary of common mechanism and probe used for the development of colorimetric sensors.

of flight (TOF), orbitrap, quadrupole-time of flight. Besides the high cost and lack of portability, the primary barrier to the use of the above-mentioned techniques is the laborious sample pre-treatment and analysis. Consequently, the development of colorimetric, fluorescent electrochemical, and surface-enhanced Raman spectroscopy (SERS)-based sensors have gained attention. This section summarizes the comprehensive overview of different sensors developed for ECs detection.

18.4.1 Electrochemical sensors for the detection of ECs

Electrochemical-based detection of ECs is mostly carried out using voltammetry and potentiometry techniques. For instance, the detection of naproxen has been carried out using modified gold digital disc (LOD: 0.08 μM) and nano-spinel zinc ferrite film (LOD 13 μM) (Bhuvaneshwari *et al.*, 2022).

So far, the lowest LOD (0.000016 nM) has been achieved using aspirin-modified screen-printed carbon (SPE) electrodes (Diouf *et al.*, 2020). Pesticide antibiotics can be detected using a molecularly imprinted polymer as the conductive probe. Electrochemical enzymatic biosensors, aptamers, and immunosensors have also been explored to detect ECs. However, an optimum method to immobilize the probe on the substrate and modifier of the probe is not yet developed. Paper-based electrochemical sensors have been investigated because they are sustainable, biocompatible, and inexpensive. For example, sulphamethoxazole (SMX) (LOD: 0.09 μM) and trimethoprim (TMP) (0.04 mM) detection has been reported on parchment paper-printed electrodes (Martins *et al.*, 2021). The primary challenges with paper-based sensors are sample evaporation, low mechanical stability, and detection getting affected due to the flow of the analyte. Though electrochemical sensors offer high sensitivity, they have not yet become popular for real-time monitoring. The use of expensive electrode substrates and the poor durability of biocompatible and paper sensors limit their widespread use.

18.4.2 Optical sensors for the detection of ECs

Colorimetric detection of ECs such as analgesics and antibiotics has been commonly achieved using AuNPs (Bhuvaneshwari *et al.*, 2022). Since nanomaterial-based colorimetric sensors show poor stability, sensitivity, and selectivity, SERS and fluorescent sensors have been explored. For example, a thiolated aptamer-based SERS system coupled with Ag dendrites enabled multiple pesticide detection. Similarly, Ag nanospheres and Raman active dye (ethyl violet and methylene blue) have been used for the detection of perfluorooctanoic acid. However, the use of dye as a probe resulted in the obscured SERS signal. Recently, several fluorescent sensors have been utilized for the detection of perfluoroalkyl substances (Ryu *et al.*, 2021). A fluorescence detection system integrated with HPLC has shown a LOD of 0.2 nM; however, the need for fluorescent receptor and solvent limit their regular application. The major challenges in using optical sensors for the detection of ECs are the poor reproducibility and the short shelf life of the probe. SERS-based tools are as sensitive as GC-MS. However, obstacles of poor repeatability, high instrument costs and field adaption still need to be resolved. Also, the accuracy of measurement in the fluorescence technique is hampered by the matrix effect, inner filtering effect and inherent fluorescence quenching features of organic matter present in wastewater.

18.5 SENSORS FOR THE DETECTION OF PATHOGENS

Monitoring for all known waterborne pathogens is impractical, so faecal content qualification is used as an indication. The World Health Organization (WHO) advises counting *Escherichia coli* (*E. coli*) as the major indication of faecal contamination in drinking water; however, this involves growing a point sample for 18–24 hours. Currently, our understanding of microbial contamination is severely limited due to the use of laboratory-based culture methods for identification. These procedures are inapplicable in remote areas or when drinking water is contaminated irregularly. As a result, research into developing an effective real-time *E. coli* sensor system for analysing potable water has gained attraction. Molecular detection methods are very selective and can detect particular strains, even in low-concentration samples. There is no need for the cultivation or isolation of the bacterium. However, pre-treatment actions are frequently required before analysis, which introduces possible sources of mistakes. When studying environmental samples, the major challenge is false-negative results. They are costly and involve significant effort and training. As a result, despite the promise, there are presently no field-deployable molecular detection sensors. Many *E. coli*-focused biosensors operate on the basis of optical and electrochemical detection, using an enzyme found in most *E. coli* strains (Gunter *et al.*, 2021). There are deployable voltammetry pH sensing systems that can identify coliform-specific enzymes and offer the potential for effective future field applications. Because of a lack of acceptable approaches, non-fluorescence-based field identification of *E. coli* is currently confined to proof-of-concept designs.

18.6 DESIGNING COLORIMETRIC SENSORS FOR WATER QUALITY MONITORING

Colorimetric sensors are a promising option for detecting various pollutants due to their advantages of visible-eye detection, easy fabrication, and high sensitivity and selectivity. The sensing of the target pollutants is commonly based on the aggregation of nanoparticles, decomposition of nanoparticles, enzymes, fluorescence on–off, ligand–receptor interaction (Figure 18.2). The type, structure, and functionalization of nanomaterial govern the aggregation-mediated colour change. To enhance the selectivity and detect small toxic pollutants, the use of aptamers-assisted nanoparticle aggregation is gaining attention. Enzyme-based colorimetric sensors exhibit high selectivity. However, the detection limit is usually at the micromolar level. Fluorescence and ligand receptor-interaction-based colorimetric sensors are intricately related to the probe molecules. The main limitation of all the above-mentioned probes is poor portability, which limits the fabrication of colorimetric products such as strips. The development of colorimetric sensing products, such as paper-based strips and microfluidic devices allows quick, practical, and on-site determination of various analytes. The most traditional method for developing paper-based devices is dip-coating, whereby the substrate is dipped in a probe and binder (Figure 18.3). The primary drawback of the method is the non-uniform coating, which in turn affects the sensitivity. Recently, micro-fluidic paper-based devices have gained attention as they allow the detection of multiple pollutants at the same time. Though the performance of the latter is highly selective and sensitive, the presence of hydrophobic ink used during the synthesis affects the structure of the substrate (paper). Consequently, diffusivity is adversely affected. Electrospinning is a mature technology used for the production of nanofibres based on colorimetric sensor strips. However, the sensing performance of nanofibre strips is usually affected by temperature, pH, loading amount of indicator, and size of the strips, which should be carefully optimized for practical measurements. In order to realize on-site detection using paper-based devices, the integration of various detection components like sample collection, receptor, output, and interpretation are essential.

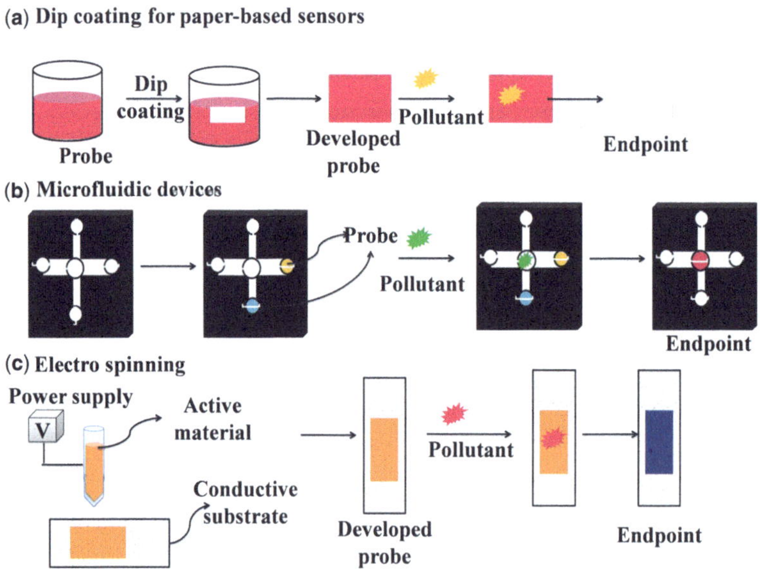

Figure 18.3 Methods for development of portable colorimetric sensors.

18.7 FIELD APPLICATION OF SENSOR TECHNOLOGY

Commercial sensors are available for the detection of nutrients and heavy metals. For instance, SUNA V2 (LOD: 42 mg N/L) (SUNVA V2) (Seabird Scientific datasheet SUNA V@ UV, 2017) and Eco Lab (Range: 0–5 mg/L) have been used for nitrate detection (Eco Lab 2). Likewise, WET Labs Cycle-PO4 (LOD: 0.023 μM) enables commercially available phosphate sensor that incorporates microfluidics with optical sensors for phosphate detection (WET Labs, 2013). It is based on the ascorbic acid method, with reagent stability of 5 months (Wet Lab's). The field-scale commercialized heavy metal detection tool mainly uses electrochemical techniques. DEP-Chip ((Biyani *et al.*, 2016) has been used for the detection of Cd (LOD: 2.6 μg/L), Zn, Pb (LOD: 4.0 μg/L), Cu (LOD: 14. 4 μg/L), and As (LOD: 15.5 μg/L). To further improve the LOD, AppliTrace (AppliTrace) devices have been developed (LOD of 1 μg/L). Portable heavy metal ion detector INE-SJB-801 (MRC) is another low-cost field-tested product. With a glassy carbon electrode, the system allows precise measurement on the spot. Nonetheless, most commercial sensors are expensive, making them impractical for large-scale environmental monitoring.

18.8 FUTURE PROSPECTS

The need for quick detection sensors will increase in the near future due to increased concentrations of pollutants like heavy metals, micro-pollutants, or pathogens. Despite the fact that many earlier studies and current research have reported on the development of electrochemical, optical, and biosensors, the instruments possessing high sensitivity and limited experimental drift have not been developed yet. The development of sensors for a target analyte and their subsequent integration into the IoT platform is another area of research that should be pursued. Long-term continuous monitoring of water quality can have far-reaching effects on aquatic ecosystems by providing spatiotemporal data sets of varied parameters and enabling the operation of water and wastewater treatment facilities in an energy-saving and cost-effective way. However, present water monitoring devices lack long-term accuracy in data gathering and processing capability. The development of innovative sensor materials is the key to long-term continuous monitoring. Thus, anticorrosive, conductive, and antimicrobial/anti-adhesive materials are needed. Also, in most cases, the performance of the developed probe is hampered due to poor stability when loaded on the portable platform. Thus, overcoming these obstacles in terms of the development, implementation, and commercialization of sensors is essential.

REFERENCES

AppliTrace. Single and multi-parameter determination of trace metals in water by online voltammetry. www.applitek.com

Biyani M., Biyani R., Tsuchihashi T., Takamura Y., Ushijima H., Tamiya E. and Biyani M. (2016). DEP-on-go for simultaneous sensing of multiple heavy metals pollutants in environmental samples. *Sensors*, **17**(1), 45, https://doi.org/10.3390/s17010045

Bhuvaneshwari M., Gobi N. and Devasena T. (2022). Alternative methods of monitoring emerging contaminants in water: a review. *Environmental Science: Processes & Impacts*, **24**, 2009–2031, https://doi.org/10.1039/D2EM00237J

Choudhary V. and Philip L. (2021). Stable paper-based colorimetric sensor for selective detection of phosphate ion in aqueous phase. *Microchemical Journal*, **171**, 106809, https://doi.org/10.1016/j.microc.2021.106809

Diouf A., Moufid M., Bouyahya D., Österlund L., El Bari N. and Bouchikhi B. (2020). An electrochemical sensor based on chitosan capped with gold nanoparticles combined with a voltammetric electronic tongue for quantitative aspirin detection in human physiological fluids and tablets. *Materials Science and Engineering*, **110**, 110665, https://doi.org/10.1016/j.msec.2020.110665

Ecolab 2. Ecolab 2 – Multi-Channel Analyzer System. Available from: http://www.labtech.com.mx/files/ecolab_2.pdf

Gan Y., Liang T., Hu Q., Zhong L., Wang X., Wan H. and Wang P. (2020). In-situ detection of cadmium with aptamer functionalized gold nanoparticles based on smartphone-based colorimetric system. *Talanta*, **208**, 120231, https://doi.org/10.1016/j.talanta.2019.120231

Gunter H., Bradley C., Hannah D. M. and Manasekind Lei Y. (2021). Recent progress in the detection of emerging contaminants PFASs. *Journal of Hazardous Materials*, **408**, 124437. Amethoxazole and trimethoprim in water samples. Journal of Electroanalytical Chemis2, https://doi.org/10.1016/j.jhazmat.2020.124437

Huang Y., Wang X., Xiang W., Wang T., Otis C., Sarge L., Lei Y. and Li B. (2022). Forward-looking roadmaps for long-term continuous water quality monitoring: bottlenecks, innovations, and prospects in a critical review. *Environmental Science and Technology*, **56**(9), pp. 5334–5354.

Mahmud M. P., Ejeian F., Azadi S., Myers M., Pejcic B., Abbassi R., Razmjou A. and Asadnia M. (2020). Recent progress in sensing nitrate, nitrite, phosphate, and ammonium in aquatic environment. *Chemosphere*, **259**, 127492, https://doi.org/10.1016/j.chemosphere.2020.127492

Martins T. S., Bott-Neto J. L., Oliveira O. N., Jr. and Machado S. A. (2021). Paper-based electrochemical sensors with reduced graphene nanoribbons for simultaneous detection of sulfamethoxazole and trimethoprim in water samples. *Journal of Electroanalytical Chemistry*, **882**, 114985, https://doi.org/10.1016/j.jelechem.2021.114985

MRC. Portable heavy metal analyzer. www.mrclab.com

Mukherjee S., Bhattacharyya S., Ghosh K., Pal S., Halder A., Naseri M., Mohammadniaei M., Sarkar S., Ghosh A. and Sun Y. (2021). Sensory development for heavy metal detection: a review on translation from conventional analysis to field-portable sensor. *Trends in Food Science & Technology*, **109**, 674–689, https://doi.org/10.1016/j.tifs.2021.01.062

Nag A., Alahi M. E. E., Feng S. and Mukhopadhyay S. C. (2019). IoT-based sensing system for phosphate detection using graphite/PDMS sensors. *Sensors and Actuators A: Physical*, **286**, 43–50, https://doi.org/10.1016/j.sna.2018.12.020

Paepae T., Bokoro P. N. and Kyamakya K. (2021). From fully physical to virtual sensing for water quality assessment: a comprehensive review of the relevant state-of-the-art. *Sensors*, **21**(21), 6971, https://doi.org/10.3390/s21216971

Ryu H., Li B., De Guise S., McCutcheon J. and Lei Y. (2021). Recent progress in the detection of emerging contaminants PFASs. *Journal of Hazardous Materials*, **408**, 124437, https://doi.org/10.1016/j.jhazmat.2020.124437

SeaBird Scientific datasheet 'SUNA V2 UV Nitrate Sensor'. 2017. Available at: http://satlantic.com/suna?qt-product_tabs=4#qt-product_tabs

Silva G. M., Campos D. F., Brasil J. A. T., Tremblay M., Mendiondo E. M. and Ghiglieno F. (2022). Advances in technological research for online and in situ water quality monitoring – a review. *Sustainability*, **14**(9), 5059, https://doi.org/10.3390/su14095059

Teodoro K. B., Shimizu F. M., Scagion V. P. and Correa D. S. (2019). Ternary nanocomposites based on cellulose nanowhiskers, silver nanoparticles and electrospun nanofibers: use in an electronic tongue for heavy metal detection. *Sensors and Actuators B: Chemical*, **290**, 387–395, https://doi.org/10.1016/j.snb.2019.03.125

Vellingiri K., Choudhary V. and Philip L. (2019). Fabrication of portable colorimetric sensor based on basic fuchsin for selective sensing of nitrite ions. *Journal of Environmental Chemical Engineering*, **7**(5), 103374, https://doi.org/10.1016/j.jece.2019.103374

WET Labs, Case Studies of WET Labs' Cycle-PO4: Simple and Reliable Nutrient Monitoring. 2013. Available at: http://wetlabs.com/news/case-studies-wet-labs-cycle-po4-simple and reliable nutrient-monitoring

doi: 10.2166/9781789063714_0213

Chapter 19

Conductivity sensors for water quality monitoring: a brief review

K. K. Tejaswini[1], B. George[1], S. C. Mukhopadhyay[2] and V. J. Kumar[1]*

[1]Department of Electrical Engineering, Indian Institute of Technology Madras, Chennai, India
[2]School of Engineering, Macquarie University, Sydney, Australia
**Corresponding author: boby@iitm.ac.in*

ABSTRACT

Conductivity of water is one of the important parameters that can, to some extent, reveal the quality of water. As water quality changes, the ability of water to pass current through it changes. This makes the conductivity measurement important. Some of the conductivity sensors are designed to use for flow-through applications while some others are immersion types. Conductivity can be measured using specially designed sensors or probes; some of those need direct contact with water. There are also sensors with no such restriction. Sensors that need direct contact with water suffer due to contamination of the measurement electrodes. Accuracy of the measurement is affected by the electrode contamination, undefined current path, noise, and non-idealities of the measurement electronics employed. There are non-contact-type conductivity sensors based on inductive techniques and capacitively coupled techniques. They have no electrode contamination problem. Undefined current path in water is a limitation for the inductive-type probe and complexity level of the capacitively coupled sensor is relatively high. To select or design a certain type of conductivity sensor, it will be advantageous if the associated details are available in a concise manner. This chapter presents an overview about the existing conductivity sensors for water. It is focussed on latest developments and a performance comparison is presented by listing the limiting factors. The chapter concludes with a discussion of future opportunities in this field.

Keywords: conductivity of water, water quality, inductive probe, capacitive probe, measurement system

19.1 INTRODUCTION

Conductivity of water is an important parameter that can, directly and indirectly, reveal valuable information about the water including its quality. Conductivity information of water is a must in water treatment or recycling plants, several process control industries, oceanographic applications, and so on. Salinity, for example, has a direct impact on aquatic life. Reliable and affordable instruments with sufficient accuracy and low maintenance requirements are needed for continuous monitoring of water conductivity. Contamination on the sensing electrodes and the corrosive nature of some of the elements that may be present in water make it difficult to use conventional contact-type conductivity measurement methods. Similarly, the current path is undefined leading to varying cell constant. Some of these problems can be overcome by using inductive non-contact-type conductivity measurement methods but the current path is still undefined. In this chapter, we present some of the latest developments in conductivity measurement methods which can be used for even corrosive water or liquids and have stable cell constant. The performance and key features of the schemes are compared with conventional methods and the results presented briefly.

19.2 WATER CONDUCTIVITY AND CONVENTIONAL MEASUREMENT METHODS

The electrical conductivity of a solution is the measure of the ability of the solution to conduct current (Ali *et al.*, 2017). This ability to conduct current depends on the concentration of ions present in the solution. Higher the ion concentration, higher is the ability to conduct current and hence higher conductivity. In natural water bodies, the ions that contribute to increased conductivity come from the minerals and salts dissolved therein.

By measuring the conductivity of a solution, certain properties of that solution can be ascertained. For example, the conductivity of seawater enables the determination of its salinity (Ramos *et al.*, 2008). The conductivity of a solution can be measured using a standard conductivity cell (Arthor, 2017). Conductivity σ is computed by multiplying the measured conductance G with the cell constant K as given in equation (19.1).

$$\sigma = GK \tag{19.1}$$

In equation (19.1), G is the conductance expressed in siemens, σ is the conductivity in S/m, and K is the cell constant which is the ratio of the length (m) and area of cross-section (m^2) of the solution or water column held in the cell used for measurement (Webster, 1999). Conductivity of the distilled water is in the range of 0.5–3 μS/cm while that of tap water is 50–800 μS/cm. Similarly, the conductivity of freshwater streams varies between 100 and 2000 μS/cm, that of industrial waste water is in the range of 10,000 μS/cm and that of sea water is in the range of 55,000 μS/cm (Fondriest, 2023). Conductivity can be measured using a contact-type electrode conductivity meter (Arthor, 2017; Ramos *et al.*, 2008) or the non-contact, inductive-type conductivity meter (Bard & Faulkner, 2001; Karbeyaz & Gencer, 2003) as shown in Figure 19.1.

The contact-type conductivity measurement can be accomplished either with two or four electrodes. Two-electrode conductivity meters have two contact-type electrodes (one is an excitation electrode and the other is a receiving electrode) placed at a distance from each other in the target solution as shown in Figure 19.1a. They are excited by a known voltage source and current through the solution is measured. By calculating the value of resistance of the path and cell constant of the meter, the

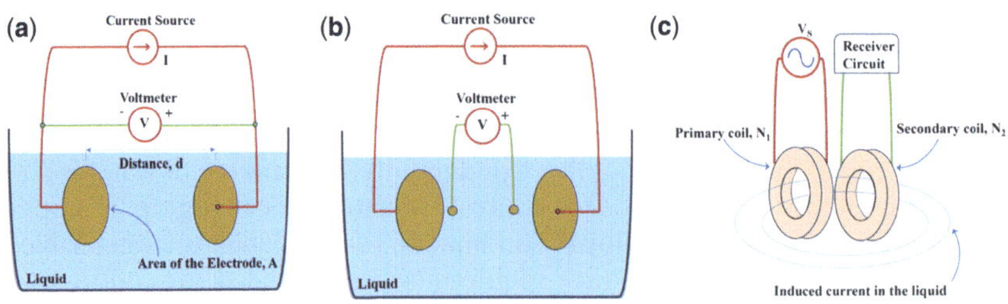

Figure 19.1 Various types of conductivity meters: (a) two-electrode conductivity meter, (b) four-electrode conductivity meter, and (c) inductive-type conductivity meter. Primary is excited using a sinusoidal excitation source V_s, and the secondary voltage is measured using a receiver circuit. (*Source*: Image source: (a), (b) https://instrumentationtools.com, (c) Karlsruhe Institute of Technology).

conductivity of the solution is determined. But the contact-type electrodes have low accuracy due to undefined current path, polarization, and contact or lead resistance (Ali *et al.*, 2017; Bard & Faulkner, 2001; Ramos *et al.*, 2008).

Four-wire electrode conductivity meters have four contact electrodes (Zhang *et al.*, 2013) as shown in Figure 19.1b. Two electrodes act as current electrodes as they inject current into the solution and the other two electrodes are potential electrodes that are placed in between current electrodes to measure the voltage drop in the solution. Using the current, voltage drop, and the cell constant of the meter, the conductivity of the solution is calculated. Four electrode meters do not suffer from the effects of lead or contact resistance or polarization but still suffer from the problem of undefined current path (Schiefelbein *et al.*, 1998). In Ramos *et al.* (2008), the ring-type electrodes make the current path confined but it suffers from error due to the assumption that half of the current injected by the transmitter electrode flows between the transmitter and receiver electrodes via the specified solution column and the other half flows via an external path. This assumption will also lead to additional errors when the solution has suspended particles or contamination of electrodes occurs. Electrode conductivity meters can only be used to measure the conductivity of solutions with unsuspended solids or non-corrosive solutions (Harrison & Roach, 1963; Relis, 1951).

Inductive-type conductivity meters are specifically useful when it comes to solutions with corrosive nature and for molten metals (Fougere, 2000; Striggow & Dankert, 1985). They have two toroidal coils (primary coil N_1 and secondary coil N_2) that are magnetically isolated from each other as shown in Figure 19.1c. They act as two transformers with the surrounding solution acting as the secondary for the first transformer and primary for the second transformer. The first transformer acts as a potential transformer and the second one acts as a current transformer. One toroid coil is excited by an AC voltage source and induces current into the surrounding solution. The second toroid coil acts as detection coil and measures the current passing through the solution which is directly proportional to the conductivity of the target solution. Unlike the electrode conductivity meter, the inductive-type conductivity meter does not suffer from polarization and electrode contamination (Fougere, 2000; Ismail & Shida, 1998; Natarajan *et al.*, 2004) since it is a non-contact way of measuring conductivity. However, the current is not confined to a predefined path in this method which leads to variations in the cell constant (Harrison & Roach, 1963).

In Groger *et al.* (1996) and Susa *et al.* (2015), an eddy current-based technique is used to measure the conductivity of the electrolytic solution. The variation in the inductance with the concentration of electrolyte in the solution is measured using an LCR (inductance (L), capacitance (C) and resistance (R)) meter. In Groger *et al.* (1996), the resonant frequency method is employed and from the inductance measurement, the unknown concentration of the electrolyte is determined with the help of a look-up table. In Susa *et al.* (2015), a quartz oscillator was used to measure the concentration of solutions, but the range of conductivity measured using this technique is very low (0.01–4.6 mS/cm) and also the latter is not determined directly. Here too, a look-up table has to be created and employed for determining the conductivity of the sample.

A non-contact inductive-type conductivity measurement method based on resonant frequency proposed earlier is quite complex and measures the conductivity indirectly (Kruipers *et al.*, 2014). The non-contact water salinity sensor proposed earlier is based on the electromagnetically induced eddy current measurement principle and is developed for desalination plants (Sonehara *et al.*, 2016). This method does not conform to the standard measurement principle of conductivity of solutions. An alternative but similar method based on eddy current measurement also suffers from non-conformity (Kleinburg *et al.*, 1989). Another method proposed earlier simultaneously measures the water level and conductivity (Yin *et al.*, 2008). Here, the inductances of two circular coils are measured and the conductivity of water is determined therefrom. In this method, conductivity measurements are affected by the material of the storage container. A brief comparison of various existing methods is given in Table 19.1.

Table 19.1 Comparison of various existing methods.

Feature	Types of the Conductivity Meter/Probe					
	Contact Type (Ali *et al.*, 2017)	Contact Type (Zhang *et al.*, 2013)	Non-Contact Type (Harrison & Roach, 1963)	Non-Contact Type (Ismail & Shida, 1998)	Non-Contact Type (Sonehara *et al.*, 2016)	Non-Contact Type (Yin *et al.*, 2008)
Sampling type or online monitoring	Online monitoring	Online monitoring	Sampling type	Sampling type	Sampling type	Sampling type
Flow-through or immersion	Both	Both	Immersion	Flow -through	Flow -through	Neither
Electrode contamination	Yes	Yes	No	No	No	No
Current path	Not defined	Not defined	Not defined	Not defined	Not defined	Not defined
External leakage current	Yes	No	Yes	Yes	No	Not applicable
Complexity of the probe	High	High	High	High	High	High
Excitation sSource	Constant current source	Constant current source	Constant voltage source	Constant voltage source	Constant voltage source	Constant voltage source
No. of electrodes	4	6	2	2	2	2
Range of conductivity measured (mS/cm)	0.5–50	0.01–50	0.8–65	2–70	0.01–50	40–100
Worst case % relative error	1.86	Not available	0.9	Not available	Not aAvailable	3

19.3 CAPACITIVELY COUPLED CONDUCTIVITY PROBE

Traditional contact-type conductivity meters for sea water have four electrodes in which two are current electrodes and the other two are potential electrodes. These potential electrodes eliminate the problem of polarization that was present in two-electrode contact-type conductivity meters. The source current electrode sends known current into the solution which is collected by the receiver current electrode. If the conductivity meter is a flow-through one, then there will be only one current path between excitation and receiver current electrodes but when the conductivity meter is of immersion type, there might be error due to the external current path between current electrodes. Although the effect of polarization of the excitation and receiver electrodes is removed, the potential electrodes need to have direct contact with the solution. This is a critical requirement for functionality but contamination of the electrodes affects the reliability of this contact. Hence, it is better if the measurement can be achieved with non-contact electrodes. To reduce this error, a four-terminal contact-type conductivity meter was proposed in which the current from the excitation current electrode was assumed to split equally between the path inside the probe and the external path in the surrounding solution (Ramos *et al.*, 2008). This assumption might lead to errors in conductivity in situations when there are suspended particles or contamination of electrodes occurs. Contact-type conductivity meters have high sensitivity and are used for solutions which are not corrosive in nature. The electrodes must be coated with platinum to prevent corrosion or contamination which makes them costly.

In this chapter, a conductivity meter which uses non-contact potential electrodes with guard ring is presented (Tejaswini *et al.*, 2019). This method uses non-contact-type potential electrodes and the

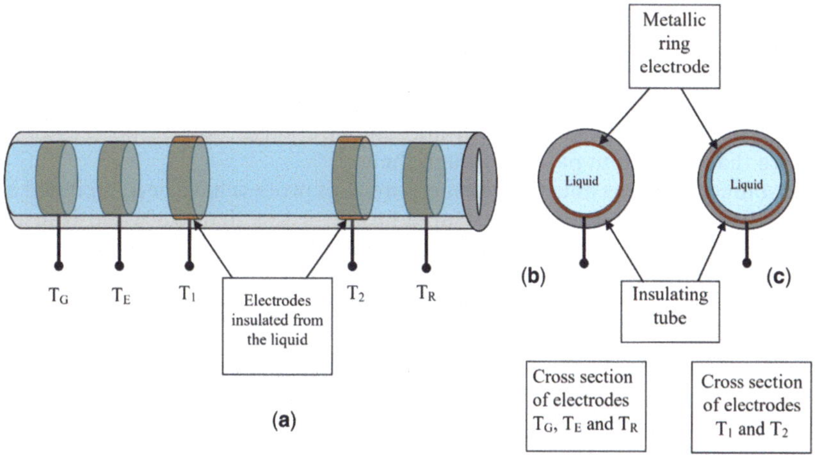

Figure 19.2 Conductivity measurement probe with five electrodes (Tejaswini *et al.*, 2019) (b) cross-section of electrodes T_G, T_E, and T_R, (c) cross-section of electrodes T_1 and T_2. © 2019 IEEE. Reprinted, with permission, from IEEE Sensors Journal.

guard ring to prevent current flow in the surrounding water; hence it has only one defined current path between the excitation and receiver current electrodes. Since there is no current flow in the surrounding water, unlike the one in Wang *et al.* (2019), it can be used in both flow-through and immersion modes. This combines the advantages of both contact-type and non-contact-type conductivity meters. An auto-balancing signal conditioning circuit was designed and implemented to obtain the resistance of the solution column which is not affected by the coupling capacitance of the non-contact potential electrodes.

A diagram of a water conductivity probe is shown in Figure 19.2. It uses capacitive coupling for the electrodes that are sensitive to the presence of contamination to measure the conductivity of the water column. It has five ring-shaped electrodes inserted inside an insulating hollow cylindrical tube as shown in Figure 19.2. The electrodes are T_E: excitation electrode, T_R: return electrode, T_G: guard ring, and T_1 and T_2: the non-contact electrodes used to sense the voltage drop in the water column under measurement.

All the electrodes T_G to T_R are connected to a measurement circuit. T_E is excited using a sinusoidal voltage source which drives a current through the T_R which is maintained at ground potential by measurement electronics. T_G ensures that the current leaving the probe is suppressed. Under this condition, the current flowing from T_E to T_R makes a voltage drop in the water column under T_1 and T_2. The electronic system used in the measurement circuit measures the voltage drop. As the voltage sensing electrodes are not in direct contact with the water column, the measurement requires a new solution which is presented in Tejaswini *et al.* (2019). The current received by the T_R is also measured by the measurement electronics. Once the voltage across T_1 and T_2, and the current through the water column are available, the resistance of the column and the conductance G can be computed. Since the probe pipe dimensions are available, the cell constant is known. With this information, the conductivity of the water column can be computed as in equation (19.1).

Based on this approach, a conductivity probe with non-contact-type potential electrodes (Tejaswini *et al.*, 2019) and a guard ring was designed and tested in the laboratory. An auto-balancing signal conditioning circuit (Tejaswini *et al.*, 2019) connected to the prototype probe to obtain the conductivity of the solution was also designed and tested. The excitation, return and guard electrodes are contact type while the potential electrodes are non-contact type. The structure of the electrodes and the guard ring ensure that the current from the excitation electrode passes through a predefined path and reaches the return electrode thus ensuring that the cell constant remains unaltered. Capacitive coupled potential electrodes eliminate errors that may arise due to contamination of contact-type potential electrodes. The balancing scheme helps to accurately trace the changes in the water column resistance under measurement, without there being any leakage current in the surrounding water. The value of reactance of coupling capacitors will not affect the measurement process as the balancing circuit makes sure that no current passes through them.

The potential sources of errors affecting the measurement process have been analysed and corrected (Tejaswini *et al.*, 2019). The conductivity measured has error less than 0.8% and is insensitive to the changes in the coupling capacitors' values. This method is less expensive, consumes less power, has less weight and uses auto-balancing circuitry to find the conductivity of the water. The probe can be used to determine conductivity of other electrolytic solutions in the desired conductivity range by making small changes in the signal conditioning circuit like changing the range of digital potentiometers and feedback resistor used in the measurement circuit (Tejaswini *et al.*, 2019). The probe developed can be used as a flow-through type or immersion type without making changes in the signal-conditioning circuit.

19.4 INDUCTIVE–CAPACITIVE PROBE

Inductive-type conductivity meters are of the non-contact type and can be used for solutions that are corrosive in nature. They are also free from polarization effect and contamination problem. However, they are prone to errors in measurement due to the undefined current path in the solution. The current in the conductivity meter may also be affected by other materials present near the probe. This leads to the cell constant value becoming different from the set value used in calculating the conductivity of the target solution and results in errors. An inductive-type conductivity meter developed on the basis of resonant frequency method portrayed earlier is quite complex and does not measure conductivity directly (Kruipers *et al.*, 2014). The non-contact-type water salinity sensor proposed earlier is based on the electromagnetic induction-based eddy current measurement principle and is developed for desalination plants (Sonehara *et al.*, 2016). This method requires high frequency of operation and also great care must be taken to make the eight-figure coil (eddy current detection coil) to reduce the error in the measurement process. An alternative but similar method based on eddy current measurement is highly prone to error due to the mismatch of placement of receiver coils symmetrical to the excitation coil (Yin *et al.*, 2008).

In this section, an inductive–capacitive conductivity probe is proposed which is a completely non-contact-type probe. The inductive part of the probe establishes a current in the solution and the capacitively coupled electrodes measure the voltage drop across the pre-defined solution column. This method has all the advantages of inductive-type conductivity meters as well as capacitively coupled conductivity meters. The solution column created inside the probe results in stable cell constant throughout the measurement process, hence eliminating the problem of undefined current path. The output is unaffected by the change in the value of coupling capacitances.

The schematic of the inductive–capacitive coupled conductivity probe is shown in Figure 19.3 (Tejaswini *et al.*, 2020). In the previous section, a capacitively coupled approach is presented but not all the electrodes are non-contact electrodes. In applications where a fully non-contact probe is needed with calculable cell constant and no undefined path problem, the solution is to combine the inductive and capacitive approaches. In the new combined approach, we use inductive approach to send a current through the water column of interest, as shown in Figure 19.3 and use the capacitively coupled approach to measure the voltage drop across the water column of interest. The current being sent through the water column is measured using inductive approach. Since the voltage across the water column in the tube under the capacitive electrode is measured, the dimensions of the water column and hence the cell constant is known, and current through the water column is measured using the inductive approach, we can compute the resistance of the water column and estimate the conductivity using equation (19.1).

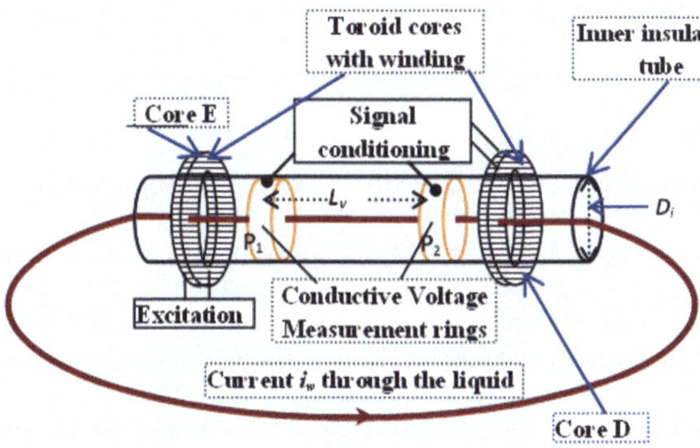

Figure 19.3 Measurement of conductivity of solutions using inductive–capacitive method. (*Source*: Tejaswini *et al.*, 2020). © 2020 IEEE. Reprinted, with permission, from IEEE Transactions on Industrial Electronics.

The probe is designed with two concentric cylindrical tubes made of non-conducting material. The ends are closed (between the inner and outer tubes) with lids made of insulating material so that water cannot enter the space between the inner and outer tubes and contaminate electrodes and electronics (Tejaswini *et al.*, 2020). Two coils were wound on the high permeability soft ferromagnetic toroid core. These are mounted on the inner insulating tube on either end. One of the coils, say coil E with NE turns, serves as the excitation coil, and induces a current in the solution. The other coil, coil D with ND turns, functions as the secondary of a bar-primary current transformer and senses this current. Electrodes P_1 and P_2 are similar to the electrode T_1 and T_2 in the previous section. A special circuit is needed to perform the measurement of voltage across P_1 and P_2 and to measure the current. The details are available in Tejaswini *et al.* (2020).

An overview of the inductive–capacitive coupled probe to measure the conductivity of the water is presented. It has all the advantages of inductive-type conductivity meter and capacitive-coupled-type conductivity meter. It measures conductivity of the solution when it is completely immersed in the water. The capacitive-coupled electrodes create a water column which makes the cell constant K stable over the conductivity range. The measurement is also less effected by the presence of external material near the probe, whether that be conductive or non-conductive. This is a major advantage over the existing inductive probes. The measurement is unaffected by the change in the value of coupling capacitances as an additional voltage source was used to eliminate that effect. The probe is completely non-contact type and can be used for solutions which are corrosive in nature.

19.5 FUTURE PERSPECTIVE

Continuous monitoring of vital water quality parameters like conductivity is becoming a must. The solution provided in this chapter helps to address the problem of contamination of electrodes used in the probe. In addition, it can provide a calibration-free instrument with high accuracy. The material required to manufacture the probe is not expensive, which helps to keep the cost low. The developed probes need further field trials and long-term tests to ascertain their long-term reliability.

Some of the possible improvements are (a) significant reduction in the overall size and weight of the probe, (b) similar reduction in power consumption and battery capacity will be valuable,

(c) development of self-powered probes will be a valuable improvement, (d) use of modern low power data transmission methods, and (e) the use of Internet of Things together with data analytics can improve the effectiveness of continuous monitoring and its impact on society.

19.6 CONCLUSIONS

Measurement of conductivity of water is important in determining quality and it serves as an early indicator of change in the water system. Total dissolved solids and salinity can be related to the conductivity data. Salinity is useful to study sea life. Due to the nature of the water depending on the source, contamination of electrodes of conductivity meter can be a major issue. Contact-type conductivity meter suffers from corrosion of electrodes in the long run. The inductive-type conductivity meter, though non-contact in nature, cannot maintain stable cell constant due to undefined current path. There is a necessity of non-contact-type conductivity meter which has stable cell constant and whose electrodes do not corrode over a period of time. The latest developments in this direction, detailed in this chapter, provide solutions to the listed problems. It is proved theoretically and experimentally (Tejaswini *et al.*, 2019) that a combination of conductive and capacitive electrodes can be used to measure the conductivity of solution in immersion and flow-through modes and the output can be made insensitive to contamination of the electrodes and current through the external water surrounding the probe. The theoretical and experimental studies conducted by Tejaswini *et al.* (2020) showed that the proposed inductive–capacitive conductivity probe provides conductivity of solution in immersion mode while being fully non-contact type and it does not have contamination problems. The solution column created by the capacitively coupled electrodes makes the cell constant stable over the given conductivity range. The conductivity measured by the proposed method is unaffected by the objects present in the surroundings of the probe.

REFERENCES

Ali A., Fadhil H., Yousif E., Hussain Z., Abdul-Wahab S. and Zageer D. (2017). An insight into measuring conductivity of solutions with ease: a review. *Research Journal of Pharmaceutical, Biological and Chemical Sciences*, **8**, 2033–2037.

Arthor A. (2017). The Electrical Conductivity of Aqueous Solutions. Andesite Press.

Bard A. J. and Faulkner L. R. (2001). Electrochemical Methods, 2nd edn. John Wiley and Sons, Hoboken, New Jersey, United States.

Fondriest. (2023). https://www.fondriest.com/environmental-measurements/parameters/water-quality/conductivity-salinity-tds/ (downloaded 6th May 2023).

Fougere A. J. (2000). New non-external field inductive conductivity sensor (NXIC) for long term deployments in biologically active regions. *Proceedings of Oceans MTS/IEEE Conference and Exhibition*, **1**, 623–630.

Groger H. P., Churchill R. J. and Kelsch J. (1996). Chemical sensor using eddy current or resonant electromagnetic circuit detection. US Patent 5 514 337.

Harrison E. and Roach P. F. (1963). An electrodeless conductivity meter for process control. *Journal of the British Institution of Radio Engineers*, **26**, 25–34.

Ismail A. B. Md and Shida K. (1998). Non-contact multi-sA: physical technique for the precise measurement of concentration of electrolytic solution. *Sensors and Actuators A: Physical*, **69**(2), 152–155, https://doi.org/10.1016/S0924-4247(98)00069-7

Karbeyaz B. and Gencer N. (2003). Electrical conductivity imaging via contactless measurements: an experimental study. *IEEE Transactions on Medical Imaging*, **22**(5), 627–635, https://doi.org/10.1109/TMI.2003.812271

Kleinburg R. L., Chew W. C. and Griffin D. D. (1989). Non-contacting electrical conductivity sensor for remote, hostile environments. *IEEE Transactions on Instrumentation and Measurement*, **38**(1), 22–26, https://doi.org/10.1109/19.19992

Kruipers J., Bruning H., Yntema D., Bakker S. and Rijnaarts H. (2014). Self-capacitance and resistive losses of saline water filled inductors. *IEEE Transactions on Industrial Electronics*, **61**(5), 2356–2361, https://doi.org/10.1109/TIE.2013.2270220

Natarajan S. P., Huffman J., Weller T. M. and Fries D. P. (2004). Contact-less toroidal fluid conductivity sensor based on RF detection. *IEEE Sensors Journal*, **1**, 304–307.

Ramos P. M., Pereira J. M. D. and Ramos H. M. G. (2008). A four terminal water-quality monitoring conductivity sensor. *IEEE Transactions on Instrumentation and Measurement*, **57**(3), 571–582, https://doi.org/10.1109/TIM.2007.911703

Relis M. J. (1951). Method and apparatus for measuring the conductivity of an electrolyte. U.S. Patent 2 542 057.

Schiefelbein S. L., Fried N. A., Rhoads K. G. and Sadoway D. R. (1998). A high-accuracy, calibration-free technique for measuring the electrical conductivity of liquids. *Review of Scientific Instruments*, **69**, 3308–3313, https://doi.org/10.1063/1.1149095

Sonehara M., Toai N. V. and Sato T. (2016). Fundamental study of non-contact water salinity sensor by using electromagnetic means for seawater desalination plants. *IEEE Transactions on Magnetics*, **52**(7), 1–4, https://doi.org/10.1109/TMAG.2016.2537921

Striggow K. and Dankert R. (1985). The exact theory of inductive conductivity sensors for oceanographic application. *IEEE Journal of Oceanic Engineering*, **OE-10**(2), 175–179, https://doi.org/10.1109/JOE.1985.1145085

Susa T., Watanabe T., Sohgawa M. and Abe T. (2015). Non-contact sensor for measurement of liquid concentration based on quartz oscillator. In: 2015 Transducers-2015 18th International Conference on Solid-State Sensors, Actuators and Microsystems (Transducers), pp. 1472–1475. IEEE.

Tejaswini K. K., George B. and Kumar V. J. (2019). Conductivity measurement using non-contact potential electrodes and a guard ring. *IEEE Sensors Journal*, **19**(12), 468–4695, https://doi.org/10.1109/JSEN.2019.2900798

Tejaswini K. K., George B. and Kumar V. J. (2020). Assay of inductive-capacitive probe for the measurement of the conductivity of liquids. *IEEE Transactions on Industrial Electronics*, **68**(9), 8911–8918, https://doi.org/10.1109/TIE.2020.3013754

Wang X., Wang Y., Leung H., Mukhopadhyay S. C., Chen S. and Cui Y. (2019). A self-adaptive and wide-range conductivity measurement method based on planar interdigital electrode array. *IEEE Access*, **7**, 173157–173165, https://doi.org/10.1109/ACCESS.2019.2956568

Webster J. (1999). The Measurement, Instrumentation, and Sensors Handbook. CRC, Boca Raton, FL.

Yin W., Peyton A. J., Zysko G. and Denno R. (2008). Simultaneous non-contact measurement of water level and conductivity. *IEEE Transactions on Instrumentation and Measurement*, **57**(11), 2665–2669, https://doi.org/10.1109/TIM.2008.926054

Zhang J., Li D., Wang C. and Ding Q. (2013). An intelligent four-electrode conductivity sensor for aquaculture. In: Computer and Computing Technologies in Agriculture VI: 6th IFIP WG 5.14 International Conference, CCTA 2012, Zhangjiajie, China, 19–21 October 2012, Revised Selected Papers, Part I 6 (pp. 398–407). Springer Berlin Heidelberg.

doi: 10.2166/9781789063714_0223

Chapter 20

Multi-analyte assessment of water quality

*Kamalesh Chaudhari, Vidhya Subramanian and Thalappil Pradeep**

Department of Chemistry, Indian Institute of Technology Madras, Chennai, India
**Corresponding author: Pradeep@iitm.ac.in*

ABSTRACT

A thorough water quality assessment requires sample testing for a set of parameters. In field measurements, it is not necessary to precisely determine water composition. Field measurements can be planned only to roughly understand the necessity of in-depth lab-based analysis. This can be done by developing methods that are reagent free, independent of high-end instrumentation, and facilitative of multiplexed detection. While using multiplexed detection techniques, it is necessary to identify hidden signals using a probabilistic/predictive approach. This can be achieved by analysing the data from the measurements using AI and ML-based prediction techniques. The spectroscopy or electrochemistry-based signals exhibit minor sample-to-sample variations, which can be processed with data analytics tools to predict multiple analytes in water. Due to advancements in the miniaturization of sensors and cloud computing-based data analysis technologies, it has become possible to realize such data analysis-based multi-analyte sensing and detection methods.

Keywords: multi-analyte assessment, sensors, water quality, portable devices, predictive analysis

20.1 WATER QUALITY ASSESSMENT

Spatiotemporal variations in water quality are common. These variations are mainly due to differences in the composition of rocks, soil, and changes in seasons across the country (Bengtsson, 2010). In the hydrological cycle, water interacts with the surrounding environment, air, and soil. These environmental properties can either be beneficial or detrimental to water quality. Different human activities and the production of waste also add to the hazardous substances in water, causing further deterioration in its quality. In rural areas, uncontrolled use of fertilizers, improper management of domestic waste, and poor agricultural practices are the sources of contamination in water. In the urban and industrial areas, discard of effluent from production processes, discard of electronic items/batteries, and chemical spills lead to the contamination of surface water with highly toxic chemicals.

A set of quality parameters to be tested in water sample varies according to the source of water and its surroundings. The most common source is the groundwater pumped from aquifers formed by underground rocks and soil. Ground water levels vary according to region, seasonal changes and the amount of rainfall. These impact and contribute to the various complex changes in water quality. Another source of water is surface water, which is available from lakes, ponds, rivers, springs in the mountains, and so on. Once this water is exposed to the open environment and sources of pollution, it becomes unsafe for drinking. Rainwater is a relatively purer source of water as compared to groundwater or surface water. If rainwater is collected directly or from a clean rooftop, it remains clear and free of contaminants. In conclusion, contaminants in water can be grouped and correlated based on water sources and regions (Singh *et al.*, 2020).

The presence of contaminants in excess of their permissible limits as set by the Bureau of Indian Standards (BIS) makes the water unsafe for drinking. Water quality can be determined by a set of physical and chemical observations. Such measurements can be performed with portable devices or a set of chemical reagents, even in field environments. In some cases, there are trace level contaminants in water which are invisible and harmful even at ultra-low concentrations (Zamora-Ledezma *et al.*, 2021). A few of these can be tested with chemical reagents, and the remaining may require sophisticated instruments. For such measurements, water samples can be collected with BIS-specified protocols and transported to professional laboratories with high-end technologies and facilities. BIS has grouped contaminants into six major groups, organoleptic and physical parameters, substances undesirable in excessive amounts, toxic substances, radioactive substances, pesticide residues and microbial contaminants. While monitoring drinking water quality, microbial contamination is a major concern. Most of the illnesses and deaths that occur from unsafe drinking water are due to the presence of microbes in water (Saravanan *et al.*, 2022).

20.1.1 Requirement of multianalyte assessment

As discussed, certain contaminants can be present in a specific region due to environmental issues or human activities. In water quality analysis, only a few crucial parameters are deemed necessary, while the remaining tests may or may not be required, depending on the historical water quality data of the region. Due to temporal variations in water quality, samples need to be tested frequently for multiple parameters. Periodic testing of water can be time consuming and cost intensive. Many of the parameters cannot be tested in the field with portable devices but samples need to be collected and sent to laboratories. The transport of samples can involve additional expenses and the physicochemical state of the sample can change due to time lapse between collection and testing. Such issues can be avoided if techniques can be developed to analyse multiple water quality parameters with the same method in a single measurement (Hara & Singh, 2021; Mao *et al.*, 2015). Also, to avoid the need for laboratory measurements, it is necessary to enable water quality measurements in field environment. For optimum accuracy in field measurements, methods which require minimal efforts are necessary. By using multianalyte assessment, the efforts required during measurements can be reduced, the consumption of chemicals can be minimized, and the obtained data can be employed for data analytics to predict the presence of contaminants that are challenging to test otherwise.

20.1.2 Water quality assessment in the field

Testing of water quality at the consumer end is a challenging task. In rural areas specifically, where there are problems such as a lack of local water quality testing labs and poor connectivity, such testing cannot be done on a regular basis. Collection of water samples from scattered locations across the region becomes a tedious job and requires trained resources. Due to the serious health issues created by contaminated water and the absence of any household potability testing device, tap water is normally considered unsafe for drinking. This leads to the installation of household water purifiers and subsequent wastage in the form of reject water. Hence, ingenious but simple-to-use designs for devices to test water quality are required. Also, to localize the variations in the level of contaminant and their trends, it is necessary that portable devices include location tagging. Several devices have been created to analyse either a single contaminant or a group of contaminants, which can be detected through small signal variations in the sensing probe.

Such portable equipment need not be as precise as lab-level equipment, but it needs to be able to roughly measure and indicate the potability of water in field environments, and generate alerts upon the detection of any abnormalities in the water samples. Despite potential interference from varying water composition and predictive data from statistical measurements, multianalyte measurements can still fulfil the water quality assessment needs at the field level.

20.2 MULTIANALYTE METHODS

As discussed in the previous section, water quality parameters can be grouped according to conditions, such as region, source, and physicochemical properties. These environmental conditions and the presence of contaminants in water exhibit high level of correlation in some cases. For instance, in regions affected by arsenic, the occurrence of iron in groundwater can be correlated. Through these correlations, the necessity for a comprehensive water analysis can be eliminated, and only a few selected parameters need to be examined. In the next step, the data obtained from the measurements can be processed for the prediction of correlated parameters which are otherwise difficult to measure in a field environment. Nonetheless, in certain instances, like with wastewater or surface water flowing across a region, the water's composition can be complex and affected by constantly changing chemistry. In such cases, correlations may not be useful.

The methods used for multianalyte assessment need to be free of high-end instrumentation. Major techniques used for the field-level assessment are chemical reagent-based colorimetric assays or potentiometric sensors. In colourimetry, a set of chemical reagents are used to develop a visible colour which can be correlated with the standard colour charts to infer the concentration of contaminants. Due to the discrete nature of colourimetric charts, digital field test kit readers (FTKs) can be used for an accurate measurement. In potentiometric measurements, water samples are exposed to ion-selective electrodes, which can generate a signal in the form of changes in current or potential difference. In techniques such as anode stripping voltammetry (ASV), samples are deposited on a working electrode and then changes in current due to the stripping of ions at different potentials can be used to infer contaminant presence in water. Many sensors based on various principles have been developed in the past. A few examples of the same are provided in Table 20.1.

20.2.1 Limitations of multianalyte methods

Although multianalyte methods seem promising for the collective measurement of multiple contaminants in water, a few drawbacks arise due to the emphasis on simplifying and lowering the cost of measurements. Due to the aforementioned points, the detection efficiency of portable devices is less as compared to lab equipment. In field environment, deconvolution of interfering components is a complex and computationally intensive task. A huge concentration of interfering ions can increase the baseline. Hence, multianalyte methods can be handled well by data analytics tools developed for the grouping of signals obtained from the measurement of samples. However, when data analytics and grouping techniques are used for the detection of contaminants, the values obtained can have error bars and they are mostly predictive in nature. Such probabilistic values can be accurately analysed only by trained resources, whereas for the common public, these readings need be converted into a weighted estimation of an index for the potability of water. Hence in conclusion, multianalyte assessment can be of quantitative use to trained resources, but for common people, it can only be a qualitative binary indication of the potability of water.

Table 20.1 Some of the sensors developed for detecting contaminants in water.

Analyte	Substrate	Sensing Method	Range	References
Copper	2,2′:5′,2″-terthiphene and MWCNTs on Nylon-6 nanofibres	Amperometry	0.65–39 ppm	Lu *et al.* (2019)
Mercury	Thiamine functionalized silver nanoparticles on filter paper	Colourimetry	>0.5 μM	Budlayan *et al.* (2022)
Silver	Modified carbon paste electrode	Potentiometry	1×10^{-5}–1×10^{-1} M	Adeli *et al.* (2022)
Cobalt	Modified poly(vinyl chloride) membrane	Potentiometry	1×10^{-6}–1×10^{-1} M	Özbek *et al.* (2022)
Cadmium and lead	Bismuth nanoparticle@laser-induced graphene on polyimide film	Electrochemical	>0.5 μg/L and >0.8 μg/L	Zhao *et al.* (2022)
Arsenic and mercury	Gold nanoparticles with dithiothreitol – 10,12-pentacosadiynoic acid (DTT-PCDA) and lysine (Lys)	Colourimetry	710–1248 μg/L and 10.77–53.86 μg/L	Motalebizadeh *et al.* (2018)
Fluoride	Cubic ceria@zirconia nanocages	Colourimetry	0.1–5 ppm	Mukherjee *et al.* (2020)
Lead	Black phosphorus (BP) integrated tilted fibre grating	Polarization	0.1–1.5 × 10^7 μg/L	Liu *et al.* (2018)
Chromium, copper	ZnO quantum dot	Fluorescence	0.0834 and 2.34 μg/L	Khan *et al.* (2020)
Lead	AuNPs/PPy/$Ti_3C_2T_x$	Electrochemistry	5×10^{-14}–1×10^{-8} M	Zhang and Karimi-maleh (2023)
Arsenic	Hydrous ferric oxide	Plasmon resonance	0.6 μg/L	Yasmin *et al.* (2022)

20.3 DATA ANALYTICS IN MULTIANALYTE ASSESSMENT

20.3.1 Clustering methods

Clustering methods are of great importance in multianalyte assessments. Conventional techniques face challenges in detecting noisy readouts and minor signal variations when trace level contaminants are present. Techniques like principal component analysis (PCA) can be utilized to detect multiple components from a single signal spectrum or to predict the presence of trace level contaminants based on the concentration of other major components in the water sample (Mahapatra *et al.*, 2012).

Cluster analysis utilizes various algorithms to process data according to user-defined criteria and form clusters. The conditions for cluster formation can be the distance between data points, and the statistical distributions or the density of data points in the observation space. Selection of the clustering algorithm depends on user interest and parameters that are used to define the cluster. A few examples of cluster analysis algorithms are: K-means method, fuzzy c-mean, self-organizing map (SOM), interval clustering, and so on (Delgado *et al.*, 2021; Vo-Van *et al.*, 2020). The effectiveness of a clustering algorithm is an optimization problem and it is controlled by factors, such as the number of clusters to be formed, function used to find distance between points, density threshold, and so on. Cluster analysis is an iterative process and depends on the selection of initial parameters followed by manual observations till a satisfactory result is obtained.

The prediction of water quality parameters using the measurements of only a few selected parameters requires data processing techniques, such as PCA. This analysis can be treated as a multidimensional problem and PCA can be conducted to find principal components that are independent and contribute to the variance of the system. PCA helps in improving the interpretability of data and thus prevents information loss. Independent principal components can be found only after the processing of large datasets of the measurements collected over a statistical number of samples. Hence, PCA is an adaptive process which can improve when more measurements are added to the data analytics. Following PCA analysis, highly correlated components enable the reduction of measurement quantities and the prediction of other parameters using independent variables (Parinet *et al.*, 2004).

20.3.2 Machine learning and AI

We are in the era of big data. Terms like data science, data analytics, machine learning (ML), and artificial intelligence (AI) are flooding the market, and data are being considered as wealth of the company. Data collected from surveys and measurements can help build predictive models. Data processing can also give more insights about business or scientific understanding (Rubinger *et al.*, 2022). This significantly impacts decisions regarding future preparations, infrastructure management, planning, or scientific problem solving. Data can reveal new or unknown patterns. To briefly distinguish data science and data analytics, data science is a broader branch that focuses on finding new methods of data analysis and the setting up of standard data processing protocols. On the contrary, data analytics is a more specific area which uses the models and data processing protocols developed by data scientists to find the desired outcome.

ML is a subset of AI which focuses on continuous improvements in generating specific outputs based on data fed into an algorithm. ML algorithms enable self-learning and correction of the model using continuous inputs from data. This helps in improving the prediction and understanding of patterns in data to obtain accurate results. Learning algorithms can be trained in various ways. There is supervised learning that uses specific inputs and outputs for training the ML algorithm. The algorithm finds the general rule which connects inputs with a specific outcome. After training, ML algorithms can take inputs from available data and generate an output based on a general rule. These algorithms are also used to filter the data based on experience or rules defined by a user. ML algorithms can be classified as classification algorithms, producing a limited set of values as outputs, based on the type of input. The outcome of a regression-type algorithm can be numerical values within a specified range. Unsupervised algorithms are employed to discover unknown patterns or groupings within the input data, without relying on any predetermined outputs. The data with no defined categories are grouped/clustered based on shared similarities. Cluster analysis is an example of unsupervised learning in which a predesignated criterion is used to classify measurements into subsets. In addition, reinforcement learning techniques are used in the area of ML where outcomes are partially random and partially under the control of a user. These are optimization processes based on dynamic programming, also termed Markov decision process (Geist *et al.*, 2019). AI are data processing algorithms which mimic the human brain in solving problems. These algorithms/codes are written to rationalize the decision-making process for machines. As compared to data science and data analytics, AI is less process intensive and its major focus is on developing predictive models for decision making and forecasting outcomes based on the data at hand.

20.4 MULTI-ANALYTE ASSESSMENT OF WATER QUALITY

20.4.1 Review of the progress so far

Various detection principles have been used for the multi-analyte assessment of water quality. Detectors are selected based on the type of signal generated and the level of accuracy required. These techniques can be label free and can depend solely upon the intrinsic properties (mass, charge, etc.) of the analytes, or they can be based on the way analyte interacts with a sensing element developed to selectively interact and transduce a signal (Magro *et al.*, 2019). Selective sensing elements can be achieved by modifying substrate surfaces with selective ligands or thin film coatings possessing desired chemical properties. Another way to achieve selectivity is by tuning the properties of the sensing element through various processing techniques, making it more responsive to different analytes under different conditions of electric potential, temperature, and so on. In some cases, the same sensing substrate can be scanned through a range of electric potential to detect different analytes. The existing literature (Biyani *et al.*, 2017; Mukherjee *et al.*, 2020) covers a wide range of developments for assessing multiple analytes in water. We have surveyed and summarized a few selected reports in Table 20.2. To combine multiple assays in a miniaturized platform is a challenging task. In many situations this has been done with the help of microfluidic techniques.

An attempt to develop microfluidic device for the detection of multiple toxins was demonstrated. The technique used antibodies relevant to the toxins with a Lab-On-A-Disc (LOAD) platform for the detection of cyclic peptide toxins in marine water (Maguire *et al.*, 2017). Immunoassays can be developed based on the changes in optical properties of different types of probes such as fluorescence or SPR. Due to tunability permitted by carbon-based nanomaterials, their selectivity towards multiple analytes has been explored extensively in the past few decades. A few examples of the same involve studies conducted by our group for the detection of arsenite and relevant groundwater analytes (manganese and iron) using electrochemically reduced graphene oxide (ERGO)-based paper strips (Jana *et al.*, 2022). Several studies have demonstrated the tunability of GO-based sensors for detecting multiple analytes, albeit not in water (Bauer *et al.*, 2021).Variability noted in the responses of sensing elements can be used to collect the signals from complex mixtures and PCA can be used to further separate the components for the sensing of multiple analytes. Such methodology has been demonstrated by Song *et al.* using an array of sensing elements based on polymer-coated nanowires modified with catalytic nanoparticles (Song & Choi, 2015). Other than these, based on intrinsic molecular properties, approaches have been developed with the combination of on-line solid-phase extraction–liquid chromatography–electrospray–tandem mass spectrometry for the detection of 20 pesticides in natural and treated waters. Such techniques have also been demonstrated for the detection of as many as 103 analytes (semivolatile organic compounds (SVOCs)) using GCMS (gas chromatography-tandem mass spectrometry). Another study performed by Hermes *et al.* has shown that LCMS can be used to quantify >150 micropollutants in aqueous samples (Hermes *et al.*, 2018).

Table 20.2 Survey of multianalyte-sensing devices.

Device	Analyte	Methodology	Detection Range / LOD	References
ANDalyze	Pb, Cu, Zn, CD	Fluorescence	LOD – Pb – 2 μg/L, Cu – 0.6 μg/L, Zn – 1 μg/L, Cd – 0.1 μg/L	ANDalyze
DEP-Chip	Cd, Pb, As, Zn, Cu	Differential pulse voltammetry	LOD – Cd – 2.6 μg/L, Pb – 4 μg/L, As – 5 μg/L, Zn –14.4 μg/L, Cu – 15.5 μg/L	Biyani *et al.* (2017)
uMED	Cd, Zn, Pb	Square Wave ASV	LOD – 4 μg/L	Nemiroski *et al.* (2014)
Vitality Plus Australia	Cd, Co, Pb, Cu, Fe, Mn, Hg, Ni, Zn		1–1000 μg/L	Heavy Metal Test Kit–Vitality Plus Australia
HF–38 Fluorimeter	Pb, U, Cu, Hg	Fluorescence	10–3000 μg/L	HF–38 Fluorimeter For Heavy Metals Testing At ACE12
Metrohm 946 Portable VA analyser (SPE)	As, Sb, Bi, Cd, Co, Pb, Cu, Fe, Mn, Hg, Ni, Zn		1–20 μg/L	Metrohm 946 Portable VA Analyzer

Data analytics and related tools have brought a whole new dimension to the information that can be inferred from multianalyte measurements. In complex composition of water, response from the analyte of interest can be separated using PCA analysis. A few examples of such studies in literature are listed here. One of the major benefits of PCA is reducing the number of parameters and sensing hardware which needs to be used for the measurement of water quality or the same required in water management infrastructure. A study from Ortuno and group has shown that PCA combined with a single ion-selective electrode can be used for multi-ion measurements (Cuartero *et al.*, 2017). Another example has shown that PCA can be useful in simplifying the number of parameters required for water resource management with the example of tropical lake system (Parinet *et al.*, 2004). PCA has also been demonstrated for prediction and evaluation of water quality and reliability of monitoring sources.

Apart from PCA-based projections of dependent water quality parameters, cluster analysis can reveal valuable information, primarily to distinguish different components contributing to changes in water quality. Vo-Van *et al.* (2020) have demonstrated a novel clustering algorithm to understand water quality which represents contribution from four different sources. The application of Grey clustering has been demonstrated in understanding the impact of mining operations on water samples collected from various points in the surrounding region (Delgado *et al.*, 2021). An artificial neural network-based technique has been developed to predict the presence of faecal coliform by Asaf *et al.* (Pras & Mamane, 2023). Cluster analysis can also facilitate addressing discrepancies in the collected data. The data collected in intervals due to various sources of errors or poor measurements can be processed using clustering methods such as interval clustering algorithms (ICA) (Wong & Hu, 2013).

20.4.2 The future of multi-analyte assessment

The development of new technologies is being driven by the need to move towards sustainability, reduce resource consumption, and minimize the impact of techniques on environmental pollution. Advancements in microfabrication techniques and a boom in data analytics with the supporting infrastructure of cloud computing are being used increasingly in the development of sensing elements. Lab-on-chip microsensors with readout interfaces from smartphones are becoming popular. An app-based operation and miniature sensing/readout devices are being developed extensively. Mobile operation with smaller form factor allows direct field application of these techniques. Smartphone-based operation and internet connectivity facilitates the application connectivity with cloud for rigorous computation and a quick predictive analysis based on PCA and advanced data analytics tools. Many new technologies developed in this direction are in a state of lab-to-field transition. Large number of such devices have been launched and commercialized in the last decade (Figure 20.1). Given this background, we envision a future where smartphone-incorporated sensors will enable users to easily perform multianalyte assessments of water quality with just a few clicks on an app, and receive an indication of the potability of the water.

20.5 SUMMARY

We started with an introduction of the different levels of technologies available for the testing of water quality parameters. Different levels of water quality testing are performed for water quality assessment from lab, water source to the end user with field assessment. All these have pros and cons in terms of the readout accuracy, transport of samples, and speed of assessment in the task to indicate whether the water sample being tested is potable or if it needs detailed investigation with high-end equipment. With these we discussed a need for the development of multianalyte-sensing technologies. Followed by this we have discussed various sensing principles which can be used to develop multianalyte sensors and how materials can be tuned to support such measurements. We have listed out the risks involved in predictability of dependent parameters and what impact they may have on the common user. Data analytics has found great applications in the area of multianalyte sensor development and inference from the collected data. PCA and clustering tools have been found useful in simplifying the required

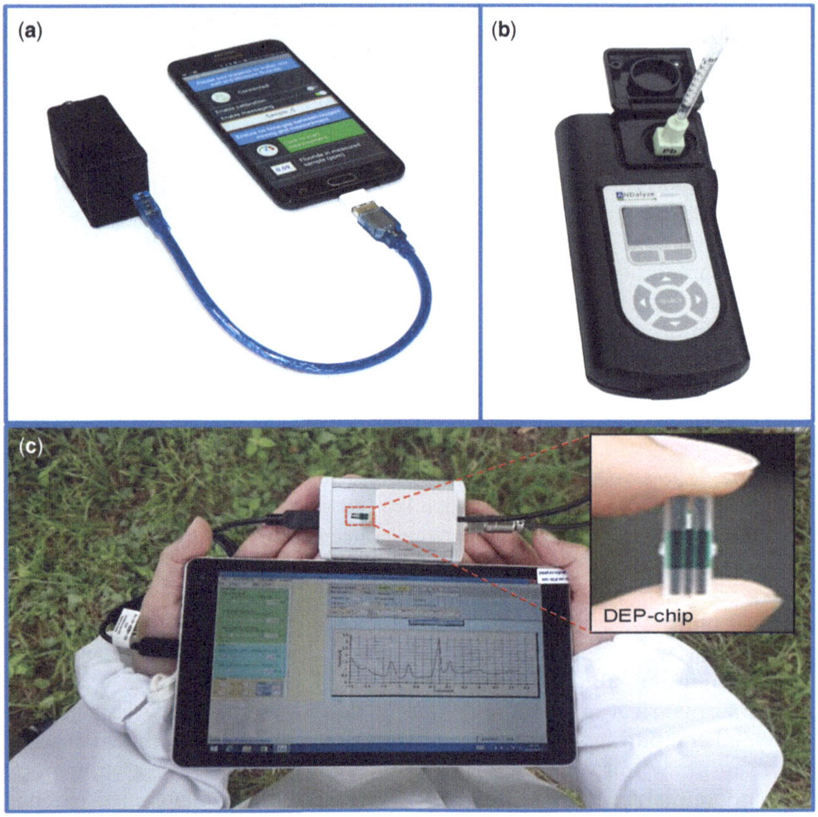

Figure 20.1 Portable devices developed for water quality analysis in field environments. (a) Smartphone based colorimeter for the testing of fluoride in groundwater (Mukherjee *et al.*, 2020). (b) AND 1100 Fluorometer utilised to determine toxicity levels of lead, copper, uranium, mercury, zinc, and cadmium (Reproduced with permission from Alpha Measurement Solutions). (c) DEP-On-Go (Biyani *et al.*, 2017).

multiparametric measurements and understanding different contributions to the water composition. Based on literature survey and an explanation of technical jargon related to area, this chapter can be concluded by looking at the possibility of smartphone-integrated sensors which will be used to indicate potability of water.

REFERENCES

Adeli A., Khoshnood R. S., Beyramabadi S. A., Pordel M. and Morsali A. (2022). Multivariate optimisation of a novel potentiometric sensor to determine silver ions in real water and pharmacological product samples. *Monatshefte Für Chemie – Chemical Monthly*, **153**(3), 227–235, https://doi.org/10.1007/s00706-022-02904-0

Bauer M., Wunderlich L., Weinzierl F., Lei Y., Duerkop A., Alshareef H. N. and Baeumner A. J. (2021). Electrochemical multi-analyte point-of-care perspiration sensors using on-chip three-dimensional graphene electrodes. *Analytical and Bioanalytical Chemistry*, **413**(3), 763–777, https://doi.org/10.1007/s00216-020-02939-4

Bengtsson L. (2010). The global atmospheric water cycle. *Environmental Research Letters*, **5**(2), 025202, https://doi.org/10.1088/1748-9326/5/2/025202

Biyani M., Biyani R., Tsuchihashi T., Takamura Y., Ushijima H., Tamiya E. and Biyani M. (2017). DEP-on-go for simultaneous sensing of multiple heavy metals pollutants in environmental samples. *Sensors*, **17**(1), 45.

Budlayan M. L., Dalagan J., Lagare-Oracion J. P., Patricio J., Arco S., Latayada F., Vales T., Baje B., Alguno A. and Capangpangan R. (2022). Detecting mercury ions in water using a low-cost colourimetric sensor derived from immobilized silver nanoparticles on a paper substrate. *Environmental Nanotechnology, Monitoring & Management*, **18**, 100736, https://doi.org/10.1016/j.enmm.2022.100736

Cuartero M., Ruiz A., Oliva D. J. and Ortuño J. A. (2017). Multianalyte detection using potentiometric ionophore-based ion-selective electrodes. *Sensors and Actuators B: Chemical*, **243**, 144–151, https://doi.org/10.1016/j.snb.2016.11.129

Delgado A., Andy J., Alfredo J., Cesar J. and Carbajal-Mancilla C. (2021). Grey clustering method for water quality assessment to determine the impact of mining company, Peru. *International Journal of Advanced Computer Science and Applications*, **12**(January), 557–564.

Geist M., Scherrer B. and Pietquin O. (2019). A theory of regularized Markov decision processes. Proceedings of the 36th International Conference on Machine Learning, K. Chaudhuri and R. Salakhutdinov (eds.), **97**, 2160–2169. Proceedings of Machine Learning Research. PMLR. https://proceedings.mlr.press/v97/geist19a.html

Hara T. O. and Singh B. (2021). Electrochemical biosensors for detection of pesticides and heavy metal toxicants in water: recent trends and progress. *ACS ES&T Water*, **1**(3), 462–478, https://doi.org/10.1021/acsestwater.0c00125

Hermes N., Jewell K. S., Wick A. and Ternes T. A. (2018). Quantification of more than 150 micropollutants including transformation products in aqueous samples by liquid chromatography-tandem mass spectrometry using scheduled multiple reaction monitoring. *Journal of Chromatography A*, **1531**, 64–73, https://doi.org/10.1016/j.chroma.2017.11.020

Jana S. K., Chaudhari K., Islam M. R., Natarajan G., Ahuja T., Som A., Paramasivam G., Raghavendra A., Sudhakar C. and Pradeep T. (2022). Selective and practical graphene-based arsenite sensor at 10 ppb. *ACS Applied Nano Materials*, **5**(8), 11876–11888, https://doi.org/10.1021/acsanm.2c02860

Khan M. M. R., Mitra T. and Sahoo D. (2020). Metal oxide QD based ultrasensitive microsphere fluorescent sensor for copper, chromium and iron ions in water. *RSC Advances*, **10**(16), 9512–9524, https://doi.org/10.1039/C9RA09985A

Liu C., Sun Z., Zhang L., Lv J., Yu X. F., Zhang L. and Chen X. (2018). Black phosphorus integrated tilted fiber grating for ultrasensitive heavy metal sensing. *Sensors and Actuators B: Chemical*, **257**, 1093–1098, https://doi.org/10.1016/j.snb.2017.11.022

Lu Y., Yu G., Wei X., Zhan C., Jeon J. W., Wang X., Jeffryes C., Guo Z., Wei S. and Wujcik E. K. (2019). Fabric/multi-walled carbon nanotube sensor for portable on-site copper detection in water. *Advanced Composites and Hybrid Materials*, **2**(4), 711–719, https://doi.org/10.1007/s42114-019-00122-7

Magro C., Mateus E. P., Paz-Garcia J. M., Sério S., Raposo M. and Ribeiro A. B. (2019). Electronic tongue coupled to an electrochemical flow reactor for emerging organic contaminants real time monitoring. *Sensors*, **19**(24), 5349, https://doi.org/10.3390/s19245349

Maguire I., Fitzgerald J., McPartlin D., Heery B., Murphy C., Nwankire C., O'Kennedy R., Ducrée J. and Regan F. (2017). A centrifugal microfluidic-based approach for multi-toxin detection for real-time marine water-quality monitoring. OCEANS 2017 – Aberdeen, 1–8, https://doi.org/10.1109/OCEANSE.2017.8084975

Mahapatra H. S., Patel R. and Panda B. (2012). Prediction of water quality using principal component analysis. *Water Quality Exposure and Health*, **4**(May), 93–104, https://doi.org/10.1007/s12403-012-0068-9

Mao S., Chang J., Zhou G. and Chen J. (2015). Nanomaterial-enabled rapid detection of water contaminants. *Small*, **11**(40), 5336–5359, https://doi.org/10.1002/smll.201500831

Motalebizadeh A., Bagheri H., Asiaei S., Fekrat N. and Afkhami A. (2018). New portable smartphone-based PDMS microfluidic kit for the simultaneous colorimetric detection of arsenic and mercury. *RSC Advances*, **8**(48), 27091–27100, https://doi.org/10.1039/C8RA04006K

Mukherjee S., Shah M., Chaudhari K., Jana A., Sudhakar C., Srikrishnarka P., Islam M. R., Philip L. and Pradeep T. (2020). Smartphone-based fluoride-specific sensor for rapid and affordable colorimetric detection and precise quantification at sub-Ppm levels for field applications. *ACS Omega*, **5**(39), 25253–25263, https://doi.org/10.1021/acsomega.0c03465

Nemiroski A., Christodouleas D. C., Hennek J. W., Kumar A. A., Maxwell E. J., Fernández-Abedul M. T. and Whitesides G. M. (2014). Universal mobile electrochemical detector designed for use in resource-limited applications. *Proceedings of the National Academy of Sciences*, **111**(33), 11984–11989, https://doi.org/10.1073/pnas.1405679111

Özbek O., Kalay E., Cetin A., Isildak O., Tokalı F. S., Aslan O. N. and Isildak I. (2022). Synthesis and potentiometric sensor applications of a hydrazone derivative molecule: fast and selective determination of cobalt(II) ions. *ChemistrySelect*, **7**(28), 202201307, https://doi.org/10.1002/slct.202201307

Parinet B., Lhote A. and Legube B. (2004). Principal component analysis: an appropriate tool for water quality evaluation and management – application to a tropical lake system. *Ecological Modelling*, **178**(3), 295–311, https://doi.org/10.1016/j.ecolmodel.2004.03.007

Pras A. and Mamane H. (2023). Nowcasting of fecal coliform presence using an artificial neural network. *Environmental Pollution*, **326**, 121484, https://doi.org/10.1016/j.envpol.2023.121484

Rubinger L., Gazendam A., Ekhtiari S. and Bhandari M. (2022). Machine learning and artificial intelligence in research and healthcare, *Injury*, **54**, S69–S73, https://doi.org/10.1016/j.injury.2022.01.046

Saravanan A., Kumar P. S., Hemavathy R. V., Jeevanantham S., Harikumar P., Priyanka G. and Devakirubai D. R. A. (2022). A comprehensive review on sources, analysis and toxicity of environmental pollutants and its removal methods from water environment. *Science of the Total Environment*, **812**, 152456, https://doi.org/10.1016/j.scitotenv.2021.152456

Singh J., Yadav P., Pal A. K. and Mishra V. (2020). Water pollutants: origin and status. In: Sensors in Water Pollutants Monitoring: Role of Material, D. Pooja, P. Kumar, P. Singh and S. Patil (eds.), Springer Singapore, Singapore, pp. 5–20.

Song E. and Choi J. W. (2015). Multi-analyte detection of chemical species using a conducting polymer nanowire-based sensor array platform. *Sensors and Actuators B: Chemical*, **215**, 99–106, https://doi.org/10.1016/j.snb.2015.03.039

Vo-Van T., Nguyen-Hai A., Tat-Hong M. V. and Nguyen-Trang T. (2020). A new clustering algorithm and its application in assessing the quality of underground water. *Scientific Programming*, 2020, 6458576, https://doi.org/10.1155/2020/6458576

Wong H. and Hu B. Q. (2013). Application of interval clustering approach to water quality evaluation. *Journal of Hydrology*, **491**, 1–12, https://doi.org/10.1016/j.jhydrol.2013.03.009

Yasmin M. K., Al-Rekabi S. H., Bakar M. H. A., Fen Y. W., Mohammed H. A., Halip N. H. M., Alresheedi M. T. and Mahdi M. A. (2022). Arsenic detection using surface plasmon resonance sensor with hydrous ferric oxide layer. *Photonic Sensors*, **12**(3), 220306, https://doi.org/10.1007/s13320-021-0643-4

Zamora-Ledezma C., Negrete-Bolagay D., Figueroa F., Zamora-Ledezma E., Ni M., Alexis F. and Guerrero V. H. (2021). Heavy metal water pollution: a fresh look about hazards, novel and conventional remediation methods. *Environmental Technology & Innovation*, **22**, 101504, https://doi.org/10.1016/j.eti.2021.101504

Zhang Z. and Karimi-maleh H. (2023). In situ synthesis of label-free electrochemical aptasensor-based sandwich-like AuNPs/PPy/Ti3C2Tx for ultrasensitive detection of lead ions as hazardous pollutants in environmental fluids. *Chemosphere*, **324**(March), 138302, https://doi.org/10.1016/j.chemosphere.2023.138302

Zhao G., Wang X., Liu G. and Thuy N. T. D. (2022). A disposable and flexible electrochemical sensor for the sensitive detection of heavy metals based on a one-step laser-induced surface modification: a new strategy for the batch fabrication of sensors. *Sensors and Actuators B: Chemical*, **350**, 130834, https://doi.org/10.1016/j.snb.2021.130834

doi: 10.2166/9781789063714_0233

Chapter 21

A point-of-use single probe multi-analyte sensor

*J. Narayanan, P. Yuvaraj, J. B. Gowthaman, M. Kalipriyadharshini, J. Raghava Rao and S. Easwaramoorthi**

Inorganic & Physical Chemistry Laboratory, CSIR-Central Leather Research Institute, Chennai, India
**Corresponding author: moorthi@clri.res.in*

ABSTRACT

Water is an essential molecule for all life on earth and it carries numerous other species including different elements and neutral molecules which necessarily decide its usability for a variety of applications. The presence of external species in water occurs either naturally or due to anthropogenic activities. Though some of them are essential nutrients for living organisms, many of them might lead to adverse health effects. Rapid industrialization and the ever-increasing human population have resulted in an increased demand for potable water. Unfortunately, the scarcity of quality potable water owing to the depletion of resources in several countries demands the identification of new resources and day-to-day water quality monitoring. Indeed, an inadvertent mix of sewage and domestic wastewater in drinking water resources also contributes to water pollution, in addition to industrial waste which is a predominant water-polluting activity. Thus, continuous water quality monitoring at the source is essential which could be achieved by the systematic analysis of the contaminants present therein, even in remote places. While the traditional analytical techniques are quite expensive and hence cannot be deployed in remote places, molecule or material-based identification, that is, sensing of specific target species has been found to be the cost-effective method. Unfortunately, most of the probes that have been developed were specific to one target, which becomes a drawback when analysing the actual sample that would probably consist of several toxic species. Thus, simultaneous analysis of different toxic species with distinctly different signalling pathways is required as a practically viable solution. This chapter will focus on the importance of water quality monitoring, methods, and sensors that operate on colorimetric, luminescent, and electrochemical principles. The toxic species present in the water will be restricted to the metal ions. Furthermore, the molecules which sense more than one or two toxic species simultaneously will be incorporated and the reports of the integrated sensor arrays will also be included.

Keywords: multianalyte sensor, lab-on-a-molecule, fluorescent sensors, colorimetric sensors, orthogonal sensing

21.1 INTRODUCTION

Water is an essential molecule for life on the planet and has the propensity to carry ions to molecules to materials in its flow. Thus, water resources across the planet are not water molecules alone and they contain numerous dissolved and suspended species, including several elements, neutral molecules, and biomolecules. The existence of external species in water occurs naturally or through anthropogenic activities, and indeed, some of them are essential nutrients for aquatic and other living organisms, and in fact, those dissolved or suspended species decide the quality of the water and its usage. While some of the dissolved species, for instance, the transition metal ions, are essential for living organisms, their presence in even slightly excess quantity combined with their continuous consumption has proven to be responsible for adverse health effects. For example, ions like arsenic and fluoride are naturally present at high levels in groundwater which is being used for drinking, cooking, and irrigation of food crops, resulting in continuous exposure, thereby leading to cancer, skin lesion, fluorosis, arthritis, infertility, hypertension, and neurotoxicological effects. Rapid industrialization and ever-increasing human population further strain all water resources by letting toxic contaminants into it, and thus depleting potable water resources to a large extent. On the contrary, the requirement for potable, clean water is continuously on the rise to meet the demands for food production, drinking water, and industrial usage. Thus, coherent action is needed for the conservation, maintenance, and remediation of water for drinking purposes.

Analytical chemistry, one of the four traditional fields of chemical science, plays a pivotal role in qualitative and quantitative analysis by virtue of instrumental and chemical analysis. Between them, chemical analysis is considered to be simple, cost-effective, and field-deployable when compared to the more expensive instrumental analysis, though the latter has the advantage of rendering precise information even at the lowest level of abundance. Thanks to the continuous efforts by several research groups, several molecules or materials referred to as 'sensors' could quantitatively or qualitatively identify the specific components in a medium with greater specificity. By definition, according to the International Union of Pure and Applied Chemistry, the chemical sensor is a device that transforms chemical information, ranging from the concentration of a specific sample component to total composition analysis, into an analytically useful signal as shown in Figure 21.1.

The chemical information may originate from a chemical reaction of the analyte or a physical property of the system investigated. Indeed, the changes driven by the single analyte are more than a century-old technique, that is, colorimetric tests, and the most popular example is the litmus test. In principle, the sensor comprises two basic functional units: a receptor part and a transducer part.

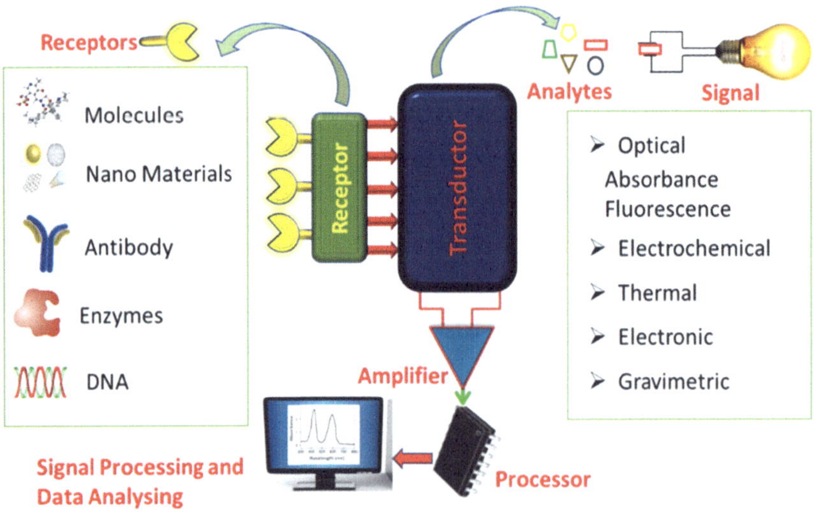

Figure 21.1 A generic sensor design based on the different principles and materials of construction.

The receptor transforms the chemical information into a form of energy that may be measured by the transducer which turns the energy into a useful analytical signal. The selectivity is defined by the receptor component, and the transducer, as an individual entity, does not show selectivity. The chemosensor represents the dynamic interaction between the analyte and the sensor, which is reversible and labile, while the chemodosimeter represents the irreversible reaction operating under kinetic control. Since the working principle is the dynamic interaction for chemosensors, the sensors designed to operate for the specified analysis under defined conditions would not necessarily respond to the same analyte under different conditions. This feature has been one of the challenging bottlenecks for the practical utility of sensors in water samples from the actual source and it might lead to misleading information. Another challenge would be that countless sensors are focused on single analytes with greater specificity and sensitivity which become an impediment to working under realistic conditions where enormous species coexist in water and might require a series of tests for individual entities to be performed to comprehend the overall water quality.

Considering the drawbacks of single analyte-specific sensors or molecular probes, chemists have developed methodologies to implement multianalyte sensing using molecular devices by virtue of constructing the sensor arrays in solution or on the surface. Several concepts in this regard such as lab-on-a-molecule, sensor arrays, displacement assay, relay assay, keypad lock and so on have been developed (Chen *et al.*, 2015). While the relay and keypad assay are found to have outstanding performance but needing sequential addition of analytes, lab-on-molecule probes respond to all the analytes in a single pot reaction, independent of the addition sequence. Nevertheless, the challenge lies in molecular design that exhibits non-interfering orthogonal and quantitative detection of two or more analytes by a single probe. The molecular design could be (1) a single transducer linked with several analyte-specific receptors, and (2) a single transducer coupled with a single receptor wherein different conditions (time, temperature) are to be adopted. It has been known that these probes allow qualitative information but fail in quantification under competitive conditions and the probes that allow quantification are conceived on multiplex sensing. One of the most common methodologies in avoiding signal interference and enabling orthogonal signalling would be aiming at different wavelengths in spectrophotometric methods, different potential ranges in electrochemical methods, and/or using both spectrophotometric and electrochemical methods together. Furthermore, the cross interferences could also be distinguished using mathematical algorithms using principal component analysis, artificial neural networks, and partial least-squares regression. A careful analysis of the published literature using Scifinder (accessed in January 2023) with the keyword of single probe multianalyte sensor yielded around 1300 reports till 2022 and the year-wise distribution is given Figure 21.2, but the numbers are intriguingly lower when compared to the one obtained with keyword 'chemosensor' or 'sensor'. This probably highlights the complexity in developing the single probe multianalyte sensors with the necessary structural and functional features. Indeed, the predominant portion of the literature retrieved with these keywords is focused on multiple parameters for the biological system rather than environmental monitoring. Nonetheless, this study describes the molecular designs, principles adopted for multianalyte, orthogonal sensing based on colorimetric, fluorescence, and electrochemical methods.

21.2 FLUORESCENT SENSORS

Fluorescence is one of the most prominent principles that offers disruptive technologies and simplified methods or instruments that does away with the expensive, traditional, instrument-intensive methods, owing to its excellent sensitivity with stronger signal at the lowest level of abundance. Most of the sensors reported to date work on the modulation of fluorescent properties such as intensity, fluorescence quantum yield, Stokes shift, fluorescence lifetime, and spectral shifts. Fluorescence intensity is an arbitrary unit, not governed by any laws unlike its complementary technique absorption, and it cannot be employed for quantification. However, ratiometric analysis obtained from two different,

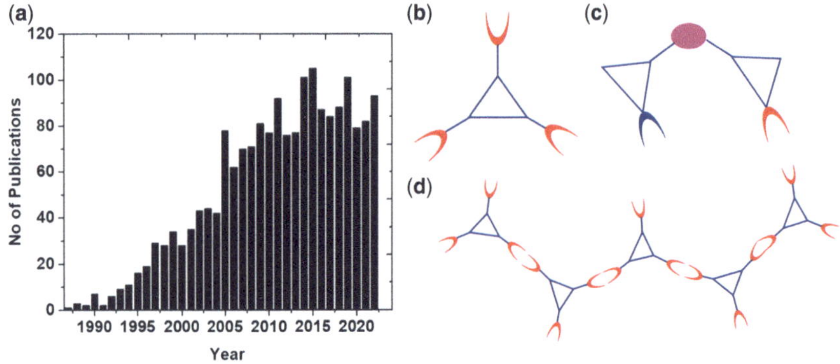

Figure 21.2 (a) Publication profile accessed from SciFindern data base (American Chemical Society) using the key word of single probe multianalyte sensor, sensor design with (b) single transducer with multiple receptors, (c) two transducers coupled with individual receptors, and (d) possible supramolecular assembly of the receptors.

interdependent fluorescence spectral bands extends its capability for quantification. In general, the fluorescent sensors can be divided into two types, turn-off, wherein the intense fluorescence of the sensor could be quenched in the presence of analytes, and turn-on sensor which describes the fluorescence intensification in the presence of analytes. The alternations in photoinduced electron transfer, resonance energy transfer (FRET or RET), charge transfer, chelation-enhanced fluorescence (CHEF), aggregation-induced emission (AIE), rigidification to reduce non-radiative decay channels are the major approaches to find the mechanistic pathways. On a molecular front, for multianalyte sensing, the strategies can be broadly classified into three categories, namely, single transducer coupled with more than two receptors, two transducers coupled through a common linker, which may contain the same or different receptors, and supramolecularly organized transducer–receptor system, which has been schematically represented in Figure 21.3.

An example of the single transducer with a gradually increasing number of receptors is represented by the triphenylamine, rhodanine-3-acetic acid donor–acceptor type molecules **1**, **2**, and **3**, respectively, called dipolar, quadrupolar, and octupolar geometries and are given in Figure 21.3. All the compounds

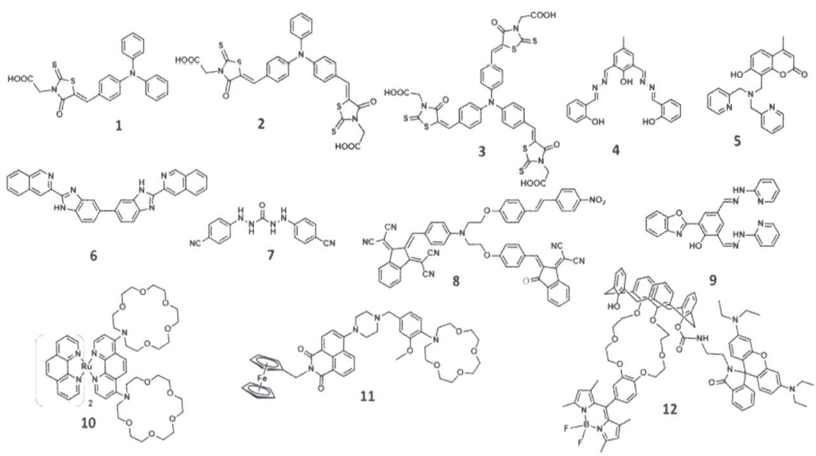

Figure 21.3 Structures of the fluorescent sensors.

exert intense absorption in the visible region owing to the intramolecular charge transfer (ICT) interactions and the rhodanine-3-acetic acid moiety acts as a receptor and its binding with the metal ions influences ICT behaviour, which could be evident from the changes in absorption and fluorescence spectra. Among the several metal ions tested, all the compounds were sensitive to Hg^{2+}, Ag^{+}, respectively, detected with colourless, and intense-red fluorescence under UV light illumination. While **2** and **3** are expected to bind with several other metal ions or exhibit different stoichiometries, owing to the free receptor groups, all of them maintain the same stoichiometry. This feature highlights that the availability of the free receptor site itself is insufficient to have multimodal binding, the successive binding using the second or third receptor is prevented by the initial complexation wherein the changes in the electronic properties of the complex do not allow the subsequent binding. This is an excellent example of negative cooperativity. However, the additional receptor does improve the probe solubility in water and complexation constants (Thamaraiselvi *et al.*, 2019). An azine-based smart probe, **4** detects the iodide with the intense-red colour in tetrahydrofuran solution, green emission for Al^{3+} ion in aqueous methanolic, and ratiometric yellow emission for Zn^{2+} in dimethyl sulphoxide (DMSO) solution. Indeed, probe **5**, a dipyridyl derivative of fluorescent coumarin, was able to bind with Cd^{2+}, Hg^{2+}, and Pb^{2+} under acidic, neutral, and basic pH conditions in an aqueous solution. Its inherent fluorescence spectral sensitivity to the surrounding pH and its insensitivity towards the largely abundant Ca^{2+} and Mg^{2+} allows the probe used to quantify multiple analytes. Indeed, **4** and **5** are the best examples to depict the role of solvent or pH in altering the target metal ions with different signals (Bowyer *et al.*, 2022; Nandi & Das, 2016). Nevertheless, it highlights the risk of expecting the same signal for the analytes under different conditions and expecting the workability of the sensing methodologies developed under the influence of partial or complete organic solvents into an aqueous solution. Probe **6**, a multiple benzimidazole and quinoline moieties with abundant nitrogen available for metal ion binding, is selective for Cu^{2+}, Hg^{2+}, and Zn^{2+} in 9:1 DMSO/H_2O (volume/volume) solution. The colourless solution turned to pale green with Cu^{2+} ion and an intense-blue fluorescence was quenched with Hg^{2+} and turned to green for Zn^{2+} (Pang *et al.*, 2019). The interaction between the Hg^{2+}, Cu^{2+} with probe **6** is the ground state complexation, thus, can also be used as a colorimetric probe, but Zn^{2+} complexation does not alter the absorption spectra, possibly due to its d^{10} electronic configuration, but rigidifies the interaction between quinoline and benzimidazole core to restrict the non-radiative decay channels with improved fluorescence intensity. Hydrazine-based chromogenic, fluorogenic, and electrochemical chemosensor **7** was able to detect, Hg^{2+}, Cu^{2+}, and F^{-} ions in a methanolic water medium. A colourless solution turns to an intense-blue solution in the presence of fluoride ion originated from the deprotonation of acidic, NH proton, and thus forms a charge transfer complex responsible for the intense absorption. Hg^{2+} and Cu^{2+} ions also turn the colourless solution to light blue and purple along with the turn-on fluorescence. The turn-on fluorescence would have been due to the rigidification of the probe after the anion or cation binding and also prevention of non-radiative photo-induced electron transfer reactions (Mondal *et al.*, 2019). A slightly different approach with the receptor containing three donor atoms, S, O, and N, having three different HSAB (hard and soft acid base) coupled with three chromogenic units, **8** responds to individual transition metal ions with different signalling mechanisms (Komatsu *et al.*, 2005). The preferential interaction of metal ions with S, O, and N depicts the direct interaction with different chromogenic units, for instance, Cu^{2+}, Fe^{3+}, Al^{3+}, Pb^{2+} ions induce varied changes with the dye attached to the nitrogen, that is, all these metal ions possess semi-selectivity. Furthermore, the comparatively stronger interaction of Pb^{2+} ion with S, and Cu^{2+} with S and O, leads to the overall signal differences; hence, the concept could be adopted for multiplex sensing. The probe **9**, a hydrazine-based dye coupled with the benzoxazole, selectively targets trivalent Al^{3+}, Cr^{3+}, and Fe^{3+} while being silent on mono- and divalent metal ions. The weakly emissive probe **9** became strongly emissive along with larger spectral shift after the binding of trivalent cations. Specifically, for Cr^{3+} and Al^{3+}, a naked eye distinguishable green fluorescence was observed with the former having nearly double fluorescence quantum yield than the latter. The intrinsic, excited state intramolecular proton transfer (ESIPT) characteristic of the probe has weaker fluorescence owing to the energy wasting isomerization around C–N bond which gets arrested when the receptor binds with the metal ion, thereby facilitating the ESIPT process (Wang *et al.*, 2014). A quadruple-channel sensing setup with UV–visible absorption, photoluminescence, electrogenerated chemiluminescence (ECL), and redox properties of bis-heteroleptic Ru(II) complex substituted with the crown ethers was demonstrated. The lab-on-a-molecule **10** was selective for Pb^{2+} through changes in the redox potential, turn-on intense orange luminescence, for Cu^{2+} turn-on red luminescence, absorption changes in the visible and NIR spectral region, and with Hg^{2+}, electrogenerated chemiluminescence was observed (Schmittel & Lin, 2007). A lab-on-a-molecule with the structural design of electron donor-fluorophore-receptor 1-receptor 2 was constructed using ferrocene, naphthalimide, pyrazine and crown ether, that is, **11**. Interestingly, it detects three chemical species, that is, H^+, Na^+, and Fe^{3+} even present simultaneously in aqueous methanol. The reduction of Fe^{3+} to Fe^{2+} by ferrocene resulted in the fluorescence enhancement, similarly the protonation of pyrazine nitrogen, and complexation of Na^+ in the crown ether alters the electron-donating behaviour, thereby responsible for the fluorescence enhancement. From the data obtained at different concentration levels of the individual analytes tested, it appears that the overall effect is additive in nature (Scerri *et al.*, 2019). An interesting molecular construct with fluorophores BODIPY and rhodamine moiety covalently linked to the calix[4]arene forms a FRET sensor **12** selective towards Ba^{2+} and Hg^{2+}. The PET from the crown ether to the BODIPY resulted in the non-fluorescent BODIPY, but the PET becomes turned off when Ba^{2+} ion complexes with the crown ether, leading to intense-green fluorescence. If Hg^{2+} is added to the Ba^{2+} containing **12**, then the solution shows intense-red fluorescence, owing to the resonance energy transfer from BODIPY to rhodamine. Incidentally, the non-fluorescent, spirocyclic rhodamine became fluorescent by spirocyclic ring-opening reaction which could be fluorescent even without the Ba^{2+} ions (Yuan *et al.*, 2008).

Supramolecularly assembled structures of the probes were found to be more promising in identifying the multiple analytes than the individual probes in the neat solution. The probes as an individual entity or in combination with the other motif form self-assembled hybrid structures with distinctly different optoelectronic properties, which expands the scope of the analytes being detected with greater accuracy. Recently, the chromophores that exert aggregation-induced emission enhancement (AIEE), called AIEgens, are considered to be the prominent emitters with diverse applications including sensing. An ionic supramolecular gel based on pyridinium functionalized-naphthalimide forms gels with n-pentanoic acid as shown in Figure 21.4a, which could detect Hg^{2+} and arginine which could be tunable to target the different metal ions as well. For example, the classical AIEgen tetraphenylene ethylene (TPE) substituted ionic liquid with the ionic liquid crosslinker forms poly(ionic liquid)photonic spheres (Lin *et al.*, 2017) as depicted in Figure 21.4b. The existence of abundant intermolecular interaction aided by the ionic liquids indicates that the single sphere could interact with the different analytes broadly to provide the multidimensional signals.

The photonic sphere is able to detect and discriminate four multianalyte systems, including natural amino acids, phosphates, metal ions, and enantiomers. It is able to detect and discriminate Al^{3+}, Ba^{2+}, Cd^{2+}, Co^{2+}, Cu^{2+}, Hg^{2+}, Ni^{2+}, Pb^{2+}, Sn^{2+}, and Zn^{2+}. The photonic spheres are versatile and can be altered with different AIEgens, different anions or cations of ionic liquid which offer the materials diverse functionalities (Zhang *et al.*, 2017). Another approach to develop AIEgen-based supramolecular assembly was reported based on pillar[5]arene (Zhang *et al.*, 2018). A host–guest interaction leads to the supramolecular assembly strengthened by π–π interactions with a blue-white AIE emission. The assembly was able to detect and separate analytes such as Fe^{3+}, Hg^{2+}, Ag^+, F^-, and Br^- with greater sensitivity and also the cations in the solution were absorbed by the gel from 90 to 99% (Liu *et al.*, 2019). Sometimes, the simple chromogenic units substituted with the long alkyl chain form supramolecular assembly through π–π stacking and hydrogen bonds and van der Waals forces and could be used for the ultrasensitive detection and separation of multianalytes in the gel state. An excellent separation of Hg^{2+} and Fe^{3+} from water was demonstrated and in fact these metallogels were also able to selectively detect CN^- and $H_2PO_4^{2-}$ in water (Yin *et al.*, 2019). The naphthylhydrazone-based organogelator, a

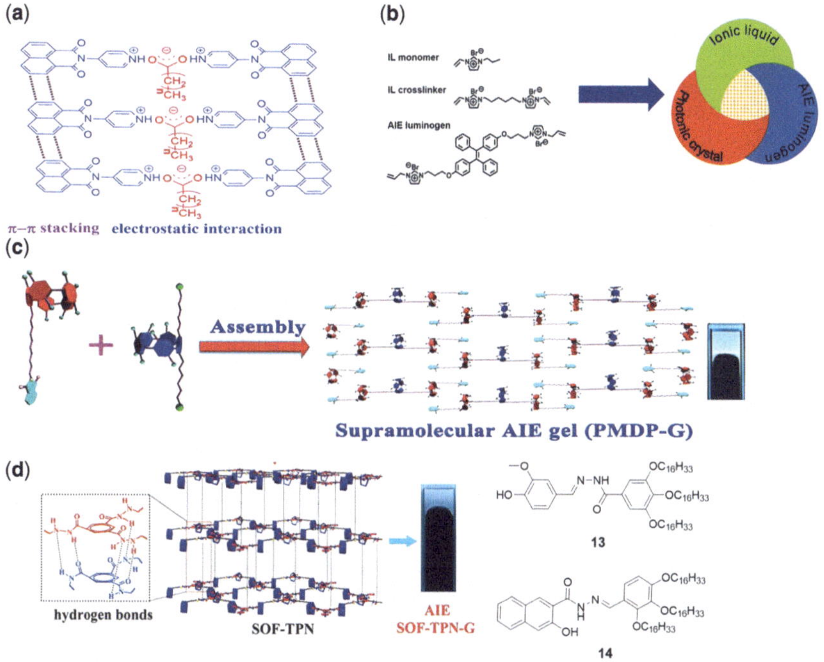

Figure 21.4 (a) supramolecular assembly of NDI with the pentanoic acid (copyright, RSC), (b) AIE-doped poly(ionic) photonic spheres (copyright, RSC), (c) supramolecular aggregation-induced emission gels based on pillar[5]arene (copyright, ACS), and (d) supramolecular organic framework (AIE SOF) gels constructed from supramolecular polymer networks based on tripodal pillar[5]arene (copyright, ACS). All the images are reproduced with permission.

supramolecular gel based 22-member sensor array, has been created and it requires only one synthesized receptor. The sensor array has been shown to accurately identify 14 kinds of important ions (F^-, Cl^-, I^-, CN^-, HSO_4^-, SCN^-, S^{2-}, OH^-, Al^{3+}, Fe^{3+}, Zn^{2+}, Hg^{2+}, Pb^{2+}, and H^+) in water. It is possible to make different metallosupramolecular assemblies with these metal ions and the resultant materials could also be used as display materials as well as security documents (Lin *et al.*, 2016). An interesting approach employed by Lin *et al.* (2016) has been the use of nano or microstructural materials comprising of three-dimensional structural colours in colloidal poly(styrene-methyl methacrylate-acrylic acid) (poly(St-MMA-AA)) nanoparticle-assembled photonic crystals (PCs). Morphological changes constructed with geometrical micromorphologies have a significant impact on the structural colour, though the mechanistic understanding is still elusive while methodology seems to be promising. The PC self-assembly is able to discriminate 14 metal cations and distinguish the 12 ground water samples with similar constitutions. Indeed, the printability of these multianalyte chips makes them ideal for the scale up to the commercial level (Huang *et al.*, 2021).

Although multianalyte sensing is also common with electrochemical sensing, more focus was on analytes of biological importance rather than the environmental monitoring. The recent effort highlights the possibility of multiple functional groups integrated on a single platform to target the multiple analytes in a one pot analysis. In this case, the graphene oxide is functionalized with thymine and carbohydrazide to target Cr^{6+} and Hg^{2+} with different signals, without any interference from each other (Jayaraman *et al.*, 2022). The limit of detection has gone down to parts per billion level and the fabrication of handheld device could be field deployable. The conceptual proof promises to have a multianalyte sensing platform with field deployability, cost effectiveness, and easy operability.

21.3 CONCLUSIONS AND PERSPECTIVE

Although a great amount of effort has been documented in developing the low-cost, easily field deployable, sensors or lab-on-a-molecule, still a lot needs to be explored for practical application. The reports on the single analyte sensors are prevalent, the ones for the multiple analytes are relatively scarce, yet there are significant reports highlighting the conceptual proof for developing those. The easy-to-use point-of-use devices are generally based on colorimetric, fluorometric, and electrochemical principles, as they offer simplified sampling protocol, results analysis, selectivity and ultra-low detection level with highest signal intensities. One of the major issues for multianalyte sensors is signal overlap or the similar signal for the different analytes, and in some cases, though the signals are different, one of the analytes masks the other. Thus, a successful multianalyte sensor might have different detection channels with orthogonal binding, reaction mechanism, and time. A careful analysis of the literature also hints that compared to the single probe in the neat solution, its supramolecular assemblies are apparently prominent to offer the versatility for multianalyte detection and ease of varying the supramolecular structure and thereby function to capture the wider analytes in the environment.

ACKNOWLEDGEMENTS

The work is financially supported by the project of Centre for Sustainable Treatment, Reuse, and Management of Water (WATER-IC for SUTRAM of Water, DST/TM/WTI/WIC/2017/82), Department of Science and Technology, India.

REFERENCES

Bowyer A. A., Mai A. D., Guo H. and New E. J. (2022). A pH-based single-sensor array for discriminating metal ions in water. *Chemistry – An Asian Journal*, **17**(10), e202200204, https://doi.org/10.1002/asia.202200204

Chen K., Shu Q. and Schmittel M. (2015). Design strategies for lab-on-a-molecule probes and orthogonal sensing. *Chemical Society Reviews*, **44**(1), 136–160, https://doi.org/10.1039/C4CS00263F

Huang Y., Liu L., Yang X., Zhang X., Yan B., Wu L. and … Li F. (2021). A diverse micromorphology of photonic crystal chips for multianalyte sensing. *Small*, **17**(12), 2006723, https://doi.org/10.1002/smll.202006723

Jayaraman N., Palani Y., Jonnalagadda R. R. and Shanmugam E. (2022). Covalently dual functionalized graphene oxide-based multiplex electrochemical sensor for Hg(II) and Cr(VI) detection. *Sensors and Actuators B: Chemical* **367**, 132165, https://doi.org/10.1016/j.snb.2022.132165

Komatsu H., Citterio D., Fujiwara Y., Minamihashi K., Araki Y., Hagiwara M. and Suzuki K. (2005). Single molecular multianalyte sensor: jewel pendant ligand. *Organic Letters*, **7**(14), 2857–2859, https://doi.org/10.1021/ol0507219

Lin Q., Lu T.-T., Zhu X., Wei T.-B., Li H. and Zhang Y.-M. (2016). Rationally introduce multi-competitive binding interactions in supramolecular gels: a simple and efficient approach to develop multi-analyte sensor array. *Chemical Science*, **7**(8), 5341–5346, https://doi.org/10.1039/C6SC00955G

Lin Q., Mao P.-P., Fan Y.-Q., Jia P.-P., Liu J., Zhang Y.-M. and … Wei T.-B. (2017). Novel multi-analyte responsive ionic supramolecular gels based on pyridinium functionalized-naphthalimide. *Soft Matter*, **13**(40), 7360–7364, https://doi.org/10.1039/C7SM01624G

Liu J., Fan Y.-Q., Song S.-S., Gong G.-F., Wang J., Guan X.-W. and … Lin Q. (2019). Aggregation-induced emission supramolecular organic framework (AIE SOF) gels constructed from supramolecular polymer networks based on tripodal pillar[5]arene for fluorescence detection and efficient removal of various analytes. *ACS Sustainable Chemistry & Engineering*, **7**(14), 11999–12007, https://doi.org/10.1021/acssuschemeng.9b00452

Mondal A., Roy Chowdhury A., Bhuyan S., Mukhopadhyay S. K. and Banerjee P. (2019). A simple urea-based multianalyte and multichannel chemosensor for the selective detection of F^-, Hg^{2+} and Cu^{2+} in solution and cells and the extraction of Hg^{2+} and Cu^{2+} from real water sources: a logic gate mimic ensemble. *Dalton Transactions*, **48**(13), 4375–4386, https://doi.org/10.1039/C8DT05097J

Nandi S. and Das D. (2016). Smart probe for multianalyte signaling: solvent dependent selective recognition of I^-. *ACS Sensors*, **1**(1), 81–87, https://doi.org/10.1021/acssensors.5b00035

Pang C.-M., Luo S.-H., Jiang K., Wang B.-W., Chen S.-H., Wang N. and Wang Z.-Y. (2019). A dual-channel sensor containing multiple nitrogen heterocycles for the selective detection of Cu^{2+}, Hg^{2+} and Zn^{2+} in same solvent system by different mechanism. *Dyes and Pigments* **170**, 107651, https://doi.org/10.1016/j.dyepig.2019.107651

Scerri G. J., Spiteri J. C., Mallia C. J. and Magri D. C. (2019). A lab-on-a-molecule with an enhanced fluorescent readout on detection of three chemical species. *Chemical Communications*, **55**(34), 4961–4964, https://doi.org/10.1039/C9CC00924H

Schmittel M. and Lin H.-W. (2007). Quadruple-channel sensing: a molecular sensor with a single type of receptor site for selective and quantitative multi-ion analysis. *Angewandte Chemie International Edition*, **46**(6), 893–896, https://doi.org/10.1002/anie.200603362

Thamaraiselvi P., Duraipandy N., Kiran M. S. and Easwaramoorthi S. (2019). Triarylamine rhodanine derivatives as red emissive sensor for discriminative detection of Ag^{+} and Hg^{2+} ions in buffer-free aqueous solutions. *ACS Sustainable Chemistry & Engineering* **7**, 9865–9874, https://doi.org/10.1021/acssuschemeng.9b00417

Wang J., Li Y., Patel N. G., Zhang G., Zhou D. and Pang Y. (2014). A single molecular probe for multi-analyte (Cr^{3+}, Al^{3+} and Fe^{3+}) detection in aqueous medium and its biological application. *Chemical Communications*, **50**(82), 12258–12261, https://doi.org/10.1039/C4CC04731A

Yin Z.-Y., Hu J.-H., Fu Q.-Q., Gui K. and Yao Y. (2019). A novel long-alkyl-chained acylhydrazone-based supramolecular polymer gel for the ultrasensitive detection and separation of multianalytes. *Soft Matter*, **15**(20), 4187–4191, https://doi.org/10.1039/C9SM00624A

Yuan M., Zhou W., Liu X., Zhu M., Li J., Yin X. and … Zhu D. (2008). A multianalyte chemosensor on a single molecule: promising structure for an integrated logic gate. *The Journal of Organic Chemistry*, **73**(13), 5008–5014, https://doi.org/10.1021/jo8005683

Zhang W., Gao N., Cui J., Wang C., Wang S., Zhang G. and … Li G. (2017). AIE-doped poly(ionic liquid) photonic spheres: a single sphere-based customizable sensing platform for the discrimination of multi-analytes. *Chemical Science*, **8**(9), 6281–6289, https://doi.org/10.1039/C7SC02409F

Zhang Y.-M., Zhu W., Huang X.-J., Qu W.-J., He J.-X., Fang H. and … Lin Q. (2018). Supramolecular aggregation-induced emission gels based on pillar[5]arene for ultrasensitive detection and separation of multianalytes. *ACS Sustainable Chemistry & Engineering*, **6**(12), 16597–16606, https://doi.org/10.1021/acssuschemeng.8b03824

Section 5
Urban Water Management

INTRODUCTION

The urban population is projected to increase by twofold; by 2050, most of the world's population will live in cities because of rapid urbanization. Therefore, water management in urban areas, from the perspectives of floods, water scarcity, and domestic wastewater management, assumes significance. Thus, the primary focus of discussion in the present section is on water distribution networks, wastewater collection, and storm drainage networks, which are significant components of urban water infrastructure. Furthermore, in this age of interconnectedness, the management of water resources can be wholesome and sustainable only if the aspect of virtual water is also addressed. Therefore, concerns related to virtual water are also discussed in this section. The present section comprises five chapters.

The chapter on **Management of floods and droughts in urban environment** discusses different causes and effects of these extreme events in different regions. Different mitigation strategies need to be adopted in different regions. This critical aspect is brought forth through illustrative cases from India and the Republic of South Korea. The failure of conventional stormwater drainage networks to effectively tackle urban flooding has given rise to the paradigm of sustainable drainage systems or low-impact development (LID). While LID measures are being implemented routinely in cities in Global North, their implementation is very recent and scanty in India. Issues pertaining to LID measures are discussed in the chapter on **Urban stormwater management through sustainable drainage systems**. The provision of water infrastructure is costly yet highly important for making cities liveable, inclusive, and environmentally sustainable. However, water infrastructure planning, design and operation are challenging because of certain factors such as unpredictable urbanization, dwindling sources, and climate change. Important issues such as water allocation, multi-functional infrastructure, integrated planning, decentralization, flexible planning, etc., are discussed in the chapter on **Urban water infrastructure: distribution and collection**.

In India, challenges to the implementation of new projects and adaptation of new technologies depend on the following factors: (1) financial resources, (2) skilled personnel for planning, designing, and operation, and (3) implementation agencies for executing the project works. Challenges also arise because of inappropriate institutional structures and the absence of facilitating government policies. In multiple scenarios, societal acceptance is also not guaranteed. These issues are discussed in detail in the chapter on **Implementation challenges for new projects and technologies**.

In recent decades, there has been a significant increase in the exchange of virtual water, defined as water used in the production of commodities, due to a rapid rise in global trade. As water is indirectly becoming a traded commodity, sustainability in water management can only be achieved if due consideration is given to virtual water management. This is increasingly becoming important because several emerging economies, such as the water-poor India, are becoming major net virtual water exporters to water-rich counties in the Global North. The last chapter addresses the **concept of virtual water and presents the role of policy intervention in virtual water management**.

doi: 10.2166/9781789063714_0245

Chapter 22

Managing floods and droughts in urban environment: case studies from South Korea and India

Mariappan Rinisha Kartheeshwari[1], Puthan Veettil Razi Sadath[1], Boosik Kang[2] and Lakshmanan Elango[3]*

[1]*Department of Geology, Anna University, Chennai, India*
[2]*Department of Civil and Environmental Engineering, Dankook University, Gyeonggi-do, The Republic of Korea*
[3]*Department of Civil Engineering, Indian Institute of Technology Madras, Chennai, India*
**Corresponding author: elango34@hotmail.com*

ABSTRACT

Floods and droughts are two contrasting extreme events that have become a major concern especially due to climate change and urbanization. These problems are due to distinct causes and effects at various places requiring different mitigation strategies. The risks due to floods and droughts cannot be eliminated but can be reduced or minimized. Transfer of the surplus water to areas of water deficit is a possible way of managing floods and droughts together. This would also help to create additional irrigational potential and promote hydropower generation to overcome regional imbalances. Other measures such as enhancing surface storage, managed aquifer recharge, interlinking of water bodies, and developing green infrastructures are practised globally under different names to combat these two extremes. This chapter presents a detailed insight into the effects and mitigation strategies of floods and droughts in general and with the case studies from the Republic of Korea and India.

Keywords: floods, droughts, climate change, urbanization, mitigating strategies

22.1 INTRODUCTION

Extreme hydroclimatic conditions such as floods and droughts are common primarily in regions that receive rainfall due to monsoons. Floods typically occur over smaller areas in association with heavy rainfall and stream flows on shorter timescales (Dhar & Nandargi, 2003; Kale, 2012; Mishra *et al.*, 2012; Sharma *et al.*, 2018). They are the most common natural hazards in terms of levels of exposure in percentage across the 1860 cities worldwide (Figure 22.1). The worldwide spatial distribution of flood events is depicted in Figure 22.2. Every year, nearly 8 million hectares of land is affected by floods in India (Ray *et al.*, 2019) and in 2020, about 15,000 ha of land were affected in Korea as per the South Korean Ministry of Safety.

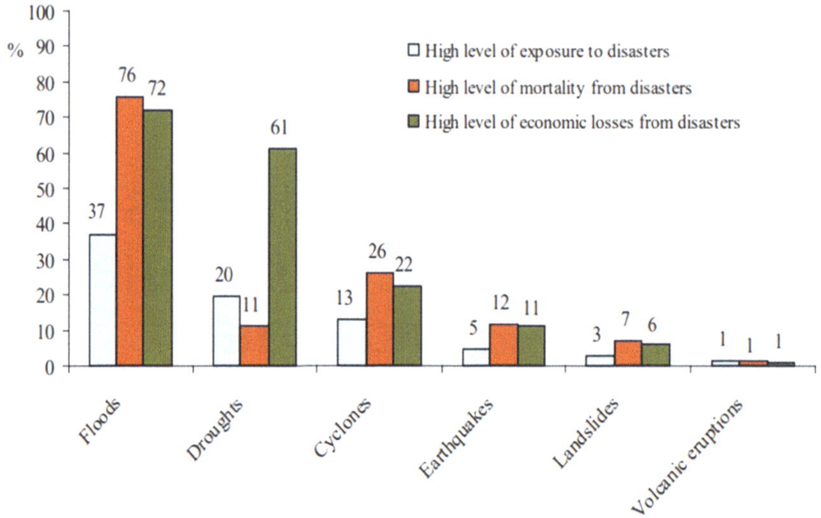

Figure 22.1 Percentage of number of cities with high level of exposure and vulnerability to six types of natural disasters. (*Source*: Gu, 2019).

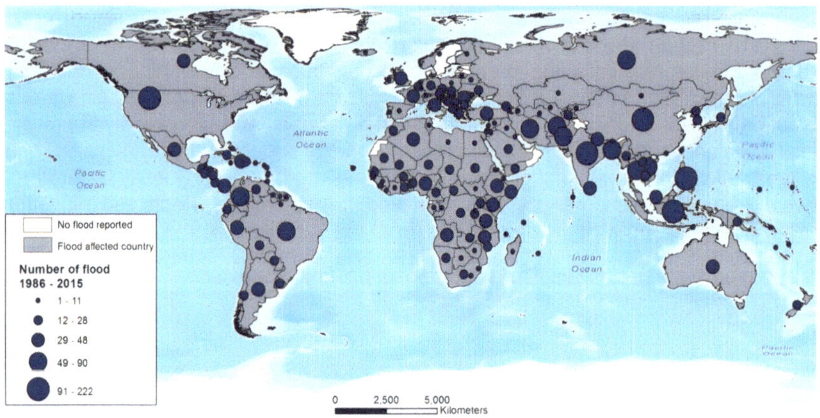

Figure 22.2 Worldwide spatial distribution of flood events for different countries between 1986 and 2015. (*Source*: Alizee Vanderveken, copyright CRED 2015 (Zhao, 2020 – Ph.D. thesis)).

Compared to floods, droughts affect a larger area and are typically associated with prolonged periods of abnormally low monsoon rainfall that can last over a season or longer and extend over large spatial scales across India (Sikka, 1999) and Korea. In India, monsoons and droughts have a considerable impact on socioeconomic activity, water supply, agriculture, and drought intensity due to their delayed progression (Bhalme & Mooley, 1980; Sikka, 1999; Swaminathan, 1987).

22.2 CLIMATE VARIABILITY TRENDS AND THEIR IMPACTS ON FLOODS AND DROUGHTS

Climate change, initially identified in the 19th century, is the long-term pattern of changes in the earth's temperature, air pressure, and precipitation caused by greenhouse gases. Due to excessive greenhouse gas emissions, the average global temperature has increased by 1.08°C from the pre-industrial era (1850–1900) (IPCC, 2021). Even a slight deviation from the earth's long-term (three decades or more) average temperature in the future can cause potentially dangerous shifts in the climate system, resulting in unusual extreme weather events. Hence, formulating concurrent climate adaptation measures has become a necessity.

According to the World Economic Forum's Global Risks Report (McLennan, 2021), the failure to mitigate and adapt to climate change is the most impactful risk faced by communities worldwide. The average global temperature is expected to rise by around 2°C from the present to 2100 even in the optimistic shared socio-economic pathway scenario which assumes that the world shifts to a more sustainable path. The projected rainfall and its intensity are likely to increase globally and more particularly over South and Southeast Asia (IPCC, 2022), resulting in higher surface runoff. These changes are already evident in India. In general, the Asian monsoon regions are more prone to extreme flooding, associated with the projected increase in the annual maximum consecutive 5-day precipitation (Almazroui *et al.*, 2021).

About 10% of the world's population (>600 million) live in coastal areas at elevations less than 10 m above sea level (low elevation coastal zone, LECZ). Nearly 40% (about 2.4 billion) live within 100 km of the coast (UN, 2019). Cities are dynamic systems that face unique climate impacts and hence their adaptation must be location-specific and tailored to local circumstances (The World Bank Group, 2011). Almost 630 million people live on the lands below the projected annual flood levels for 2100 (Kulp & Strauss, 2019). Asia has many of the world's largest metropolitan cities in the floodplains of major rivers that consist of cyclone-prone coastal areas (Huq *et al.*, 2007). Human influence has increased the chance of compound extreme events such as floods and droughts since the 1950s. Combining all this information, all the coastal urban cities are in a crucial period to have their suitable climate adaptation strategy.

22.3 MANAGEMENT MEASURES

According to the prediction made by the climate models, rising greenhouse gas concentrations will lead to frequent floods and droughts. To combat this, management measures are essential tools for mitigating the impact of extreme weather events on communities and ecosystems. The management measures (Figure 22.3), both structural and non-structural, are generally framed based on the return period of extreme events. Some basic measures are: (1) mapping the location affected by past floods and droughts; (2) mapping areas that are susceptible to future floods and droughts based on their geography; (3) comprehensive flood and drought risk assessment, and (4) sophisticated modelling. The global data indicate that structural measures such as levees along with early warning system substantially reduced the number of fatalities. Furthermore, the intensity–duration–frequency (IDF) curves are used to evaluate the return period of floods and the intensity of rainfall events. Extreme rainfall occurrences are anticipated to change due to climate change; hence, the IDF curves must be modified to account for the climate change scenarios (Singh *et al.*, 2016).

According to the National Research Development Corporation analysis of eight U.S. cities, if a city could capture all the rain that falls each year, that bounty would meet 21%–75% of its annual water needs. Capturing a portion for reuse could make an additional difference. Rainwater harvesting through cisterns for large commercial buildings can hold thousands of gallons of rainwater that can be used for irrigation, toilets, cleaning, and even firefighting. Further, replacing impermeable with permeable and porous pavement with green options allows rainwater to seep, infiltrate, and store water instead of running over. It is a great solution for parking lots, driveways, and sidewalks and has a substantial effect on flood and drought reduction. The expansion of permeable surfaces, sometimes called low-impact development (LID) or sustainable draining systems (SuDS) is now widely accepted as an effective, low-cost way to reduce flash flooding during heavy rainfall events. In addition to aestheticism, the green roofs, bioswales, and rain gardens covered with plants can

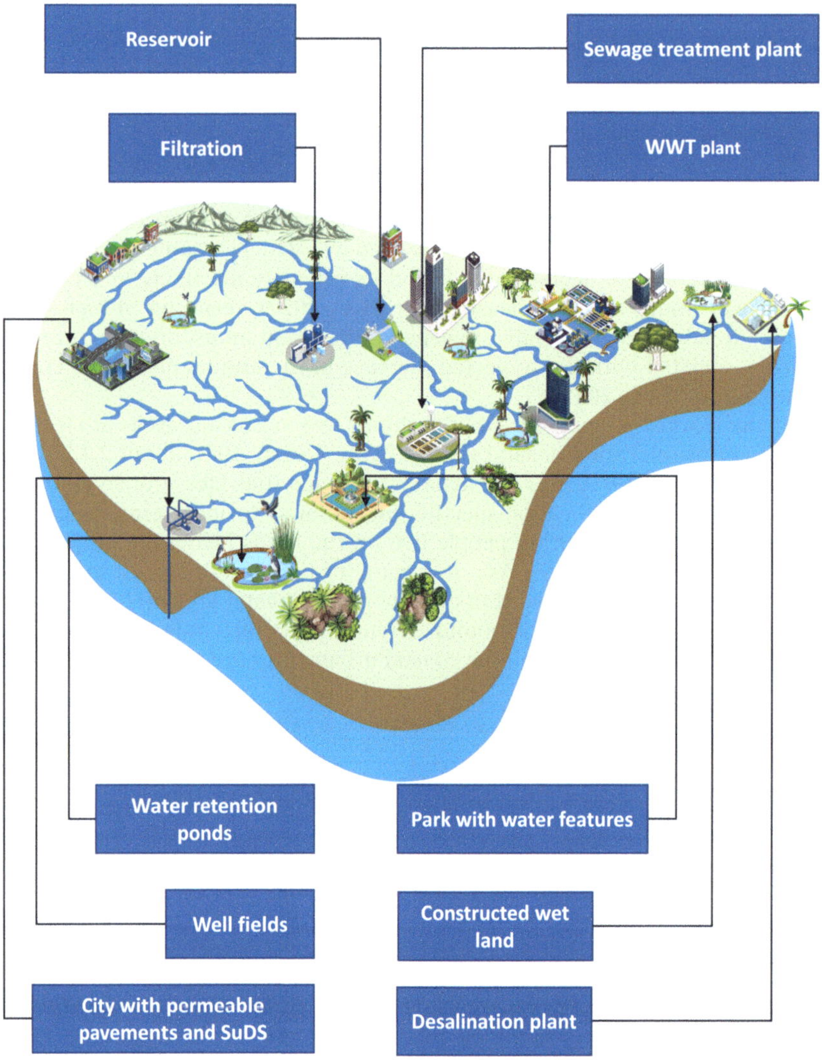

Figure 22.3 Schematic diagram illustrating different flood and drought management strategies for a watershed.

absorb up to 80% of the rain. This substantially reduces a certain percentage of runoff. Contrarily, in summer, they help to keep the environment cool.

To reduce flood risks, new development on wetlands and floodplains has to be designed as effectively as possible. If there is a need to construct in flood plains, the area should be properly zoned and must have followed the standards when building in flood plains. The flood plains should be left unused to hold the water during floods and in dry periods, they can be used for sports and leisure. The water-retention ponds can substantially reduce the impact of both floods and droughts. The elevated area of the watersheds should be used for water infiltration and the lower-elevation spaces should be used for water retention. For instance, in 2003, Seoul dismantled a ten-lane motorway and elevated highway to restore the Cheonggyecheon River that ran beneath it, providing enhanced flood protection and creating a new urban park that attracts more than 50,000 visitors daily, boosting the local economy, improving air quality and public transit ridership (Lee, 2006). Similarly, Rotterdam has built 'water squares' and transferred flat roofs to green roofs to hold excess rainwater, Bangkok opened a new 'amphibious' public park in 2018 to collect floodwater, Chennai restored hundreds of water bodies to manage water stress and heavy rainfall, Rio de Janeiro's Praça Niterói reservoir has three large underground wells with a total floodwater storage capacity of 58 million litres. (Cities 100 report, 2019). Copenhagen's Enghaveparken also received a makeover in 2019 to enable the space to hold excess water (CPH 2025 Climate plan, 2012).

22.4 CASE STUDIES

The Republic of Korea (ROK) and India are two nations that have used a variety of structural and non-structural flood and drought management techniques to lessen the effects of these climate extremities.

22.4.1 The Republic of Korea

ROK, a small East Asian country bordered by the East Sea and Yellow Sea to the west and has a diverse climate ranging from humid continental to subtropical. However, it is prone to floods and droughts which have significant impact on the economy, environment, and society, exacerbated by climate change. According to the Korean Meteorological Administration (KMA) and the National Institute of Environmental Research (NIER), the frequency and intensity of flood and drought events have increased in recent years.

K-water, a government-owned organization, is responsible for managing the country's water resources infrastructures, including dams, treatment plants, and others. It also operates a flood forecasting and warning system to alert authorities and residents before floods occur. To combat droughts, K-water has built water storage, diversion canals, rainwater harvesting, and desalination. K-water is also preparing the country for future flood and drought events.

22.4.1.1 Flood management in ROK

ROK is vulnerable to floods due to its high population density, large areas of urbanization in flood-prone areas, heavy rains, typhoons, snowmelt, and mountainous terrain. In recent years, ROK has experienced some historical flood damage. For example, in 2020 and 2022, a series of heavy rains caused widespread flooding in the country, resulting in a significant loss of lives and causing damage to infrastructure, homes, and businesses.

The country has implemented a comprehensive flood management strategy that includes the Four Rivers Project which aims to restore the natural river functions disrupted by natural or industrial activities. The project has five key objectives: (1) to implement comprehensive flood control, (2) to secure abundant water resources and safeguard against potential water scarcity, (3) to improve the water quality and restore the ecosystems in and around the rivers, (4) to create multi-use spaces for residents, and (5) to prepare for further revitalization of these river systems under regional authorities in the future. The rivers involved in the Four Rivers project are the Han River, Nakdong River, Geum River, and Youngsan River. The project includes dredging and installing 16 multi-purpose weirs to increase the cross-sectional area for flood conveyance and stabilization of river flow. Further, to deal with coastal floods due to the reverse flow of flood tides, estuary barrages were used in Nakdong, Geum, and Youngsan rivers. The Han River was provided with submerged weirs. Other initiatives under the Four Rivers project are the installation of discharge gates in addition to the existing gates to facilitate quick drain flow, and reduction of flood water level and reengineering along flood banks. Approximately 620 km of structurally degraded riverbank has been reengineered, but they will not be very effective since the elevation of the bund top remains the same. The project resulted in a reduction in flood levels during the 2011 summer flood season which produced record-high precipitation in Korea.

Before the project, many sections of the national-level rivers did not reach the freeboard height required by the river design criteria. After the project, all four rivers became the safety area. The project also remodelled and raised some of the country's 17,600 agricultural dams, gaining an additional flood control capacity of 220 Mm^3, and a further 96 agricultural dams were remodelled. The project has shown a positive impact on flood mitigation in ROK, but it is important to continue to monitor and improve the flood management strategy in response to changing weather patterns and population growth.

22.4.1.2 Drought management in ROK

Droughts in ROK are caused by a combination of factors, including low precipitation, high evapotranspiration, and high temperatures. According to the Korean Meteorological Administration (KMA), ROK has been experiencing more frequent and severe droughts in recent years, particularly in the country's southern regions. Droughts can significantly impact agriculture and water supply in the country, leading to crop failures and water shortages.

To mitigate the impact of droughts, ROK has implemented several water conservation measures such as low-flow fixtures, water transfer canals, rainwater harvesting, and greywater reuse that reduces the water demand during droughts. The country also has a system of water storage facilities and transfer canals to provide a source of water during droughts. Additionally, ROK has implemented a comprehensive drought management plan that includes all the above measures to reduce the impact of droughts in the country. Furthermore, ROK has incorporated climate adaptation strategies into its drought management plan to prepare for future droughts and reduce their effects.

22.4.1.3 Water management in Nakdong River

The Nakdong River is the longest in ROK, flows through the cities of Busan, Daegu, and Ulsan over a distance of around 514 km. It is a vital source of water for the region's agriculture and industry. The river basin is a significant economic and cultural hub and is home to nearly half of ROK's population. The river supplies water for numerous sectors, including steel and shipbuilding, agriculture, and hydroelectric power production. Though the river has numerous benefits, it is also prone to floods and droughts which can significantly impact the surrounding communities. Since 1991, the river has started getting polluted by the release of tons of phenols from the electronic industry complex. The sanitation process with chlorine controlled the issue. Later, the pollution was controlled by the optimal operation of the water releases (Kang *et al.*, 2013). Further, to predict runoff changes in the South Nakdong river basin caused by climate change, the SWAT model was used. It was predicted that summer runoff will increase by 45%. In 2012, the Four Rivers projects was completed which widened the Nakdong River and dredged approximately 40.9 Mm^3 of sediments. Using stream tubes and sediment transport equations, a hydraulic model GSTARS was used to analyse the riverbed variation due to climate change.

ROK has built water storage facilities along the Nakdong River to control droughts and has a system of water transfer canals and pipelines to move water from surplus to deficit areas. They also use water conservation measures like low-flow fixtures, rainwater harvesting, and greywater reuse to increase water supply during droughts. To ensure the effectiveness of these measures, ROK regularly monitors and evaluates its management strategies. The country promotes partnerships and community involvement in managing floods and droughts. However, pollution, deforestation, and overbuilding in the catchment area remain challenges that increase vulnerability to these disasters. Therefore, it is crucial to monitor and evaluate existing measures and implement new ones that address the substantial causes of the problem.

22.4.2 Chennai, India

India faces challenges with water resource management due to its large population and limited land and water resources. Due to hydrological, topographical and geological constraints, conventional storage and diversion structures can only use 690 km^3 of the country's average annual water resource potential out of 1869 km^3. Floods and droughts are common occurrences in India due to varying climatic and rainfall patterns in different regions. In 1980, the National Commission on Floods, also known as Rashtriya Barh Ayog, evaluated that 40 million hectares of land in the country were susceptible to flooding, making up one-eighth of the country's total geographical area. Meanwhile, droughts affect half of the country every year. India has experienced 22 major droughts during the last 131 years and the 2002 drought was one of the worst. It affected 56% of the country's geographical area and the lives of 300 million people and 150 million cattle in 18 states.

Chennai, a fast-growing metropolitan city in India, frequently experiences floods and droughts due to heavy rainfall and urbanization. The city receives an average annual rainfall of around 1200 mm, most of which occurs during the monsoon season between October and December. Major floods have occurred in the city during 2005, 2007, 2008, 2010, 2015, and 2021. In 2015, the city faced its worst flood in over a century which caused widespread damage and loss of life. Urbanization has led to concrete paving over of natural drainage systems, reducing the city's ability to absorb heavy rainfall and increasing the risk of flooding. Climate change also contributes to floods in Chennai due to the rise in sea levels caused by global warming and the increased frequency and severity of extreme weather events. A predominant increasing trend in rainfall is generally witnessed in the coastal regions of Chennai (Anandharuban & Elango 2019; Ramachandran & Anushiya 2015). The rainfall projections for Chennai city also confirm that extreme events would increase in the future (Ramachandran *et al.*, 2019).

The Water Resources Department of the state operates several dams and reservoirs to prevent flooding. Furthermore, this organization maintains a network of canals and channels to drain excess water away from populated areas and into natural storage areas. Chennai Metropolitan Water Supply and Sewerage Board (abbreviated as Metro Water) operates several water treatment plants that ensure a reliable water supply for the city and management of the sewerage systems, ensuring that wastewater is treated before it is released back into the environment. Metro Water also promotes rainwater harvesting systems to store rainwater for later use which helps to mitigate the impact of drought. The organization also works closely with the community to educate them about the importance of water conservation and encourages residents to use water responsibly. By working closely with the community, Metro Water helps to mitigate the impact of floods and droughts and ensures that the city has a sustainable water supply.

22.4.2.1 Floods in Chennai city

The city has four major reservoirs: Poondi, Cholavaram, Redhills, and Chembarambakkam (Figure 22.4), which store excess water during high precipitation periods and release it gradually during low precipitation periods to prevent flooding as a flood management measure. The fifth one, called Kannankottai Thervoykandigai, has also recently been added to the list. The reservoirs play a critical role in providing a reliable water supply for the city, though they were initially conceived for meeting irrigational demand. The city also maintains a network of canals, channels, embankments, and levees to drain excess water away from populated areas and to protect low-lying areas from flooding. Chennai has also implemented non-structural flood management measures such as land-use planning, rainwater harvesting, and emergency response planning.

Overall, Chennai has taken significant steps towards flood management and water scarcity. Several studies related to flood mitigation measures already in practice and yet to be practised in Chennai are discussed further.

22.4.2.1.1 Chembarambakkam reservoir

The controlled reservoir operation using a box-model approach under extreme conditions was developed by Anandharuban *et al.* (2019). The model was used to estimate the volume of water to be released to achieve flood mitigation and to delay the time to peak. Three extreme rainfall events that occurred in Chennai in 2005, 2008, and 2015 were used for the analysis. As per this study, reducing

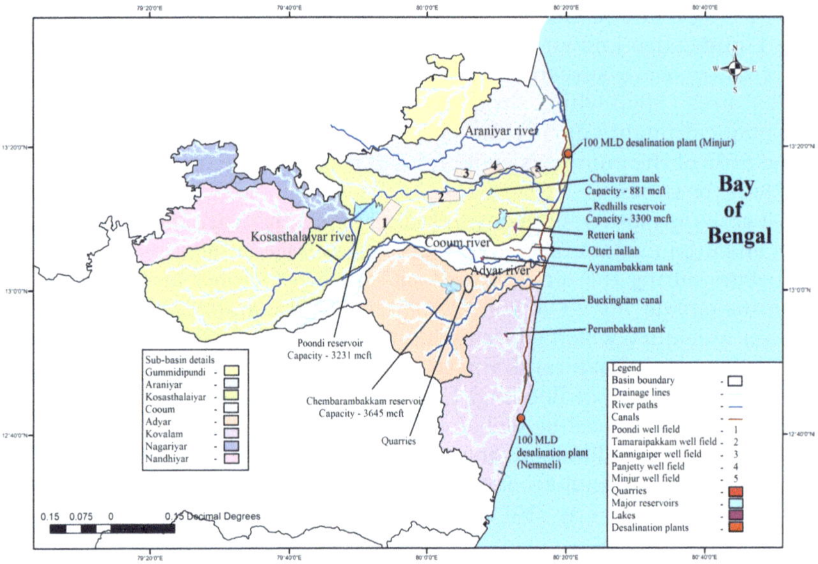

Figure 22.4 Basin-wise water resources available in Chennai.

the initial storage of the reservoir to 50% and releasing water at the beginning of the flood event could reduce the outflow volume from 9% to 37% and delay the time to peak from 1 to 6 h. The process could not completely reduce the flood but can reduce the impact to a significant extent. This type of simplified approach can be used for Chennai's hydrological conditions by practising engineers for better reservoir and flood management.

22.4.2.1.2 Flood management using interventions

The abandoned rock quarries located in and around Chennai and close to reservoirs can be used for additional water storage (Kartheeshwari & Elango, 2022). The diversions and links between the reservoirs or rivers and the quarries with planned reservoir operations will help to mitigate floods. The stored water can be used to meet the daily water demand of the city. Restoring the city's existing drains and redesigning of the culverts can increase the carrying capacity of major roads and railway lines and also increase the temporary water storage structures to mitigate floods and improve water storage in Chennai city.

22.4.2.1.3 Flood and groundwater quality

The majority of solid waste in the city ends up in water bodies and deteriorates the water quality of major rivers like Adyar and Cooum which previously carried fresh water. Further, the city's waterways are fed by several sewage outfalls. In 2015, these two rivers were flooded due to the heavy rainfall in the city's western outskirts. The impact of floods on Chennai's surface and groundwater quality in Chennai during this flood event was assessed (Gowrisankar *et al.*, 2017). The wells inundated by floods led to the entry of sewage-contaminated river water into the aquifers. The inadequacy of sewer lines has also augmented the problem of inundation in several localities, leading to the mixing of sewage in stormwater drains. This has led to increased bacterial counts in most flood-affected areas in groundwater. Brindha *et al.* (2014) also highlighted groundwater contamination in Chennai due to surface water and groundwater interaction. The provision of more screening nets at regular intervals along the rivers and flood-proofing of wells located near the riverbank can reduce the impact of the deterioration of water quality due to floods. Ecological restoration of the waterways could further improve surface and groundwater quality. The problem of seawater intrusion in the north of Chennai can be mitigated by interlinking the Araniyar and Korattalaiyar Rivers (Rajaveni *et al.*, 2021).

22.4.2.2 Water security concerns of Chennai city

Chennai gets the majority of its domestic water supply from reservoirs and well fields (Figure 22.4). The city is prone to water scarcity during the years of failure of monsoons or prolonged periods of low rainfall. Paul and Elango (2018) carried out a study using water evaluation and planning tool to quantify various water management strategies and identify the supply–demand gap for Chennai. This study has considered a new desalination plant, a new reservoir and the reuse of wastewater. The water supply and demand gap was estimated to be improved by 25%, 38%, and 50% by the independent usage of the desalination plant, reservoir and wastewater reuse alone, and the mix of the three steps could help in better water availability. Recently, the government took the initiative to set up a new desalination plant to increase the quantum of desalinated water to about 750 MLD, a new reservoir and commencement of tertiary-treated ultra-filtrated domestic sewage storage in lakes to recharge the groundwater and to supply it for domestic purposes. The initiative has commenced at Porur Lake and will soon be extended to Retteri, Ayanambakkam, Nanmangalam, Keelkattalai, and Narayanapuram lakes. These strategies to improve the water supply and its distribution in the city promote conservation of water and efficient water usage.

22.5 SUMMARY

Flood and drought management are challenges, and they require a comprehensive approach that includes both structural and non-structural measures. Both India and Korea have implemented a variety of flood and drought management strategies, including the construction of dams, levees, and other flood control infrastructure, water storage facilities, water conservation measures, and early warning systems. Both India and South Korea have faced challenges in implementing these flood and drought management strategies such as lack of funding, public awareness, and coordination between different government agencies. To overcome these challenges, promoting public-private partnerships, community participation and regular monitoring and evaluation of flood and drought management strategies is necessary. It is important to monitor the existing situation and evaluate the effectiveness of the measures in place and continue to implement new measures that can address the root causes of the problem. This can be achieved through a comprehensive approach that takes into account the social, economic, and environmental factors contributing to floods and droughts.

REFERENCES

Almazroui M., Saeed F., Saeed S., Ismail M., Ehsan M. A., Islam M. N. *et al.* (2021). Projected changes in climate extremes using CMIP6 simulations over SREX regions. *Earth Systems and Environment*, **5**(3), 481–497. https://doi.org/10.1007/s41748-021-00250-5

Anandharuban P., La Rocca M. and Elango L. (2019). A box-model approach for reservoir operation during extreme rainfall events: a case study. *Journal of Earth System Science*, **128**(229), 1–14, https://link.springer.com/article/10.1007/s12040-019-1258-7

Bhalme H. N. and Mooley D. A. (1980). Large-scale droughts/floods and monsoon circulation. *Monthly Weather Review*, **108**, 1197–1211, https://doi.org/10.1175/1520-0493(1980)108<1197:LSDAMC>2.0.CO;2

Brindha K., Neena Vaman K. V., Srinivasan K., Sathis Babu M. and Elango L. (2014). Identification of surface water-groundwater interaction by hydrogeochemical indicators and assessing its suitability for drinking and irrigational purposes in Chennai, Southern India. *Applied Water Science*, **4**, 159–174, https://doi.org/10.1007/s13201-013-0138-6

Cities 100 report (2019). Accessed 2 December 2022, https://realdania.dk/projekter/cities100/

CPH 2025 Climate Plan (2012). The City of Copenhagen, The Technical and Environmental Administration, City of Copenhagen. Accessed 3 November 2022, https://kk.sites.itera.dk/apps/kk_pub2/pdf/931_e0pg1K8O8G.pdf

Dhar O. N. and Nandargi S. (2003). Hydrometeorological aspects of floods in India. *Natural Hazards*, **28**(1), 1.

Gowrisankar G., Chelliah R., Ramakrishnan S. R., Elumalai V., Dhanamadhavan S., Brindha K. and Elango L. (2017). Chemical, microbial and antibiotic susceptibility analyses of groundwater after a major flood event in Chennai. *Scientific Data*, **4**(1), 1, https://doi.org/10.1038/sdata.2017.135

Gu D. (2019). Exposure and Vulnerability to Natural Disasters for World's Cities. Technical Paper 2019/4. Population Division, Department of Economic and Social Affairs, United Nations.

Huq S., Kovats S., Reid H. and Satterthwaite D. (2007). Reducing risks to cities from disasters and climate change. *Environment and Urbanization*, **19**(1), 3–15, https://doi.org/10.1177/0956247807078058

IPCC (2001). Summary for policymakers: climate change 2001: impacts, adaptation, and vulnerability. A report of working group II of the intergovernmental panel on climate change (IPCC), Gland, Switzerland.

Kale V. (2012). On the link between extreme floods and excess monsoon epochs in South Asia. *Climate Dynamics*, **39**(5), 1107–1122, https://doi.org/10.1007/s00382-011-1251-6

Kang J. Y., Kim J. M., Kim Y. D. and Kang B. S. (2013). Effect of climate change on water quality in Seonakdong River experimental catchment. *Journal of Korean Society of Water and Wastewater*, **27**(2), 197–206, https://doi.org/10.11001/jksww.2013.27.2.197

Kartheeshwari M. R. and Elango L. (2022). 2021 Chennai floods – an overview. *Journal of the Geological Society of India*, **98**(6), 865–866, https://doi.org/10.1007/s12594-022-2079-x

Kulp S. A. and Strauss B. H. (2019). New elevation data triple estimates of global vulnerability to sea-level rise and coastal flooding. *Nature Communications*, **10**(1), 1, https://doi.org/10.1038/s41467-018-07882-8

Lee I-K. (2006). Cheong Gye Cheon Restoration Project. Cheong Gye Cheon, Urban Revitalization and Future Vision International Symposium on the 1st Anniversary of Cheong Gye Cheon Restoration Seoul, Korea.

McLennan M. (2021). The Global Risks Report 2021, 16th edn, World Economic Forum, Cologny, Switzerland. Accessed 3 November 2022.

Mishra V., Dominguez F. and Lettenmaier D. P. (2012). Urban precipitation extremes: how reliable are regional climate models? *Geophysical Research Letters*, **39**(3), n/a–n/a, https://doi.org/10.1029/2011GL050658

Paul N. and Elango L. (2018). Predicting future water supply-demand gap with a new reservoir, desalination plant and waste water reuse by water evaluation and planning model for Chennai megacity, India. *Groundwater for Sustainable Development* **7**, 8–19, https://doi.org/10.1016/j.gsd.2018.02.005

Rajaveni S. P., Indu S. N., Bhola P. K., Zabe A., Monninkhoff B. and Elango L. (2021). Identification of management options to mitigate seawater intrusion in an overexploited multi-layered coastal aquifer by integrated rainfall–runoff, surface water and density-dependent groundwater flow modelling. *Environmental Earth Sciences*, **80**, 617, https://doi.org/10.1007/s12665-021-09836-8

Ramachandran A. and Anushiya J. (2015). Long-term rainfall trends of Indian urban station and its variation in different phases and seasons. *International Journal of Global Warming*, **7**(3), 307, https://doi.org/10.1504/IJGW.2015.069364

Ramachandran A., Palanivelu K., Mudgal B. V., Jeganathan A., Guganesh S., Abinaya B. and Elangovan A. (2019). Climate change impact on fluvial flooding in the Indian sub-basin: a case study on the Adyar sub-basin. *PLoS One*, **14**, e0216461, https://doi.org/10.1371/journal.pone.0216461

Ray K., Pandey P., Pandey C., Dimri A. P. and Kishore K. (2019). On the recent floods in India. *Current Science*, **117**(2), 204, https://doi.org/10.18520/cs/v117/i2/204-218

Sharma A., Wasko C. and Lettenmaier D. P. (2018). If precipitation extremes are increasing, why aren't floods? *Water Resources Research*, **54**(11), 8545–8551, https://doi.org/10.1029/2018WR023749

Sikka D. R. (1999). Monsoon drought in India (Issue 2). Center for Ocean-Land-Atmosphere Studies, Center for the Application of Research on the Environment. Accessed 5 December 2022.

Singh R., Arya D. S., Taxak A. K. and Vojinovic Z. (2016). Potential impact of climate change on rainfall intensity-duration-frequency curves in Roorkee, India. *Water Resources Management*, **30**(13), 4603, http://link.springer.com/10.1007/s11269-016-1441-4

Swaminathan M. S. (1987). Abnormal monsoons and economic consequences: the Indian experiment. In: Monsoons, J. S. Fein and P. L. Stephens (eds.), John Wiley, Washington, DC, pp. 121–134.

The World Bank Group (2011). Guide to climate change adaptation in cities. Accessed 31 December 2022, http://siteresources.worldbank.org/INTURBANDEVELOPMENT/Resources/336387-1318995974398/GuideClimChangeAdaptCities.pdf

UN Department of Economic and Social Affairs (2019). World urbanization prospects: The 2018 Revision. Department of Economic and Social Affairs. https://doi.org/10.18356/b9e995fe-en

Zhao, G. (2020): A sub-model approach for fast large-scale high-resolution two-dimensional urban surface flood modelling, PhD thesis, April 2020, Department of Geosciences and Natural Resource Management, University of Copenhagen, Frederiksberg.

doi: 10.2166/9781789063714_0255

Chapter 23

Sustainable urban drainage systems

Balaji Narasimhan[1], Sreethu S.[1], Krushil Modi[1], Arun R. S.[1], Renato Anelli[2], Maryam Imani[3] and S. Murty Bhallamudi[1]*

[1]*Department of Civil Engineering, Indian Institute of Technology Madras, Chennai, India*
[2]*Mackenzi University, Sao Paulo, Brazil*
[3]*Anglia Ruskin University, Chelmsford, United Kingdom*
**Corresponding author: nbalaji@civil.iitm.ac.in*

ABSTRACT

Changes in the hydrological processes within watersheds due to rapid urbanization along with climate change have been causing frequent pluvial floods in urban areas. Conventional stormwater drainage systems are unable to cope up with the situation. Therefore, latest practices rely on 'source control techniques', 'permeable conveyance systems' and 'end of pipe systems' for storm water management in urban areas. These measures are variously known as sustainable drainage systems in the United Kingdom, low impact development (LID) techniques in the US, water-sensitive urban designs in Australia, and sponge city approaches in China. In this chapter, we review various LID measures, status of their practice, and implementation challenges faced in India. We also outline further research needs in this area.

Keywords: sustainable urban drainage systems, LID measures, LID modelling, basin-scale LID

23.1 INTRODUCTION

Population growth and changing lifestyles of humankind have paved the way for urbanization. In India, it is estimated that 14.4% of the total area classified as 'statutory town' in the census accounts for 26.3% of the population and is built up in nature. The impact of urbanization on the hydrologic behaviour of catchments has been well-documented in the literature related to the field. Changes in the hydrological processes within watersheds, along with uncertain precipitation patterns due to climate change, have caused urban pluvial floods. Urban areas, being regions of social and economic importance, need to be made more resilient to frequently occurring urban flash floods.

The conventional approach for storm water management focuses on transporting the storm water away from the site in the shortest possible time. This approach not only deprives the catchment from its natural processes such as infiltration, evaporation, and transpiration, but also causes floods during precipitation events in which storm sewer design capacities are exceeded. On the other hand, the sustainable approach for water management, known as sustainable drainage systems (SuDS), relies on 'source control techniques' to reduce the generation of runoff from source sites, 'permeable conveyance systems', which reduce the velocity of runoff water to facilitate settlement filtration as well as infiltration, and 'end of pipe systems' to provide passive treatment to collected storm water. The measures adopted for sustainable storm water management are variously known as SuDS in the United Kingdom, low impact development (LID) in the United States of America, water-sensitive urban designs (WSUD) in Australia, and sponge city approaches in China. In this chapter, we use the terms SuDs, LID, and WSUD interchangeably.

In this chapter, we review the current measures available for sustainable urban drainage in India and other countries in the Global South, status of its practice, challenges that are being faced for its implementation, and further research needs in this area.

23.2 LID MEASURES

23.2.1 Green roofs

Green roofs are typically constructed by partially or completely filling the rooftops with soil media that can support plant growth. This soil media, along with vegetation growth, can help reduce the runoff generated from rooftops. Many studies have estimated the effectiveness of LIDs in regulating runoff generation. Experiments have shown that when providing a 200 mm depth for green roofs, runoff can be reduced between 42.8% and 60.8% (Lee *et al.*, 2015). However, the runoff reduction potential of LIDs decreases as the depth of soil media decreases. Other factors that determine the potential of green roofs in reducing runoff include: the type and depth of soil media used, hydraulic parameters of the soil media, type of vegetation grown on the surface, the area occupied by the LID and so on.

In urban areas with limited open space availability, green roofs can be an effective sustainable alternative to reduce runoff and also to mitigate urban heat island effects. However, in arid regions, green roofs may not be viable because maintaining plant growth during water scarcity months may add extra water stress.

23.2.2 Permeable pavements

Permeable pavements are designed with a pervious pavement layer to enhance infiltration and facilitate temporary rainwater storage in the bottom layers consisting of coarse grains. Permeable pavements such as porous concrete and interlocking pavements can reduce peak runoff and increase the time to peak when compared to conventional pavements. These pavements can be laid in lawns and parking spaces in residential as well as non-residential building areas, and also along sidewalks. They have to be designed by considering the climate, moisture conditions, and rainfall characteristics of the region. Permeable pavements in cold regions may have their performance impeded by sediments or frost compared to those in semi-arid or arid regions. The thickness of various pavement layers, porosity, and size distribution of particles can also affect how these structures respond to various rainfall events.

In cities where open space availability is a constraint for implementing other LID measures, retrofitting permeable pavements in the existing impermeable surfaces can help tackle floods to some extent.

23.2.3 Swales and infiltration trenches

Grass-lined conveyance structures that facilitate storage and infiltration of storm water runoff, along with pollution capture, are referred to as swales. Swales can aid in reducing the runoff by allowing infiltration (between 9% and 100%) and can also provide passive treatment to the runoff before disposing it into storm sewer networks (Ahiablame *et al.*, 2012; Yousef *et al.*, 1987). Swales reduce the runoff velocity and thus cause an increase in the time to peak. Factors contributing to the performance of swales include: (a) slope of the conveyance structure; (b) nature of the vegetation grown on it; (c) permeability of the native soil layer; (d) groundwater levels at the site, and (e) provisions for underdrain pipes. Swales can contribute to improving the resilience towards pluvial floods, provided they are effectively designed and placed. Infiltration trenches bear resemblance to swales. They are channels with stones and gravels filled in the ground excavations carried out in a linear pattern. The functionality of trenches differs from swales. Infiltration trenches are implemented to facilitate infiltration and storage of water, rather than to convey the runoff.

23.2.4 Retention ponds and detention basins

Lakes and ponds that exist naturally in the watersheds are classic examples of regional-level LIDs that can aid in runoff regulation. These water bodies have multiple ecosystem functions, and they also act as an additional source of water. However, urban expansion and encroachments have caused a decline in the number of these natural water bodies. Retention ponds try to emulate the runoff regulation functionalities of natural ponds by allowing the retention of the runoff during flooding events. Detention basins, on the contrary, are designed to detain water during floods. The water from the basins will be eventually released through regulated outflow structures. These structures can be adopted as a part of regional-level LID control measures and have been found effective in regulating volume and peak of runoff, especially for storm events of smaller return periods.

23.2.5 Recharge shafts

Exploitation of groundwater levels due to increased demand is a major concern in densely populated regions. Recharge shafts can be implemented in these areas to enhance groundwater recharge to the aquifers. The low space requirements of recharge shafts make them suitable for regions with space constraints. A study conducted in the city of Aurangabad, India, showed that by implementing sufficient number of recharge shafts, the water availability in the region could be significantly increased (Aher *et al.*, 2015). In most of the LID measures discussed earlier, there is a loss of water through evapotranspiration. Hence, there is reduction in the quantity of water available for infiltration and groundwater recharge. However, in the case of recharge shafts, the water loss is minimal as they are constructed deep into the ground. A detailed study on the soil layer stratification, aquifer material, and thickness is required prior to the implementation of recharge shafts.

23.2.6 Bio-retention cells

Bio-retention cells are constructed by replacing the natural soil to certain depths with a porous soil medium that has relatively higher hydraulic conductivity and can support plant growth. The potential of bio-retention cells in reducing runoff increases as the extent of their implantation increases. With higher LID implementation ratios, bio-retention cells can contribute substantially to runoff reduction and augmentation of groundwater levels (Zhang & Chui, 2020). Where the native soil layer has low conductivity, bio-retention cells can be provided with underdrain pipes which would take the flow to storm water networks during high-intensity storm events.

23.2.7 Rain barrels

Rain barrels are similar to conventional rainwater harvesting structures used for capturing water from a roof within the premises of a house. The water held in rain barrels can be used for several non-potable purposes by households. Additionally, capturing rainwater from rooftops reduces the flow in downstream storm water drains, thus reducing the necessary capacity. Rainwater harvesting has been made mandatory in many cities in India, such as Chennai, for tackling water scarcity problems. It is to be noted that rain barrels help in reducing peak flows during storms.

23.3 CURRENT STATUS OF LID PRACTICE

23.3.1 Global North

Urban water management transition framework (Figure 23.1) suggested by Brown *et al.* (2008) indicates that cities evolve in the order: water supply city – sewered city – drained city – waterways city – water cycle city – water-sensitive city, as they attempt to cope with water supply access and security; public health protection; flood protection; environmental protection; constraints on natural sources; intergenerational equity, and resilience to climate change. Figure 23.1 indicates why historically the basic ideas for SuDS originated in the Global North.

Seeds for LID measures were sown in the USA in the 1970s. The first implementation of LID in the USA dates back to its practice in Maryland in 1999. Radcliffe (2019) provides a comprehensive history of LID adoption worldwide. Currently, many cities worldwide, especially in the Global North, have already implemented LID measures. LID practices have recently been adopted by Kansas City, Missouri, Philadelphia, Pennsylvania and New York as primary measures to address the problems caused by old combined sewer systems (Weiss *et al.*, 2019).

From 2014 to 2016, 30 cities including Beijing and Shanghai, were designated as sponge cities in China. China follows a top-down approach for implementing the sponge city program which is expected to control 85% of annual runoff in Beijing through sponge city implementation. Lashford *et al.* (2019) provide an in-depth comparison of the practice of SuDS in the UK with the sponge cities program in China.

In the European Union, the Water Framework Directive requires member states to manage their water resources sustainably. Thishas encouraged the adoption of SuDS and other sustainable water management practices in the continent. Hence, many cities across Europe have implemented SuDS and WSUD in recent years, demonstrating the potential for these approaches to be successful in a variety of contexts.

In the UK, the implementation of SuDS was initiated in 1980 and it has led to a wide range of local, regional and national guidelines, and legislation mandating the incorporation of SuDS into new development plans wherever possible. Schedule 3 of the Flood and Water Management Act (FWMA) was introduced in 2010 for regulation of new development and compliance with the Act.

In Australia, the ACT (Australian Capital Territory) Government introduced the Water-Sensitive Urban Design General Code with a review in 2014 and substantial revisions supported by Practice Guidelines in 2019. The General Code is designed to embrace the SuDS principles by encouraging the reduced use of mains water, improving water quality, and managing stormwater flows in urban areas. There have been more than 300 sites where WSUD projects were implemented in South Australia alone until 2016 (Sharma *et al.*, 2016).

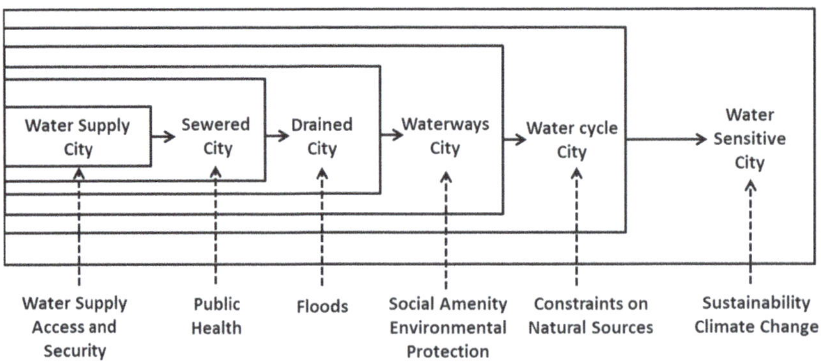

Figure 23.1 Urban water management transition framework. (*Source*: Brown *et al.*, 2008).

23.3.2 Global South

The implementation of LID measures in Global South cities has only recently started. In the Brazilian context, the city of Porto Alegre, in the south of the country, was a pioneer in using low-impact LID-type systems in its Urban Drainage Master Plan prepared in 2003. In 2006, the federal Ministry of Cities proposed national regulation through the Sustainable Urban Drainage Program, encouraging municipalities to use LID techniques in their macro-drainage plans. However, the regulation was not mandatory, and the initiative was not successful in increasing the implementation of LID or equivalent systems. The practices of accelerating the flow through channelling in reinforced concrete remained the main principle applied in urban drainage in the 21st century.

The failure of the 2006 national regulation to introduce LID systems on a large scale in Brazil can be attributed to the lack of managerial capacity to bring together different fields of public management needed for their implementation. Integrated action by municipal departments in São Paulo faced several difficulties including legal instruments, implementation of sanitation actions (clean stream program), creation of a linear parks system along the rivers ('100 Parks in São Paulo' program), and removal and allocation of informal poor settlements in areas at risk of flooding and landslides (Municipal Housing Plan). This sort of problem is absent in wealthy Global North cities but occurs frequently in Global South cities. Besides, the Master Plan of São Paulo underwent a major revision in 2014, which may actually aggravate the risks of flooding. The application of nature based solutions (NBS) and SuDS depends on the existence of areas that are open or likely to be expropriated for their implementation. The analysis of the master plans reveals relevant limits for overcoming the grey infrastructure paradigm. Nevertheless, from 2000 onwards, greater use of large retention ponds can be observed in major cities of Brazil, distributed in a network planned throughout the catchment areas. Initially, multiple uses of these ponds were sought, divided into compartments according to the time of recurrence of the design rains. Leisure and sports squares would occupy the ponds during the dry period, being cleaned after the floods, and returned for community use. However, untreated sewage released into most water bodies made these urban integration spaces unhealthy, which triggered public resistance to having such devices nearby.

In contrast, in South Africa, another major country in the Global South, guidelines for LIDs were formulated almost a decade ago in 2013. Gajjar *et al.* (2021) report on the best management practices, based on environment-based adaptation (EbA) and NBS, adopted by the cities of Cape Town and Durban for disaster risk reduction. These cities have been responding adequately to climate change in terms of appropriate policy, planning, and action frameworks. A comparative study carried out by Gajjar *et al.* (2021) for cities in South Africa and Kenya showed that sustainable water management needs the involvement of a wide range of stakeholders and institutions as well as intra- and cross-organizational collaboration. Practitioners and decision-makers should be well-informed of the latest knowledge so that these practices are adopted. It is also important to involve civil society organizations and NGOs who have the knowledge of local constraints for the practices to be effective.

23.3.3 India

The effectiveness of the LID practices implemented is highly dependent on the hydraulic parameters of the native soil, topography of the region, groundwater conditions at the site, rainfall characteristics as well as overall land use and drainage practices in the region. This necessitates the development of planning and implementation policies that are region-specific. While many countries such as UK, US, Australia, China, and Singapore have developed generic guidelines for LID development plans and have advanced in integrating LID measures into urban landscapes, LID philosophies are only at incubation stage in India. Although there is an increase in the awareness and willingness at institutional level and among the practitioners to adapt to sustainable measures, the availability of directives and guidelines to steer the process is still limited.

Recently, the Centre for Science and Environment (CSE) and the Ministry of Housing and Urban Affairs (MoHUA), Government of India, have developed the 'Water Sensitive Urban Design and

Planning – Practitioner's Guide' to assist practitioners in incorporating sustainable water management strategies into urban planning (Rohilla *et al*., 2017). The 'Guide for Green Infrastructure', which was developed in conjunction with this, attempts to bring water management and green infrastructure practices together for different geographical settings. MoHUA has also provided guidelines regarding the ratio of built-up to open area that has to be maintained for different land-use types. It suggests that 25–35% of a city's area should be made available as open space. However, the present data indicate that in most of the prominent Indian cities, the percentage of open space available falls much below international norms. From an environmental protection perspective, India has norms and guidelines recommended at the institutional level for the protection of open spaces, water bodies and environmentally sensitive areas. These recommendations tend to emphasize the need for development that is sustainable.

The manual prepared by the Centre for Public Health and Environment Engineering Organization (CPHEEO) emphasizes that the water-sensitive drainage design should be such that it ensures water quality, provides protection against flooding, reduces the runoff from catchments, enhances infiltration, minimizes the dependency on artificial drainage networks, and reduces the changes in the water balance in the natural systems. Most of the guidelines that presently exist in India provide design philosophies and approaches required for planning LIDs. Considering the impact that local site and climatic conditions have on the performance of LID techniques, it is imperative to frame rules for designing LIDs suitable for a region. Currently, manuals and guidelines do not elaborate on how site-specific designing can be carried out.

State governments and local authorities have come up with instructions for implementing water-sensitive drainage systems in Orissa, Delhi, and Uttar Pradesh in collaboration with the Centre for Science and Environment (CSE). The rainwater harvesting (RWH) movement was launched in Tamil Nadu in 2001, making it mandatory to have RWH structures in all newly constructed buildings by bringing amendments in the building codes. Research by CSE has shown that there is significant scope for RWH in the cities of Chandigarh and Noida. CSE has also provided plans for WSUD for Dwarka area in Delhi and Medinipur town in West Bengal (Rohilla *et al*., 2017). They have suggested: (a) ponds and wetlands; (b) infiltration and retention basins; (c) filter strips; (d) swales; (e) bio-retention; (f) filter drains, and (g) canals and rills for open spaces. Filter strips, swales, bio-retention, and filter drains are recommended for roads. Acknowledging the role of water bodies in providing resilience towards water-related extremities, the Ministry of Jal Shakti has launched a scheme aimed at repairing, renovating, and restoring water bodies in various states. However, given the size of India and the number of urban areas, significantly more efforts are needed to implement LID measures.

23.4 CHALLENGES FOR LID IMPLEMENTATION IN INDIA

Given the urban expansion that has already taken place on the natural watersheds, one of the major challenges is to incorporate LID measures in the existing limited open spaces in urban areas to enhance flood resiliency. Many major Indian cities were developed without allocating sufficient space for open areas and environmental conservation, resulting in haphazard development in environmentally sensitive zones. Retrofitting LIDs in the existing spaces would require detailed study and analysis. Although separate sewerage systems are designed and implemented in Indian cities, due to unplanned or poorly planned urban expansion, the sanitary sewer capacities are often exceeded and hence the existing storm water drains in most Indian cities carry both storm water and sewage. LID implementation, along with storm water drains in such situations, has to be carefully planned to prevent the inflow of sewage to LIDs. Inflow of sewage to LIDs would not only clog the system but also raise issues of health and hygiene by becoming breeding grounds for mosquitoes and insects. For places that receive rainfall only during the monsoon months, the LID objectives should also encompass water conservation to prevent acute shortages during the summer. Thus, LID planning should be done comprehensively, taking into account the regional climate, social, and economic conditions.

Another major challenge comes in the form of willingness of people to shift from conventional drainage solutions to sustainable solutions. The concepts of sustainability may not be popular among the policy makers and community members alike, making adaptation challenging. Installing LID measures might involve taking up private property spaces that would bring in legal challenges. Overcoming these challenges involves creating awareness among the various stakeholders regarding the necessity and benefits of LIDs.

The economics of adopting LID practices are a major aspect that determines the success rate of implementation. Currently, the economic analyses that exist have been performed for countries with different economic and social statuses. Considering the living conditions and economic–social status of the end-user communities in Indian cities, cost–benefit studies need to be performed.

The performance of LIDs during their design life is dependent on how the structures are maintained. Currently, in India, regulations regarding LID maintenance have not been established. While governing bodies can undertake monitoring and maintenance of regional level LIDs such as detention basins, lakes, and ponds, the responsibility for maintaining plot-scale LIDs implemented in the premises of residential and commercial/non-residential buildings will be upon the respective property owners. During the planning stage, identifying and allocating responsibilities among various stakeholders should also take place.

Often decision makers are stuck in the conventional technical paradigm for drainage projects, although the risks and damages are clear. The conventional paradigm may persist in many cities of the Global South because it is intrinsically related to powerful production chains. The first is the production chain of large infrastructure contractors, totally dependent on heavy reinforced-concrete technologies. LID practices challenge established practices, and hence there is a reaction and reaffirmation of consolidated practices. The second is the productive chain of the real estate market, mainly in the allotment sector, which always looks for the highest yield of the land. Areas belonging to rivers (meanders, floodplains, and protection of slopes and springs) are necessary to implement SuDS patterns. But these are often ignored by real estate projects resulting in end-to-end development without adequate buffer or open spaces.

23.5 NEED FOR ENHANCEMENT OF ENABLING METHODS AND TOOLS

Several methods and tools are available which aid in the decision making for implementation of LID measures. They help in the selection of appropriate type of LID measure and its design based on local site conditions (native soil characteristics, groundwater level, topography, and space constraints), rainfall characteristics, and the specific goal to be achieved. In this section of this chapter, we discuss the need for enhancement of existing methods and tools.

23.5.1 Modelling of LID components

To assess the performance of different types of LID measures against different site and storm conditions, it is crucial to represent LID techniques in hydrological models. The storm water management model (SWMM) tool comes with discrete LID modules for bio-retention cells, rain gardens, green roofs, permeable pavements, block pavers, rain barrels, and swales. However, the model is most appropriate for smaller catchments and its overtly simplified representation of the catchment limits its applicability for watersheds with heterogeneous land uses. The model for urban sewers (MOUSE) is a spatially distributed model that can model urban drainage systems. The model has provisions to represent LID measures such as ponds and wetlands, detention tanks, and swales explicitly. It also allows for implicit modelling of other LID measures. However, LID modelling in the software might experience limitations as it is less sensitive towards the soil and rainfall conditions at site (Broekhuizen *et al.*, 2019). Furthermore, the application of both the models in simulating LID processes for large complex watersheds or river basins with different land-use patterns and corresponding hydrological responses may have limitations. Thus, for the modelling of LIDs, it is essential to incorporate process-based modelling of LIDs into river basin-scale models that can adequately capture the complexities of a watershed.

The soil and water assessment tool (SWAT) is a model widely used for simulating the hydrological processes for large river basins. Recent developments in the software have enabled its applicability in storm water management system modelling. SWAT is capable of modelling multiple soil layers, a step which is essential for LID modelling. Recently, the integration of LIDs such as green roof, rain garden, cistern, and porous pavement has opened the potential of SWAT model in aiding sustainable storm water management development plans. Even though SWAT model is effective for hydrological modelling at river basin scales, it needs to be improved with integration of more LID modules to extend its applicability for sustainable watershed management. Thus, it can be inferred that expanding the strengths of river basin-scale models such as SWAT to develop and integrate various types of LID modules will facilitate formulation of comprehensive sustainable drainage solutions for complex river basins.

23.5.2 Multi-criteria decision making for LID planning

The integration of LID into the urban water cycle requires a detailed and systematic design and planning framework at various spatial scales to restore the hydrological conditions to the pre-development level. In this context, the multi-criteria decision making (MCDM) methods can be used to help decision-makers evaluate and select among various potential LID designs based on numerous competing criteria. MCDM methods provide a structured and systematic approach for assessing the importance of criteria and determining the most suitable LID design for a particular site. Over the years, a range of decision support tools, optimization techniques, and site suitability analysis methods have been developed and refined over time to assist with the selection of appropriate LID design (Gogate & Rawal, 2015; Kuller *et al.*, 2019; Martin-Mikle *et al.*, 2015). Most techniques aimed at developing optimal LID interventions do not consider urban planning considerations or stakeholder preferences. Urban biophysical environments and technologies simulator (UrbanBEATS) serves the purpose of integrating an urban planning module and an LID planning module to provide a variety of LID options based on the challenges and opportunities in an urban area (Bach *et al.*, 2020).

Various decision-support systems for implementing blue-green infrastructure measures exist. However, none of them holistically address the increasing monsoon deluge and the alarming exigency for water conservation. This highlights the demand for a comprehensive framework that perceives this issue as a water resources management problem to integrate LIDs. Such a framework, coupled with educational initiatives to define the multifaceted principles and frequently tacit agreements between stakeholders and governments on water management, aimed at raising community awareness of socio-economic and environmental benefits, is likely to have a positive impact and reinforce the acceptance of the LIDs. So, it is important to study and understand the feasibility of LID components using comprehensive site suitability analysis for the development of an integrated planning framework for a watershed at a regional and local scale.

The remote sensing datasets and geographic information systems (GIS) are often integrated with multi-criteria analysis (MCA) methods to develop geospatial models to identify viable sites for any target resource or activity. LID-related spatial planning studies are few and far between. Existing spatial planning models have been prepared using a few parameters that influence the suitability of LID elements. The existing studies are based on ranking the factors or thematic layers influencing the target resource or activity. This sometimes creates an inherent cognitive bias in decision making. Instead, the potential of existing hydrological models can be utilized to analyse and rank the factors considered. A comprehensive framework, which embeds the MCA method as well as an adequate physical process-based model of LIDs, is needed to improve storm water management.

23.5.3 Basin-scale LID implementation

The cumulative effect of LID practices on a watershed is a research problem that has been relatively less explored. Although there is a well-adhered to belief that locale scale LIDs can impact the watershed flow regimes, uncertainties regarding the translation of LIDs to broader spatial scales are still unattended. While flood control is the primary objective for implementing LIDs, they can also facilitate groundwater recharge. This is especially critical in the Global South where groundwater forms a major source of water and must be augmented during the few rainy days of the monsoon season. However, an assessment of the surface – subsurface flow interactions due to LIDs has not yet been achieved. Therefore, an assessment of the impact of LIDs on the hydrological processes of a watershed conducted at a regional scale can be a starting point for a holistic water management practice for river basins.

23.6 SUMMARY

The integration of LIDs as part of managing the urban water cycle requires a thorough design and planning framework at different spatial scales to replicate the pre-development hydrology. Planning frameworks at different scales facilitate the development of site plans tailored to natural topographic constraints while preserving hydrologic functions and offering aesthetically pleasing and cost-effective stormwater management controls. Over the years, several challenges have emerged that hinder the implementation of LIDs. The absence of a reference framework and institutional barriers to guide the creation of a sustainable urban water management strategy are significant obstacles to progress. Future LID advancements should focus on developing a comprehensive design, planning, and development framework at various spatial scales to enhance the watershed's resilience to changing climates and address pluvial flooding and water scarcity. By integrating LIDs, we can transition from a water supply-based watershed to a water-sensitive watershed.

REFERENCES

Aher K. R., Patil S. M. and Mane V. P. (2015). Recharge trench cum recharge shaft new concept for groundwater recharge for sustainability of source: a case study. *International Journal of Current Medical and Applied Sciences*, **6**(1), 17–21.

Ahiablame L. M., Engel B. A. and Chaubey I. (2012). Effectiveness of low impact development practices: literature review and suggestions for future research. *Water, Air, & Soil Pollution*, **223**(7), 4253–4273, https://doi.org/10.1007/s11270-012-1189-2

Bach P. M., Kuller M., Mccarthy D. T. and Deletic A. (2020). A spatial planning-support system for generating decentralised urban stormwater management schemes. *Science of the Total Environment*, **726**, 138282, https://doi.org/10.1016/j.scitotenv.2020.138282

Broekhuizen I., Muthanna T. M., Leonhardt G. and Viklander M. (2019). Urban drainage models for green areas: structural differences and their effects on simulated runoff. *Journal of Hydrology X*, **5**, 100044.

Brown R. R., Keath N. and Wong T. (2008). Transitioning to water sensitive cities: historical current and future transition states. In International Conference on Urban Drainage 2008 (pp. CD-Rom). IWA Publishing.

Gajjar S. P., Wendo H., Polgar A. and Hofemeier A. (2021). Ecosystem-based flood management: A comparative study report of the cities of Cape Town and Durban (South Africa), Nairobi and Mombasa (Kenya). PlanAdapt Collaborative. Climate and Development Knowledge Network (CDKN), Berlin, pp. 1–52. Accessed 4 September 2023. https://www.plan-adapt.org

Gogate N. G. and Rawal P. M. (2015). Identification of potential stormwater recharge zones in dense urban context: a case study from Pune city. *International Journal of Environmental Research*, **9**(4), 1259–1268.

Kuller M., Bach P. M., Roberts S., Browne D. and Deletic A. (2019). A planning-support tool for spatial suitability assessment of green urban stormwater infrastructure. *Science of the Total Environment*, **686**, 856–868, https://doi.org/10.1016/j.scitotenv.2019.06.051

Lashford C., Rubinato M., Cai Y., Hou J., Abolfathi S., Coupe S., Charlesworth S. and Tait S. (2019). SuDS & sponge cities: a comparative analysis of the implementation of pluvial flood management in the UK and China. *Sustainability*, **11**(11), 213, https://doi.org/10.3390/su11010213

Lee J. Y., Lee M. J. and Han M. (2015). A pilot study to evaluate runoff quantity from green roofs. *Journal of Environmental Management* **152**, 171–176, https://doi.org/10.1016/j.jenvman.2015.01.028

Martin-Mikle C. J., de Beurs K. M., Julian J. P. and Mayer P. M. (2015). Identifying priority sites for low impact development (LID) in a mixed-use watershed. *Landscape and Urban Planning*, **140**, 29–41, https://doi.org/10.1016/j.landurbplan.2015.04.002

Radcliffe J. C. (2019). History of water sensitive urban design/low impact development adoption in Australia and internationally. In: Approaches to Water Sensitive Urban Design, A. K. Sharma, T. Gardner, D. Begbie (eds), Woodhead Publishing, pp. 1–24, https://doi.org/10.1016/B978-0-12-812843-5.00001-0.

Rohilla S. K., Matto M., Jainer S. and Sharda C. (2017). Water-Sensitive Urban Design and Planning: A Practitioner's Guide. Centre for Science and Environment, New Delhi. ISBN: 978-81-86906-17-0

Sharma A. K., Pezzaniti D., Myers B., Cook S., Tjandraatmadja G., Chacko P., Chavoshi S., Kemp D., Leonard R., Koth B. and Walton A. (2016). Water sensitive urban design: an investigation of current systems, implementation drivers, community perceptions and potential to supplement urban water services. *Water*, **8**(7), 272, https://doi.org/10.3390/w8070272

Weiss P. T., Gulliver J. S. and Ebrahimian A. (2019). US version of water-wise cities: low impact development. In: Water-Wise Cities and Sustainable Water Systems: Concepts, Technologies, and Applications, X. C. Wang; G. Fu (eds), IWA Publishing, London, pp. 77–130.

Yousef Y. A., Hvitved-Jacobsen T., Wanielista M. P. and Harper H. H. (1987). Removal of contaminants in highway runoff flowing through swales. *Science of the Total Environment*, **59**, 391–399, https://doi.org/10.1016/0048-9697(87)90462-1

Zhang K. and Chui T. F. M. (2020). Assessing the impact of spatial allocation of bioretention cells on shallow groundwater – an integrated surface-subsurface catchment-scale analysis with SWMM-MODFLOW. *Journal of Hydrology*, **586**, 124910, https://doi.org/10.1016/j.jhydrol.2020.124910

doi: 10.2166/9781789063714_0265

Chapter 24

Urban water infrastructure: distribution and collection

Sridhar Kumar Narasimhan[1], Shankar Narasimhan[1], Timo C. Dilly[2], Amin Bakhshipour[2], Ulrich Dittmer[2] and S. Murty Bhallamudi[3]*

[1]Department of Chemical Engineering, Indian Institute of Technology Madras, India
[2]RPTU University of Kaiserslautern-Landau, Kaiserslautern, Germany
[3]Department of Civil Engineering, Indian Institute of Technology Madras, India
*Corresponding author: sridharkrn@iitm.ac.in

ABSTRACT

Significant urbanization has occurred in the last few decades, especially in the Global South. The provision of sustainable water infrastructure is crucial for making these urban areas liveable, inclusive, and environmentally sustainable. Planning, designing, and operating water infrastructure is highly challenging because of the unpredictability of the growth of urban agglomerations, dwindling water resources, competing needs, transforming governance structures, limited finances, and the impact of climate change. This chapter focuses on issues involved in the planning and operation of water distribution and wastewater and stormwater collection systems in rapidly growing cities, especially in the Global South.

Keywords: water supply systems, sewerage systems, storm drainage systems, integrated planning, flexible planning

24.1 INTRODUCTION

It is projected that by 2025, more than 50% of the population in the Global South will be living in urban areas, accounting for a population of 3.75 billion. The provision of sustainable water infrastructure (water supply, sewerage, and stormwater drainage systems) is crucial for making these urban areas liveable, inclusive, and environmentally sustainable. Piped water distribution networks (WDNs) supply treated water, sourced from dams, lakes, ponds, aquifers, and so on, to the consumers through a network of pipes, valves, pumps, and storage tanks. The main purpose of the water distribution system is to deliver water to consumers in the required quantity, quality, and adequate pressure. Any inefficiency in operating WDNs, such as wastage of water and degradation of water quality, is akin to poor utilization of the already dwindling freshwater resources.

In many urban areas, except where onsite and decentralized sanitation systems are adopted, the wastewater generated in residential and commercial areas needs to be collected through a system of interconnected sewers for treatment before disposal. Discharge of untreated wastewater leads to an unhygienic environment, public health problems, and contamination of sources. The third major component of the water infrastructure is the stormwater drainage system. As discussed in Chapter 23 on 'Sustainable Urban Drainage Systems', the provision of an appropriate stormwater drainage system is essential for avoiding urban pluvial flooding. In the past, many cities in India (e.g. Chennai), other Asian countries (e.g. Antananarivo), and in Africa (Douglas *et al.*, 2008) incurred significant damages due to both fluvial and pluvial flooding. Efficient drainage systems, based on urban design concepts that consider the need for flood protection, can help to mitigate the damage. Proper management of stormwater also protects water resources from pollution by urban surface runoff.

In this chapter, we discuss the issues involved in planning, designing, and operating water infrastructures in rapidly growing cities, especially those in the Global South. In this chapter we do not discuss the related important questions of water and wastewater treatment, and water quality issues since they are covered in other chapters of this book.

24.2 CHALLENGES TO WATER INFRASTRUCTURE PLANNING

Planning, designing, and operating water infrastructure is highly challenging because of the unpredictability of the growth of urban agglomerations, dwindling water resources, competing needs, transforming governance structures, limited finances, and the impact of climate change.

24.2.1 Unpredictable urbanization

Water distribution and drainage infrastructures are usually planned and designed to last 50–75 years into the future because of the cost and effort involved. The design of the infrastructure is based on the needs of the future population, which is estimated using conventional (e.g., statistical extrapolation) forecast methods. In the case of cities in the Global South, there is a significant uncertainty in the estimation of future total population as well as spatial variation in the population density. It is difficult to predict in which direction the city will be expanding. This, in turn, has a cascading effect on the sizing of distribution and collection systems. Errors in the estimation of future land use and land cover lead to errors in the estimation of design storm runoff, making it difficult to appropriately size the stormwater drainage system.

24.2.2 Multiple stakeholders

There are multiple uses of water in an urban area. While the demand for domestic water is high in the core city area, the need for agricultural water is generally high in the peri-urban areas. The proportion of the demand in different sectors continuously changes as peri-urban areas transition into urban areas. There is also a demand for water from industrial establishments. Water bodies provide crucial eco-services and should not be exploited beyond a tipping point. Therefore, the planning of water infrastructure should be based on the sustainable allocation of water among different competing sectors. Social and macroeconomic problems could arise if water is not distributed equitably among all the stakeholders with varying income levels. The way different sections of society respond to a drought or a flood and the coping mechanisms they adopt are complex, making it challenging to plan a resilient system.

24.2.3 Increasing gap between demand and supply

The gap between demand and supply in urban areas is ever increasing for the following reasons: (1) increase in the population, (2) aggregation due to urbanization at locations away from sources, (3) changing lifestyles, (4) industrialization, (5) climate change, and (6) dwindling usable water sources due to overuse by other sectors and increasing pollution. For example, as the city of Chennai has expanded its peripheries over the last few decades, many water bodies have disappeared due to encroachment, which otherwise could have been used for water storage. There is a significant gap between the amount of sewage generated and the existing capacity for sewage treatment in many cities in India. The discharge of untreated wastewater into the environment is deteriorating the water quality in many sources, making them unfit for use. Over-exploitation of aquifers in coastal areas is inducing seawater intrusion, making many aquifers saline.

24.2.4 Climate change

Climate change will affect all parts of the hydrologic cycle. Warming climate will: (a) increase the evaporation losses, (b) alter the spatial and temporal distribution of precipitation, (c) alter the frequency of occurrence and intensity of extreme events, (d) change the surface run-off and stream flow, (e) increase the seawater intrusion, (f) change water quality characteristics, and (g) change the water demand. There is a likelihood of more variability in climate in the future as compared to the present. This will have to be factored into planning robust, flexible, and resilient water infrastructure systems. Climate change will have a significant effect on intensity–duration–frequency curves (Chandra *et al.*, 2015), which are basic inputs for planning stormwater drainage systems. It is important to note that the uncertainty in the quantification of climate change effect increases with decreasing scale in time and space, posing challenges to planners. For example, the uncertainty of local design storms (a few hours) is much higher than long-term water balances.

24.2.5 Other challenges

Most urban agglomerations in the Global South are characterized by fragmented, top-down decision-making processes, and governance in silos. Many a time, there is little coordination between different governmental organizations, which are independently responsible for the planning of water supply, sewerage, and storm drainage systems. Typically, the planning and implementation of different components are carried out at different times. In contrast, the interdependent systems perspective is needed to establish circular material flows. Governance practices should enable the participation of all stakeholders and enhance the capacity to respond in an adaptive manner to changes.

24.3 WATER INFRASTRUCTURE PLANNING: GLOBAL SOUTH VS GLOBAL NORTH

Planning of water infrastructure differs from country to country because of: (a) economic conditions, (b) social–cultural–political structures, (c) data availability, (d) the current state of infrastructures, and (e) development dynamics. These differences are especially perceptible between countries belonging to Global North and Global South. Herein, this important issue is brought forth through a brief discussion on how the planning of water infrastructure in India differs from that in Germany (Dilly *et al.*, 2020).

Data scarcity poses the biggest challenge for planners in India. For example, sub-hourly rainfall data and high-resolution digital elevation models, essential for urban drainage system planning, are not available for most Indian cities. The planning process, involving stakeholders in decision making and governance is more structured in Germany, as compared to that in India. A basic input required for planning either a water supply or a sewerage system is the spatial distribution of the water supply demand or wastewater generation in the future. In Germany, the population in several areas is expected to decrease in the future, as opposed to a rapid increase in the urban population in India. The difference in planning objectives is also obvious from the way 'Smart Cities' goals are articulated differently in the two countries.

The current state of water infrastructure development also affects the planning process. At present, many cities in India neither have proper sewerage nor a stormwater drainage system. The Government of India plans to invest significantly in this sector, and it has mandated that the municipalities should go for separate systems. Germany, by contrast, has already achieved 97% coverage, and the majority of the systems are combined systems. Thus, measures like sewer network control to reduce the frequency of combined water overflows are not an issue in India. Recycling of treated wastewater for domestic use is encouraged in India to bridge the gap between supply and demand, unlike Germany. The main drivers for the adoption of sustainable drainage systems in India include the reduction of peak storm flows and enhancing the water supply. However, low impact development measures (LIDs) in Germany are designed to minimize the effect on the urban hydrologic cycle. Also, differing socio-economic conditions introduce differences in planning processes in both countries.

24.4 WATER INFRASTRUCTURE PLANNING AND DESIGN: IMPORTANT ISSUES

Sustainable water allocation, consideration of multifunctional infrastructure, integrated planning of water infrastructure systems, the introduction of decentralized systems, and adaptive planning are important issues to address water infrastructure planning challenges.

24.4.1 Sustainable water allocation

Sustainable water allocation among different stakeholders is a critical prerequisite to the planning and design of water infrastructure. This task can become complex given: (a) the interdependency of the systems, (b) differences in the required water quality depending on the type of use, (c) availability of water from different types of sources having varying quality, (d) geographical distribution of sources, (e) seasonal variation in water availability, and (f) uncertainty in the estimation of future demand. Also, there could be a water scarcity situation, making equity in water allocation a prime issue.

The difficult task of sustainable water allocation may be facilitated by mathematical models. For example, the software WEAP is a popular tool used for integrated water resources planning. It facilitates the allocation of limited water resources from different sources to different users, based on the scenarios (set of rules) specified by the planner (https://www.weap21.org/; downloaded on 05-04-2023), using an embedded mass balance model with link-node architecture. It has in-built simulators to calculate water demand in different sectors, supply, components of the hydrologic cycle, storage, pollution generation, treatment, discharge, and in-stream water quality under varying hydrologic and policy scenarios. However, WEAP cannot be used for optimal allocation based on specified objectives and constraints. Recently, Dilly *et al.* (2022a) developed a linear programming-based tool for holistic decision making for planning water supply, urban drainage, wastewater treatment, and water reuse based on the urban water mass balance

24.4.2 Multi-functional infrastructure

The paradigm of multi-functional infrastructure is becoming popular in urban planning, especially in the context of the evolution of eco-cities due to factors, such as climate change, the necessity for maintaining ecological balance, resource crunch, and a greater appreciation of the water–food–energy nexus. For example, improper implementation of transportation infrastructure interferes with local urban hydrology and has a bearing on urban flooding (Narasimhan *et al.*, 2016). Multi-functional planning of the transportation infrastructure can be based on the function of roads as transportation corridors as well as drainage routes during heavy rainfall events. Similarly, objectives of urban drainage networks can include: (i) minimization of flooding, (ii) provision of water during scarcity, and (iii) creation of water bodies for recreation and reduction of the heat island effect (Dilly *et al.*, 2022b). Sewerage systems with the tertiary treatment of wastewater can be planned for providing treated wastewater for domestic and industrial water supply. In the context of water-sensitive urban design, the entire water infrastructure may be treated as one single system with multiple functions.

24.4.3 Integrated planning of water infrastructure systems

The implementation of water circularity will have a significant impact on how the water infrastructure is planned, designed, and operated. On-site and decentralized treatment of grey water and wastewater for reuse purposes would reduce the amount of flow into the existing sewerage system, resulting in increased maintenance problems. Thus, for retrofitting existing water supply and sewerage systems to incorporate treated wastewater reuse, the total cost of the system should be minimized to obtain an optimal solution. The total cost consists of the capital cost of different components, the operational cost of the wastewater treatment plants, the maintenance cost of the sewerage system, and the cost of supplying fresh water through the existing system (Dev *et al.*, 2021). Recently, Zhang *et al.* (2023a) developed an optimization tool that provides pareto-optimal solutions for sewerage system design based on minimizing capital cost, minimizing energy consumption, and maximizing the water reuse capacity. In the case of new water infrastructure, significant cost saving can be achieved if the reuse of wastewater for potable purposes is factored upfront and the planning and design of both systems are carried out simultaneously. The cost reduction occurs due to a reduction in the required sizes and capacities.

For the management of stormwater runoff, there is an opportunity for employing LID measures for simultaneously tackling spatially and temporally varying water excess as well as the risks of water scarcity (Yang *et al.*, 2020). For example, Chennai city has implemented a policy of rainwater harvesting (RWH) to tackle the chronic problem of water scarcity. However, these spatially distributed RWH measures can be factored in while designing the stormwater drainage system. Similarly, Chennai Metropolitan and Sewerage Board is implementing projects wherein treated wastewater would be stored in rejuvenated ponds and abandoned stone quarries for indirect potable use. It is important to note that many of these water bodies can also be utilized as flood detention and recharge structures, underscoring the importance of integrated planning and design of all components of water infrastructure.

24.4.4 Decentralized systems

Conventional water infrastructure systems are centralized because they have been designed to perform single functions, and the constraints have been manageable. For example, domestic water supply was sourced from one or two large surface reservoirs and distributed in the entire city. However, the necessity to look for alternative and sustainable sources such as tertiary treated wastewater, spatially distributed groundwater pumping, and so on, makes it imperative to plan a water supply system as a decentralized system. Similarly, in recent years there has been a push to go for decentralized sewerage and stormwater drainage systems because of factors such as the topographical conditions, spatial variation in population density, rapidly transforming peri-urban areas, effect on the environment, the necessity for treated water recycling, and so on. Zahediasl *et al.* (2021) developed an optimization algorithm for designing a decentralized sewerage system that is cost-effective and sustainable. Hesarkazzazi *et al.* (2022) presented a novel framework to investigate the impact of adding redundant flow paths on resilience for optimal centralized versus decentralized urban stormwater networks.

24.4.5 Flexible planning

Planning of expensive water infrastructure in rapidly expanding urban agglomerations in the Global South is complex because of uncertainties associated with the prediction of urbanization, lack of land area, transforming governance structures, limited finances, and so on. It is essential to adopt a flexible design framework to consider system changes, growth, multi-period constructions, new technological developments, and other uncertainties (Cunha *et al.*, 2019). In a flexible design approach, although initial planning and design are based on predictions for the future, spanning the entire lifetime of the project, the project is implemented in multiple stages. At the beginning of each stage, predictions for the remaining lifetime are carried out again, and the design of the system is revised if needed. Zhang *et al.* (2023b) have developed an innovative multi-stage planning framework for hybrid low-impact development and grey infrastructure urban drainage systems in response to land-use changes. One can also explore the possibilities of applying model predictive control concepts (Herman *et al.*, 2020).

24.5 OPERATION OF WDNS

24.5.1 Leakage management

The magnitude of water loss in a WDN is typically quantified by 'unaccounted-for-water (UFW)', which is defined as the ratio of the difference in the quantity of water supplied and delivered to the total amount of water supplied. This UFW includes water lost through leaks in WDNs and unauthorized use of water. It is estimated that the average water loss in many countries in Global South could be as high as 40–50%. The UFW rates are higher in the Global South, where WDNs are poorly monitored for leakage due to limited instrumentation and improper operation. Besides the loss of precious resources, leakages also lead to longer pumping times, higher energy consumption, and water quality degradation.

Techniques based on transient analysis of pressure, acoustic, or magnetic flux signals caused by leaks in pipelines are possible (Colombo *et al.*, 2009). Another approach called the inverse analysis uses hydraulic models, simulators, and measurements to estimate the location of the leak. Machine-learning methods are also becoming popular. Network theory and water balances have also been used to identify and isolate leaks (Rajeswaran *et al.*, 2018).

24.5.2 Pressure management

The pressure at any point in a WDN should be above a minimum value so that water can be withdrawn from the system by the customers at an appropriate rate. At the same time, pressures should not be excessively high as to minimize the background leakage through joints. Originally, the concept of district metered areas (DMAs) was introduced as a convenient way to carry out continuous water auditing and easy detection of water-leaking areas by measuring flow rate using strategically located bulk flow meters (Khoa Bui *et al.*, 2020). A DMA may be defined as a discrete area of a WDN created either by temporarily closing boundary valves or permanently disconnecting pipes from neighbouring areas. Recently, Zhang *et al.* (2021) combined a graph theory algorithm, community detection algorithm, and two-objective optimization framework for creating DMAs, which minimizes the number of flow meters and cumulative pressure differences compared with critical points in each DMA.

24.5.3 Equitable water supply

WDNs are typically designed for continuous operation. However, the WDNs that are originally designed for continuous operation cannot be operated continuously in cases when either water is not available, or it is not possible to ensure appropriate pressures. Hence, the operators are forced to supply water intermittently to manage the resulting stress based on some heuristics and consumption patterns. In most intermittent WDNs, the supply policies are often inadequate and inequitable (Bhave & Gupta, 2006). Although intermittent operation is common in the Global South, it is not as well studied. Minimization of pressure variations across the network for achieving equitable distribution is discussed in Vairavamoorthy *et al.* (2008). Equitable distribution and prevention of excess withdrawal are the focus of operational planning in intermittent networks, whereas, in 24/7 systems, it is mostly operation cost and quality of water supplied.

24.5.4 Energy management

Pumping costs are a significant portion of the operating expenses of a water utility operator and hence optimal pumping strategies can reduce energy consumption. The efficient operation of WDNs has gained a lot of interest in recent decades, and this topic has been recently reviewed by Mala-Jetmarova *et al.* (2017). These studies propose operational methods for minimizing pumping cost as a primary goal for the efficient network by pump scheduling. A common assumption in most of these studies, originating in the Global North, is the availability of a sufficient amount of water. Recently, an approach based on a non-linear model predictive control for optimal operation in water-deficient conditions was developed by Sankar *et al.* (2015). Kurian *et al.* (2018) have addressed the problem of optimal operation of WDNs with intermediate storage facilities.

24.5.5 Internet of Things and smart systems

Smart WDN monitoring using the Internet of Things (IoT) has received wide attention due to the geographical spread of water networks, buried pipelines, and difficulties in accessing vital components of the network. While cellular communication for voice and data has become ubiquitous, it is still expensive and energy-intensive for monitoring water networks. Low-power, short-range radios such as IEEE 80215.4 with multi-hop routing have been used. However, it increases the complexity of the network. PipeNet (Stoianov *et al.*, 2007), among others, is an early example of the use of wireless sensor networks (WSNs). Recently developed low-power wide area networks (LPWANs) based on SigfoxTM and SemtechTM are attractive alternatives to conventional cellular technology. LoRa-based communication can achieve 20 km or higher in range with line of sight (LOS) and about 2 km without LOS. It uses an unlicensed spectrum and can be deployed in each area without any additional subscription, whereas Sigfox is entirely operator managed. Although several works use LoRa for monitoring applications, relatively fewer works have reported the use of LoRa for the remote operation of valves (Chinnusamy *et al.*, 2018).

24.6 OPERATION AND MAINTENANCE OF WATER AND WASTEWATER INFRASTRUCTURE

The water distribution and sewerage collection systems are typically designed to operate for several decades. Unless these are well maintained, they will not provide the level of service expected of them. The reality is that continuous monitoring and maintenance of the infrastructure is not practised in most developing countries. The Central Public Health and Environmental Engineering Organisation (CPHEEO) of the Government of India has published manuals on the operation and maintenance (O&M) of water supply and sewerage systems. Broadly, the O&M activities can be grouped under the following areas:

(i) **Public health and safety:** To ensure that the supplied water meets the standards of water quality, it is necessary to monitor the water sources as well as the treated water. Water sources can be monitored by taking samples at appropriate intervals. Appropriate online instruments can be employed for continuous monitoring of the various water quality parameters. Although usually disinfection using chlorine is carried out in the treatment plant, water may get contaminated during transportation. Therefore, the residual chlorine at the delivery end should also be monitored. Models for the optimal dosing strategy to ensure minimum residual chlorine have been developed (Karaderek *et al.*, 2016).

(ii) **Environmental impact:** Direct and indirect emissions of greenhouse gases such as methane and nitrous oxide from sewerage systems, wastewater treatment facilities (WWTFs), and receiving waters must be monitored and minimized. Of recent concern is emerging contaminants such as pharmaceuticals and microplastics. The sludge from WWTFs can either be disposed of in landfills, in which case the impact of the sludge on surface and groundwater contamination must be monitored.

(iii) **Asset maintenance:** The equipment (pumps, pipelines, valves, sensors, controllers, filters, sedimentation tanks, and advanced reactors) used in the infrastructure for water supply and wastewater management need to be continuously monitored to assess their health and performance and take appropriate preventive maintenance strategies. With the advances made in sensing, communication, and control, it is possible to monitor these facilities and take prompt remedial action to prevent the breakdown of this equipment even centrally. The need for good record-keeping cannot be overemphasized. Also, it is imperative that the infrastructure be digitized and maintained up to date.

(iv) **Operational efficiency:** In water supply systems, 30–50% of the operating cost is on energy consumption. Similarly, in WWTFs, the cost of energy is 10–30% of the total annual operating costs. Advanced control strategies can be used to reduce energy consumption as well as the quantity of chemicals used. This requires several quality parameters to be monitored online using strategically placed sensors in the system.

24.7 NEED FOR NEW PLANNING TOOLS AND SOFTWARE

Several software packages are available for the design and simulation of WDNs, sewerage, and stormwater drainage systems. EPANET1 and WATERGEMS are public domain open-source and commercial software systems, respectively, for hydraulic simulation in WDNs. WATERGEMS can also be used for design. BRANCH is an optimization tool by the World Bank that attempts to minimize pipe costs for branched pipe networks with a single water source. Jaltantra is an open-source tool for designing branched rural water supply schemes developed by IIT Bombay. OpenFlows Sewer GEMS is commercial software for the design and operation of urban sanitary and combined sewer systems. The software can be used for optimal urban sewer planning. Software ++SYSTEMS of Tandler.com can also be used for this purpose. SWMM is public domain open-source software for the analysis of stormwater drainage management. It includes modules for analysing LID measures too.

SCADA systems (e.g., Aquis, Sedaru) used for monitoring and control typically do not include hydraulic models. Conventional and novel thermodynamic-based pump monitoring systems have been developed by Riventa. Sensus Analytics from Xylem can be used for meter management, billing, and leak detection. Several commercially available software offers real-time monitoring and operation packages for water network management.

There is a need to develop new tools for planning and operation utilizing state-of-art computational resources (hardware and open-source software). Presently available software tools do not model or account for network designs and operational practices in many countries in the Global South. For example, reservoirs in India are typically fill-and-draw type, which is not supported natively by standard tools. Further, in many countries in the Global South, water networks are operated intermittently even though this was not the design intent. Mathematical model-based optimal operation can be applied to obtain significant energy savings and improvements in supply (Kurian *et al.*, 2018). These are not available in standard software packages or management systems. Also, easy-to-use and computationally efficient software tools are not yet available for (i) integrated planning and design of all water infrastructure, (ii) adaptive planning, and (iii) planning of hybrid centralized–decentralized systems, as the research on these aspects is of recent origin.

24.8 SUMMARY

In this chapter, the focus of the discussion has been the planning of urban water infrastructure, a topic of paramount importance in contemporary urban planning. A comprehensive overview of the challenges that may arise during the planning process, as well as the necessity of context-based planning, have been carefully analysed. Critical factors and issues that must be considered during the planning stage have been thoroughly examined, highlighting the significance of integrated and flexible planning methodologies. The indispensable role of multi-functional infrastructure and decentralized systems in achieving efficient and sustainable water management has been discussed. The chapter concludes with an examination of the operational and maintenance issues that are fundamental to the successful implementation of any urban water infrastructure plan.

REFERENCES

Bhave P. R. and Gupta R. (2006). Analysis of Water Distribution Networks. Alpha Science International, Oxford, United Kingdom.

Chandra R., Saha U. and Mujumdar P. P. (2015). Model and parameter uncertainty in IDF relationships under climate change. *Advances in Water Resources*, **79**, 127–139, https://doi.org/10.1016/j.advwatres.2015.02.011

Chinnusamy S., Mohandoss P., Paul P., Rohit R., Murali N., Bhallamudi S. M., Narasimhan S. (2018). IoT enabled monitoring and control of water distribution network. 1st International WDSA/CCWI Joint Conference, Kingston, ON.

Colombo A. F., Lee P. and Karney B. W. (2009). A selective literature review of transient-based leak detection methods. *Journal of Hydro-Environment Research*, **2**(4), 212–227, https://doi.org/10.1016/j.jher.2009.02.003

Cunha M., Marques J., Creaco E. and Savić D. (2019). A dynamic adaptive approach for water distribution network design. *Journal of Water Resources Planning and Management*, **145**(7), 04019026, https://doi.org/10.1061/(ASCE)WR.1943-5452.0001085

Dev A., Dilly T. C., Bakhshipour A. E., Dittmer U. and Bhallamudi S. M. (2021). Optimal implementation of wastewater reuse in existing sewerage systems to improve resilience and sustainability in water supply systems. *Water*, **13**(15), 2004, https://doi.org/10.3390/w13152004

Dilly T. C., Schmitt T. G. and Dittmer U. (2020). Water infrastructure for smart cities: developing tools for integrated planning for Germany and India, Gewässerschutz – Wasser – Abwasser, Aachen, 2020, ISBN 978-3-938996-58-4.

Dilly T. C., Schmitt T. G., Dittmer U. and Bhallamudi S. M. (2022a). Holistic decision-making for planning water supply, urban drainage, wastewater treatment and water reuse through linear optimization by using the urban water mass balance. IWA World Water Congress. Water for smart liveable cities, 11–15 September 2022, Kopenhagen.

Dilly T. C., Sedki K., Habermehl R., Dittmer U. and Bhallamudi S. M. (2022b). Sustainable stormwater management: a holistic planning approach for water sensitive cities. *Journal of Hydraulic Structures*, **8**(2), 40–50.

Douglas I., Alam K., Maghenda M., Mcdonnell Y., McLean L. and Campbell J. (2008). Unjust waters: climate change, flooding and the urban poor in Africa. *Environment and Urbanization*, **20**(1), 187–205, https://doi.org/10.1177/0956247808089156

Herman J. D., Quinn J. D., Steinschneider S., Giuliani M. and Fletcher S. (2020). Climate adaptation as a control problem: review and perspectives on dynamic water resources planning under uncertainty. *Water Resources Research*, **56**, e24389, https://doi.org/10.1029/2019WR025502

Hesarkazzazi S., Bakhshipour A. E., Hajibabaei M., Dittmer U., Haghighi A. and Sitzenfrei R. (2022). Battle of centralized and decentralized urban stormwater networks: from redundancy perspective. *Water Research*, **222**, 118910, https://doi.org/10.1016/j.watres.2022.118910

Karaderek I. E., Kara S., Muhammetoglu A., Muhammetoglu H. and Soyupak S. (2016). Management of chlorine dosing rates in urban water distribution networks using online continuous monitoring and modeling. *Urban Water J.*, **13**, 245–359.

Khoa Bui X. S., Marlim M. and Kang D. (2020). Water network partitioning into district metered areas: a state-of-the-art review. *Water*, **12**(4), 1002, https://doi.org/10.3390/w12041002

Kurian V., Chinnusamy S., Natarajan A., Narasimhan S. and Narasimhan S. (2018). Optimal operation of water distribution networks with intermediate storage facilities. *Computers & Chemical Engineering*, **119**, 215–227, https://doi.org/10.1016/j.compchemeng.2018.04.017

Mala-Jetmarova H., Sultanova N. and Savic D. (2017). Lost in optimisation of water distribution systems? A literature review of system operation. *Journal of Environmental Modelling & Software*, **93**(Issue C), 209–254, https://doi.org/10.1016/j.envsoft.2017.02.009

Narasimhan B., Bhallamudi S. M., Mondal A., Ghosh S. and Mujumdar P. (2016). Chennai floods 2015: a rapid assessment. Interdisciplinary Centre for Water Research Rep., Indian Institute of Science, Bangalore.

Rajeswaran A., Narasimhan S. and Narasimhan S. (2018). A graph partitioning algorithm for leak detection in water distribution networks. *Computers and Chemical Engineering*, **108**, 11–23, https://doi.org/10.1016/j.compchemeng.2017.08.007

Sankar G. S., Kumar S. M., Narasimhan S., Narasimhan S. and Bhallamudi S. M. (2015). Optimal control of water distribution networks with storage facilities. *Journal of Process Control*, **32**, 127–137, https://doi.org/10.1016/j.jprocont.2015.04.007

Stoianov I., Nachman L., Madden S., Tokmouline T. and Csail M. (2007). PIPENET: a wireless sensor network for pipeline monitoring. Information Processing in Sensor Networks, IPSN 2007, 6th International Symposium, pp. 264–273.

Vairavamoorthy K., Gorantiwar S. D. and Pathirana A. (2008). Managing urban water supplies in developing countries – climate change and water scarcity scenarios. *Physics and Chemistry of the Earth, Parts A/B/C*, **33**(5), 330–339, https://doi.org/10.1016/j.pce.2008.02.008

Yang W., Brüggemann K., Seguya K. D., Ahmed E., Kaeseberg T., Daie H., Hua P., Zhang J. and Krebs P. (2020). Measuring performance of low impact development practices for the surface runoff management. *Environmental Science and Technology*, **1**, 100010.

Zahediasl A. E., Bakhshipour A., Dittmer U. and Haghighi A. (2021). Toward decentralised sanitary sewage collection systems: a multiobjective approach for cost-effective and resilient designs. *Water*, **13**(14), 1886, https://doi.org/10.3390/w13141886

Zhang T., Yao H., Chu S., Yu T. and Shao Y. (2021). Optimized DMA partition to reduce background leakage rate in water distribution networks. *Journal of Water Resources Planning and Management*, **147**(10), 04021071, https://doi.org/10.1061/(ASCE)WR.1943-5452.0001465

Zhang D., Dong X., Zeng S., Wang X., Gong D. and Mo L. (2023a). Wastewater reuse and energy saving require a more decentralized urban wastewater system? Evidence from multi-objective optimal design at the city scale. *Water Research*, **235**, 119923, https://doi.org/10.1016/j.watres.2023.119923

Zhang Y., Wang M., Zhang D., Lu Z., Bakhshipour A. E., Liu M., Jiang Z., Li J. and Tan S. K. (2023b). Multi-stage planning of LID-GREI urban drainage systems in response to land-use changes. *Science of the Total Environment*, **859**, 160214, https://doi.org/10.1016/j.scitotenv.2022.160214

doi: 10.2166/9781789063714_0275

Chapter 25

Implementation challenges for new projects and technologies

Ashok Natarajan[1] and S. Murty Bhallamudi[2]*

[1]National Task Force Member Ministry of Housing & Urban Affairs (CPHEEO)/Consultant and CEO (retired) of Tamil Nadu Water Investment Company, Chennai, India
[2]Department of Civil Engineering, Indian Institute of Technology Madras, Chennai, India
**Corresponding author: ashoknatarajan59@gmail.com*

ABSTRACT

Implementations of new water infrastructure projects must overcome several challenges. Many difficulties are faced while introducing new and emerging technologies for water and wastewater treatment. These challenges stem from the irregular availability of (1) financial resources, (2) skilled personnel for planning, design, and operation, and (3) implementation agencies for executing the project works. Challenges are also faced because of inadequate institutional structures and the absence of facilitating government policies. Quite often societal acceptance is not guaranteed. In this chapter, we discuss various challenges one can expect while implementing new water infrastructure and adopting new technologies for water and wastewater treatment. We discuss infrastructure financing, resources required for implementation, and institutional and policy framework. The importance of social acceptance is also stressed.

Keywords: new technology adaptation, water infrastructure financing, water governance, water policy, social acceptance

25.1 INTRODUCTION

There are many challenges to the implementation of new water infrastructure and the adaptation of new technologies for water and wastewater treatment. Water infrastructure requires significant capital investment and a continuous flow of funds throughout their lifetime for operation and maintenance (O&M). This makes it necessary to have an appropriate finance model which not only considers capital investment but also the running costs. Planning and design of sustainable water infrastructure whose design life usually spans several decades is a complex process. Well-trained and knowledgeable personnel are required to carry out these tasks. This requirement becomes crucial, especially in the context of adopting emerging technologies for resource recovery and optimal planning, design, and operation of integrated water utilities. There is always a fear of the unknown which prevents many decision makers from adopting new technologies for water and wastewater treatment. Innovative technologies may be expensive as well. Highly skilled personnel are also required for managing all the complexly interlinked processes and tasks involved in the implementation and operation. The availability of implementation agencies is also crucial.

The nature and scale of water infrastructure have been changing over the last decade because of the aim to achieve sustainable development goals (SDGs). The Government of India has initiated many large mission modes projects, such as Swachh Bharat Mission (SBM) and Jal Jeevan Mission (JJM) for providing sanitation and water supply services to all the people in India. Implementation and sustainability of these projects require a change in the institutional structure and governance. Water supply services and environmental protection are highly emotional matters as they intimately affect the lives of all people. Technically sound and economically viable projects may be unacceptable to the people due to various reasons, and governments may find it very difficult to implement them, without delays.

25.2 INFRASTRUCTURE FINANCING

Water infrastructure projects in India are either implemented completely by the governments or through public–private partnerships (PPP). These projects are generally financed using project finance arrangements.

25.2.1 Public financing through the government

Until the recent past, the government used to be solely responsible for the implementation of water infrastructure projects through the utilities. Funds for the implementation, operation, and maintenance of the infrastructure by the utilities are provided by the state governments as part of their annual budgets. In the recent past, there has been a significant increase in the grants provided by the Central Government under schemes such as the Jawaharlal Nehru National Urban Renewal Mission (JNNURM), Atal Mission for Rejuvenation and Urban Transformation (AMRUT) scheme, Clean Ganga Mission, and so on, considering all these under mission mode so that the project implementation takes place efficiently and on time.

Part of the water infrastructure financing also comes from tariffs. However, in most of the cases in India, especially in publicly owned and operated utilities, the tariff is not based on the high cost of providing and maintaining water infrastructure. Most of the utilities do not have meters to measure the amount of water supplied, except in the case of bulk supply. There is no metering of wastewater collected from households. Therefore, there is only a nominal tariff for domestic users in most of the cities. Financing of stormwater drainage systems is done entirely through the taxes collected. Thus, water infrastructure falls under the immature asset class from the perspective of investment. Although the investment requirement for the water infrastructure is estimated to be USD 129 billion (Sethi, 2019), expenditure on water infrastructure in India is approximately 6% of its GDP, which is less than that in many other countries in the region.

As the requirement for investment in water infrastructure has been growing significantly due to overall economic development and rapid urbanization, it is becoming increasingly challenging for the government to adequately fund the utilities. Humphreys *et al.* (2018) have identified the two following problems with public financing, especially in the case of small towns: (a) the limited ability to generate local resources, and (b) the manner in which limited resources are allocated among competing projects. Water, including treated water, is perceived as a free or minimally priced commodity to be supplied by the Government or relevant agencies. Local bodies often give more importance to health, education, and agriculture. Among the water infrastructure facilities, potable water supply is given priority, followed by the sewerage system. Although the focus has shifted to PPP in recent years, public financing of water infrastructure is expected to continue because many towns may not be able to attract investment from the private sector. It is important to note that existing models for water infrastructure financing do not consider the ground realities of small towns (Humphreys *et al.*, 2018).

In publicly financed water infrastructure projects, there is a general unwillingness to adopt new technologies because the decision makers tend to be averse to taking risks. This is especially so if the technology is emerging and has not been tried earlier at a scale comparable to their own project and in similar situations. There is always a worry about accountability. The introduction of new technology in general follows a gradual scaling up. For example, Chennai city recently implemented a 10 MLD scheme for recycling treated wastewater for potable use only after the successful implementation of a 4 MLD scheme in IIT Madras campus for five years. Successful implementation of this scheme will help in further scaling up not only in Tamil Nadu but also elsewhere.

25.2.2 Public–private partnerships

For long, water was considered a social good, and hence water infrastructure became an essential public service that should be handled by the public authorities. The context changed in the early 1990s mainly because of the people's frustration with the poor service quality offered by public utility providers and their difficulties in managing the huge investments required due to rapid urbanization. The estimated investments exceed the government's resources, underscoring the necessity of private sector participation to address the existing gap in water infrastructure development.

The Department of Economic Affairs, Ministry of Finance, Government of India, defines PPP as: *A PPP means an arrangement between Government or statutory entity or Government owned entity on one side and a private sector entity on the other, for the provision of public assets and/or related services for public benefit, through investments being made by and/or management undertaken by the private sector entity for a specified period of time, where there is a well-defined allocation of risk between the private sector and the public entity and the private entity receives performance linked payments that conform to specified and pre-determined performance standards, measurable by the public entity or its representative.*

Typically, a PPP project addresses two main issues: (a) inadequate financial resources and (b) inadequate managerial and operational know-how. To address these issues and different challenges, across cities, the PPP projects act to remedy one of the challenges. The typology of PPPs can be divided into three main categories: (i) PPPs to build infrastructure and/or operate (build-operate-transfer (BOT)-type models), (ii) PPPs to improve operational performance (concession contracts; O&M contracts, service and management contracts, performance-based contracts), and (iii) other types of PPPS (output-based contracts, PPPs with small local private entrepreneurs).

A detailed financial analysis is one of the most critical tasks to determine the financial viability of the PPP project based on the costs involved and the expected revenues. PPP projects rely on the cash flow generated by the project to repay the loans and earn a return on the investments. Funding by loan can be obtained from sources such as: (1) commercial banks and (2) public financing through multilateral or national development banks, and (3) institutions which may provide preferential treatment in relation to repayment and limit the political risk. Tariff setting, which will often determine the financial feasibility of the project, is also the most politically sensitive factor of a PPP contract. The general principle for setting the tariff is that charges to customers should be sufficient to allow an efficient company to operate its existing assets, fund new obligations, and provide a reasonable return on equity and debt capital. Regularly planned tariff increases, that at least follow the inflation, are key to avoiding delays in the investment scheme that will downgrade the quality of the service. Many PPPs have failed because of a lack of appropriate communication with the public, especially when the project induces tariff increases without immediate improvements to the service delivery.

Over the years, many treatment plants have been constructed under DBO (design, build, and operate) schemes. The technology provider intervenes as a general contractor and must assume the operation during some initial years after the commissioning of the works. In the DBO model, the public entity takes full financial responsibility. In the BOT model, the private player actively participates in the financing of the works. The private entity designs, finances, constructs, and operates the facility or system commercially for the project period, after which the facility/system is transferred to the authority. Build–own–operate (BOO), build–own–operate–transfer (BOOT), and build–operate–train–transfer (BOTT) models have been derived from the BOT model. Obstacles to the implementation of water infrastructure projects in India will likely remain because water remains highly subsidized and a non-lucrative proposition. Political opposition to private participation at various stages including ownership and control of water is strong.

Major financing sources for PPP projects are: (a) viability gap funding (VGF) scheme, (b) India Infrastructure Project Development Fund (IIPDF), (c) equity contributions, (d) long-term/short-term debt contributions, (e) India Infrastructure Finance Company Limited (IIFCL), (f) National Bank for Financing Infrastructure and Development (NaBFID), (g) bond/capital markets financing – municipal bonds, and (h) mezzanine/subordinated contributions. Usually, an SPV (special purpose vehicle) company is set up by the proposer solely to implement the PPP project.

The possibility of implementing newly developed technologies for sustainable water management is higher under the BOT models because private agencies are more likely to take risks associated with new partially tested technologies to increase the financial viability, as compared to public utilities which are averse to risk taking. BOT models also encourage new technologies (since it is out of normal purview) in the hope of bringing in efficiency and overall cost reduction.

25.2.3 International financing

With the help of the Government, utilities try to get funding from international development agencies such as the World Bank, the Asia Development Bank (ADB), the Japan International Cooperation Agency (JICA), and so on. ADB has invested in water supply improvement in Karnataka, to privately owned systems on an experimental basis in various cities. In recent years, World Bank has provided finances for efficient water supply and sanitation services in Amritsar and Ludhiana, and peri-urban areas in Uttarakhand and Shimla among others. Several case studies by Head (2006) revealed that the multilateral development banks (MDBs) could finance water infrastructure projects through the following three channels: (a) direct commercial loans to the project company, (b) indirect concessionary loans through the government, and (c) taking minority shareholding to assist financing. The last route of financing motivates the participation of other shareholders. The equity/debt ratio of projects is usually 30/70. To handle the risk of devaluation, MDBs can provide finance in local currencies as has been done in the case of the Allain-Duhangan project in Himachal Pradesh, India. It is important to note that the host government should take the responsibility for determining the most appropriate project financing model.

25.3 RESOURCES

Resources other than finances must be available for the successful implementation of sustainable water infrastructure projects, as discussed in this section.

25.3.1 Trained personnel for planning and design

In October 2021, the Government of India launched the AMRUT scheme for universal piped water supply coverage in 500 cities. One of the goals of this scheme is to transition from the present intermittent water supply scheme to a 24×7 supply scheme to achieve qualitative, quantitative, equitable, and sustainable water supply in urban areas. Innovative technologies for water and wastewater treatment are being developed for resource recovery and for recycling treated wastewater for both potable and non-potable uses. Processes must be designed and tailor-made based on the water quality requirements for specific water use. Similarly, in recent years, there has been a paradigm shift in stormwater management towards sustainable drainage systems for reducing flood risk in urban areas. Also, there will be a need for integrated implementation of water supply, sewerage, and stormwater drainage systems in order to increase the resilience of water supply projects, reduce the impact of floods and provide maximum protection to the environment. For example, Chennai city has already started introducing the recycling of treated wastewater for potable use. Similarly, Mahagenco in Maharashtra reuses the treated wastewater from the City of Nagpur in Koradi Thermal Power Station (TPS). The State of Gujarat's Policy for Reuse of Treated Waste Water (2018) aims for the full reuse of treated wastewater by 2030.

The planning and design tasks for water infrastructure require in-depth and updated domain knowledge in the area of water and wastewater treatment technology, the use of GIS, and the latest mathematical modelling tools for simulating hydraulics of flow, consideration of water loss reduction programmes, the inclusion of district metering areas, inclusion of low impact development measures, and so on. However, there is a scarce availability of trained personnel to carry out these tasks. At present, only a few higher educational institutions and top consulting firms are capable of undertaking these works in India. A major drawback of engaging external consultants for this purpose is their lack of knowledge about local ground realities and inadequate sensitivity to sociocultural issues. The importance of considering sociocultural issues during the planning, design, and implementation stages on the sustainable implementation of a water infrastructure project cannot be overemphasized. Also, the existence of trained and knowledgeable in-house personnel will facilitate the utilities to make informed decisions regarding the adaptation of new technologies.

JJM launched by the Government of India in 2019 aims to provide functional household tap connections by 2024 in every village in India. JJM is being implemented as a decentralized, demand-driven, and community-managed programme. It is envisaged to empower the Gram Panchayat and/or its sub-committee/user group, that is, Village Water & Sanitation Committee (VWSC)/Pani Samiti to plan, implement, manage, operate, and maintain in-village water supply systems. Thus, the successful implementation of programmes such as JJM crucially depends on capacity building at all levels. Training and capacity building of Pani Panchayats is reliant on: (a) community action planning, (b) O&M, (c) community-based water quality monitoring, and (d) catchment protection and source sustainability. The Ministry of Jal Shakti, Government of India, has identified several key resource centers (KRCs) to carry out capacity building.

25.3.2 Availability of implementation agencies

It is obvious that contracting firms should be available for the successful implementation of water infrastructure projects at different scales. Several construction companies are capable of implementing large-scale water resources projects. However, the availability of local implementation agencies poses a significant challenge to the implementation of projects in rural areas. For many of the large and experienced contractors, it may not be profitable to undertake rural infrastructure projects due to inefficient mobility of equipment and labour. Also, they may not be willing to take the risks arising out of the local socio-political atmosphere.

Aditya *et al.* (2017) have studied the delays and cost overruns in many of the infrastructure projects in India and have found that the non-performance of contractors is one of the main reasons. Non-performance of the contractor occurs because of: (i) the safeguarding of self-interests, (ii) inadequate resource mobilization due to inefficient management, (iii) improper use of advance payments, and (iv) improper human resources management. Contractor internal coordination deficiencies, including cash flows, the poor performance of sub-contractors, and labour non-availability have been identified as risk factors in the implementation of infrastructure projects (Makarand, 2001; Xu *et al.*, 2010). This underscores the importance of the availability of good contractors with high professional ethics and managerial skills along with good reciprocation by water boards.

Non-availability of local implementation agencies with the necessary skills also delays the adaptation of new technologies. For example, in many cases of water supply projects, the adaptation of HDPE pipes was discouraged because of the non-availability of skilled labour for pipe laying and jointing. Nature-based solutions for wastewater treatments in rural and peri-urban areas are suggested because not only are they economically viable but they can also be implemented by local contractors.

25.3.3 Trained personnel for O&M

In general, difficulties in the sustainability of water infrastructure arise due to O&M and cost recovery. Many projects in India and elsewhere have deteriorated within a few years after the completion of the project due to poor O&M. Dillon (2023) suggests careful planning of O&M as the external implementation agencies plan the withdrawal and hand it over to local owners. Among other factors, the sustainability of O&M highly depends upon the availability of trained personnel. It is not rare to find a wastewater treatment plant being inefficiently operated and badly maintained due to poor training, absence of opportunities for frequent upgradation of skills, and lack of domain knowledge on the part of plant operators. In India, several wastewater treatment plants employ electrical and mechanical engineers for O&M who have no knowledge of biological and chemical processes involved in wastewater treatment. There have been many instances where local bodies have delayed the acceptance of facilities from the implementing agencies because of the non-availability of trained personnel.

Non-availability of trained personnel for O&M prevents the decision makers in many urban utilities from adopting new efficient technologies. For example, even now many of the decision makers in charge of sewerage and storm drainage systems in India like to go with an entire gravity system and discourage systems with pumping citing difficulties in O&M, even when a system with pumping could be more efficient.

25.3.4 Funds for O&M

Many of the water infrastructure projects in India and other developing countries are poorly operated and maintained because of a lack of funds. In general, planners and decision makers focus on technical aspects of planning, design, and financing of capital investment. However, it is important to plan a provision for financing O&M even before the project starts. Emphasis on post-construction activities should be as much as that on the design and construction (Brikke, 2000). Issues related to O&M should be made integral to planning and design because the aim of O&M is to ensure efficiency, effective service delivery, and sustainability (Castro *et al.*, 2009). Selection of a project alternative should be based on life-cycle assessment, and the project finance model should explicitly consider the provision for O&M.

25.4 INSTITUTIONAL STRUCTURE

Water supply and sewerage services are municipal responsibilities in India, following the enactment of the 74th Constitution Amendment Act. However, the national government continues to have a significant role in the urban water sector. The Ministry of Housing and Urban Affairs (MoHUA) and its Technical Wing, the CPHEEO are responsible for developing the overall policy framework and guidelines, defining technical standards and norms, and financing the largest amount of capital. Additionally, The Central Ground Water Board and Central Pollution Control Board have monitoring and regulatory responsibilities.

Urban local bodies continue to face 'administrative and legislative' constraints based on their level of urbanization. While many cities have fully functional water supply departments, in some states, public health engineering departments or parastatal agencies (in the form of corporations, boards, authorities, etc.) are involved in capital works, provision of services, or both. Revenue functions are typically a municipal responsibility. In certain cases, such as metropolitan cities, special city-specific water/sanitation agencies exist and these are legally and financially separate from the local government. Typically, a state-level agency oversees planning and investment, while the local government (ULBs) oversees operation and maintenance.

Despite decentralization, ULBs remain dependent on capital subsidies from state governments. Tariffs are also often set by state governments that often even subsidize operating costs. Furthermore, when no separate utility exists, there is no separation of accounts for different activities within a municipality. This creates a vicious cycle of increased involvement of various actors within the territory of urban local bodies and lesser autonomy. This is especially complicated for smaller ULBs, where connected parastatal and state authorities have different roles to play as compared to megacities and large ULBs, which have operational independence and encourage financial support from various sources to manage service and infrastructure levels.

25.4.1 Existing institutional arrangements

In a single agency/water supply institution, a single central agency is responsible for the provision of water supply in the urban area. This might be the city municipal body or a state level department/agency, or a specially set up water board. The agency is responsible for planning, designing, and constructing water schemes, maintaining bulk assets, and internal city-wide distribution as well as O&M with or without external support. There are also institutions that single-handedly manage the water supply provisions in urban, corporations, and ULBs. For instance, the Rural Drinking Water and Sanitation Department (RDWSD) is responsible for ensuring sanitation and providing clean drinking water to the rural areas of entire Karnataka. However, BWSSB (Bengaluru Water Supply and Sewerage Board) is responsible for the complete water supply and sewerage of Bengaluru city and extended areas.

In multiple agencies/water supply institutions, multiple agencies manage urban water supply systems within the city. The agencies either have separate responsibilities for planning, design, and construction of water schemes, maintenance of bulk assets, and internal city-wide distribution and O&M, or overlapping roles. This model is seen in most of the cities. In many cases, the agencies are different for water supply and sewerage provision. In decentralized system-through parastatal, water supply provision is decentralized. While the central/state urban department or water supply and sanitation department is responsible for planning, administrative and technical approvals, scheme funding, setting performance standards, the implementation and O&M of the projects, and so on is the responsibility of the parastatal. Such an operational model is seen in Indian cities of Shimla and Bhubaneswar. For instance, the Shimla Jal Prabhandan Nigam Limited (SJPNL) and Water Corporation of Odisha (WATCO) are responsible for the end-to-end delivery of water supply and sewerage services in their respective city areas.

Peri-urban areas have low/no coverage by municipal services and because of their administrative setup, may be served by multiple agencies with no single accountability. In most of the states, the Department of Industries and Commerce (DIC) is responsible for planning the supply of raw water to large-scale industries by liaising with the water resources department (WRD). Under this arrangement, large-scale industries and water-intensive industries can directly source water from the river basins (through relevant permissions and approvals from WRD). In many cities, the smaller industries/industrial estates which fall within the urban areas get their water supply from the city water utilities either through direct application or via industrial areas development board/respective industries corporations. The private sector and SPVs are typically seen in cities that are experimenting with PPPs.

Apart from the above formal institutional structures, various non-government organizations (NGOs) and civil society organizations (CSOs) are involved in the water sector. In many cities, NGOs and CSOs have been demonstrating community-led processes, advocacy, and capacity building in the water sector. In the present context, strong and capacitated community structures have huge potential to contribute significantly to achieving functional goals as well as to providing feedback on the progress of different government initiatives. With due training, they can be involved in service delivery operations and monitoring as well.

25.5 SOCIETAL ACCEPTANCE

It is very important to have societal acceptance for the successful implementation and sustainability of new projects. Societal acceptance is also crucial for adopting new technologies. Citizens are one of the main stakeholders in any water infrastructure project and their opinions should be sought via the stakeholder consultation process and should be duly considered in the planning. Many times, inputs from stakeholders have improved the design of the processes. In a project jointly implemented by IIT Madras and the University of Guelph for improving the drinking water quality in a low-income peri-urban community in Chennai, input from stakeholders had a significant effect on the final design of point-of-use treatment method for potable water supply. Non-consultation or ineffective consultation with stakeholders has resulted in delays in the implementation of several 24 × 7 water supply projects in India in the last decade. Citizens take a confrontational attitude and resort to either street demonstrations or stalling of the project through lawsuits. For example, the employees of Delhi Jal Board made a massive sit-in protest to oppose the 'one zone-one operator' policy of the Delhi Government to privatize water utilities in Delhi. Similarly, protests have erupted in the past in other cities such as Khandwa in Madhya Pradesh, Vijayawada in Andhra Pradesh, and Coimbatore in Tamil Nadu. Many complaints are being made about the 24 × 7 project in Nagpur city. Recently, Langsdale and Cardwell (2022) have argued that appropriate stakeholder engagement results in improved decisions, savings in time and money, and improved relationships between the Government and people. This is especially so in the context of increasing conflicts due to population growth, climate change, and the ageing of existing infrastructure.

25.6 GOVERNMENT POLICIES AND INCENTIVES

Successful and sustainable implementation of water infrastructure is critically dependent on government policies and incentives. In India, the first national water policy was adopted in 1987, which was reviewed and updated in 2002 and 2012. The key points which are directly relevant to urban water infrastructure are: (a) recycling for providing maximum availability, (b) due consideration of the impact of projects on human settlements and the environment, (c) regulation of groundwater extraction, (d) first priority to drinking water while allocating scarce resources. In this latest version of policy, government is responsible for facilitating the service and the government need not be a direct service provider. The emphasis to treat water as economic good paved the way for increased participation of the private sector in water infrastructure projects. The policy on recycling has been driving many cities to go for recycling treated wastewater for potable and non-potable use. Many wastewater treatment technologies which were considered too costly and unnecessary earlier are becoming acceptable because they are able to meet more stringent discharge standards. Decentralized water supply schemes which were not adopted earlier for urban areas, will soon become popular because of the adaptation of recycling. The policy has also provided impetus for adopting decentralized greywater recycling using nature-based solutions in rural areas. The last priority for industrial water supply in the water policy is driving the industry towards adaptation of zero liquid discharge (ZLD) technologies. The government is encouraging the industry to procure either primary or secondary treated wastewater and treat it to suit their specific requirement. These factors in turn are propelling the adaptation of innovative tailor-made technologies for water and wastewater treatment.

There has been some criticism regarding the water policy in India on the grounds that: (i) water as an economic good may lead to inequity, (ii) the rich are not deterred from misuse of water, and (iii) incentives for effluent treatment is causing society to move away from the natural law of polluter pays. A study conducted by Ahmed and Araral (2019) indicates that water governance has shown improvement in the past few years in eight states as per their evaluation. They attribute this improvement to the National Rural Drinking Water Program, Accelerated Urban Water Supply Program (AUWSP), Water Framework Law of India 2016, and Namame-Gange and National Water Policy. All the states are required to appraise the NITI Aayog, which is an apex body directly reporting to the Prime Minister and is responsible for coordination among all stakeholders involved in the implementation of SDGs. Ahmed and Araral (2019) opine that all the above factors have contributed to the reported improvement in water governance.

25.7 SUMMARY

There are many challenges to the sustainable implementation of new projects and adaptation of new technologies. These are affected by a lack of (i) adequate investment funds, (ii) skilled human resources over the entire gamut of decision-making, planning, design, implementation, and O&M, (iii) proper institutional structure, and (iv) government policy. The lack of adequate funds for public finance can be overcome by adopting the PPP mode of financing, without jeopardizing the equitability concerns. Central and state governments have to invest heavily in skill development and capacity building in order to make the water infrastructure projects sustainable beyond a few years after their implementation. Although several changes are being brought about in water governance in India through changes in institutional structure and policies, more such efforts are called for. To summarize, the adaptation of new technology, innovative financing, capacity building, and effective involvement of all stakeholders in decision making are all equally critical for the implementation of water infrastructure projects sustainably.

REFERENCES

Aditya A. K., Douglass D. A. and Bhattacharya M. (2017). Urban infrastructure development works in India: delay and difficulties in implementation with reference to a water supply project. *Journal of the Institution of Engineers (India): Series A*, **98**, 349–354, https://doi.org/10.1007/s40030-017-0214-2

Ahmed M. and Araral E. (2019). Water governance in India: evidence on water law, policy, and administration from eight Indian states. *Water*, **11**(10), 2071, https://doi.org/10.3390/w11102071

Brikke F. (2000). Operation and Maintenance of Rural Water Supply and Sanitation Systems. A Training Package for Managers and Planners. IRC International Water and Sanitation Centre and World Health Organisation, Malta.

Castro V., Msuya N. and Makoye C. (2009). Sustainable Community Management of Urban Water and Sanitation Schemes (A Training Manual). Water and Sanitation Program-Africa, World Bank, Nairobi.

Dillon L. B. (2023). https://sswm.info/sswm-solutions-bop-markets/inclusive-innovation-and-service-delivery/labour-intensive-technologies/operation-and-maintenance (accessed 26 January 2023).

Head C. (2006). The financing of water infrastructure: a review of case studies, Report No. 59070, Bank-Netherlands Water Partnership Program.

Humphreys E., van der Kerk A. and Fonseca C. (2018). Public finance for water infrastructure development and its practical challenges for small towns. *Water Policy*, **20**(S1), 100–111, https://doi.org/10.2166/wp.2018.007

Langsdale S. M. and Cardwell H. E. (2022). Stakeholder engagement for sustainable water supply management: what does the future hold? *AQUA – Water Infrastructure, Ecosystems and Society*, **71**(10), 1095–1104.

Makarand H. (2001). Risk factors affecting management and maintenance cost of urban infrastructure. *Journal of Infrastructure Systems*, **7**(2), 67–76, https://doi.org/10.1061/(ASCE)1076-0342(2001)7:2(67)

Sethi S. (2019). Financing water infrastructure: bolstering the foundation today for a better tomorrow!. *Water-Digest*, **02401**, 26–32.

Xu Y., Yeung J. F., Chan A. P., Chan D. W., Wang S. Q. and Ke Y. (2010). Developing a risk assessment model for PPP projects in China – a fuzzy synthetic evaluation approach. *Automation in Construction*, **19**(7), 929–943, https://doi.org/10.1016/j.autcon.2010.06.006

doi: 10.2166/9781789063714_0285

Chapter 26
Virtual water and policy implications

S. Ragasri[1] and Thalappil Pradeep[2*]

[1]International Centre for Clean Water (ICCW)-IITM, Chennai, India
[2]DST Unit of Nanoscience (DST UNS) and Thematic Unit of Excellence, Department of Chemistry, Indian Institute of Technology Madras, Chennai, India
*Corresponding author: pradeep@iitm.ac.in

ABSTRACT

Management of water resources is vital for sustainable development. However, management of water resources can be wholesome only if the aspect of virtual water is also addressed. Virtual water is the water that is required for the production of commodities. With the rise in global trade volume, there is an increase in virtual water exchange. Therefore, water is indirectly becoming a traded commodity. Initially, it was predicted that global market forces would direct the virtual water flows from relatively wet to relatively dry regions. However, currently, water-poor countries like India are becoming the major net virtual water exporters and water-rich counties like the UK are becoming net virtual water importers. It has become important to find management methods for virtual water for water-stressed countries. Policy interventions can play a key role in such management. This chapter aims to address the concept of virtual water and presents the role of policy intervention in virtual water management. Case studies of countries with successful management of virtual water are presented and evaluated critically. Lastly, recommendations for efficient virtual water management will be presented for the Indian scenario.

Keywords: virtual water exporter, virtual water importer, sustainable development, policy interventions

26.1 INTRODUCTION

The large-scale water abstraction for human needs is putting great pressure on global water resources. Approximately 70% of the water extracted from freshwater systems is used for irrigation and approximately 40% of our food is grown on irrigated lands. Without major policy interventions, human water usage and areas under irrigation are expected to skyrocket by 2050 due to population growth and rising food demands. Water scarcity is expected to worsen in the coming decades (Pastor *et al.*, 2019). Therefore, there is great emphasis on proper planning and management to attain water sustainability.

Virtual water trade has emerged as one of the measures to conserve global water resources. It is seen as a tool to share fresh water resources and manage water resources in a better way (Nishad & Kumar, 2022). It is estimated that virtual green water exports and virtual blue water exports will more than triple from 905 billion m^3 and 56 billion m^3 in 2010 to more than 3200 billion m^3 and 170 billion m^3, respectively, by the end of the century in response to increase in population and the resultant demand increase (Harris *et al.*, 2020). Even though the concept of virtual water was once seen as a potential way to attain water sustainability, currently, export of virtual water is causing huge harm to a nation's water sustainability resulting in irreplaceable losses.

India is emerging as one of the lead players in the global trade market. With agricultural and industrial products, India has been exporting its water resources in the form of virtual water. India's virtual water exports have increased 28 times between 1986 and 2013 (SreeVidhya & Elango, 2019). They are predicted to increase further. On the contrary, India is experiencing extreme weather changes and is water stressed. According to the ongoing trend, there will be acute shortage of water for domestic production. Therefore, there should be corrective measures that should be taken in order to keep a check on virtual water trade flows.

One of the most significant ways of addressing the virtual water trade flow is to introduce stringent and science inclusive measures and policies. These policies should be able to tackle international and interstate virtual water flow. This chapter aims to explain the concept of virtual water and elucidate how virtual water flows between different regions. Furthermore, this chapter aims to discuss the virtual water trade situation of India. Finally, recommendations for inclusive policy have been made.

26.2 VIRTUAL WATER: DEFINITION

Virtual water is defined as the water that is 'inbuilt/embedded' in a commodity. The total volume of water consumed or polluted in the complete chain of production of the commodity. This concept was first defined by Allan (1998). A precise quantification depends on two approaches. One approach quantifies virtual water as the total water in reality used to produce a commodity. The second approach defines the water *that would have been* required to produce a certain commodity. This concept can be further extended to understanding the water footprint and consumption patterns.

Virtual water concept was formulated to achieve water security and efficient water usage. It was first formulated to draw attention to the idea that severe local water shortages can be effectively alleviated by global economic processes. The whole idea behind this concept was that water-rich countries can export water-intensive products to water-scare countries. Therefore, initially, virtual water planning directed attention to regions like the Middle East and North Africa, or MENA, as they were singled out as regions with significant food imports (Delbourg & Dinar, 2020).

26.3 VIRTUAL WATER: TRADE

Virtual water trade is the idea that virtual water also gets exchanged with goods and commodities in between two countries. When a country exports a water-intensive commodity, water in the virtual form is exported. It is also seen as a way to support various countries to meet their water needs. This can be an attractive strategy for attaining water security. Water-scare countries can import high water-demanding products and water-abundant countries can attain profits by exporting such commodities. The saved water can be used for producing other high value commodities (Aldaya *et al.*, 2010). For example, it was estimated that agricultural export from USA to Japan helped Mexico save global water usage by 11% (Mekonnen & Hoekstra, 2012). Even though trade of real water between water-rich and water-poor regions is generally impossible due to the large distances and associated costs, trade in water-intensive products (virtual water trade) is realistic (Hoekstra & Hung, 2002).

Virtual water trade can be an important tool to improve global water security amongst countries and continents. Trade plays an important role in redistributing water resources. It was estimated that water exported virtually in international trade accounts to one-third of the global water withdrawal (Zhan-Ming & Chen, 2013). Most of the water that is virtually exported is due to agricultural products. It accounts for around 90% of the total virtual water displacement meant for human consumption (Vallino *et al.*, 2021).

26.3.1 Virtual water trade: agriculture

Virtual water calculations over 20–30 years for crop-related trade have been reported by various literatures. It was estimated that global crop-related trade of virtual water for 1995–1999 was 695 Gm^3/year. The global water withdrawal for agriculture (water use for irrigation) was about 2500 Gm^3/year in 1995 and 2600 Gm^3/year in 2000. That means 13% of the global water for crop production was exported virtually than being used for domestic consumption. However, this percentage is different for various countries. Therefore, for the period from 1995 to 1999, the highest virtual water importer was Sri Lanka. It was followed by Japan, Netherlands, Republic of Korea, and China (Hoekstra & Hung, 2002). Figure 26.1 shows the virtual water content of various agricultural products.

Wu *et al.* (2022) conducted a study for the virtual water trade of important staple crops such as wheat, rice, and maize for 2008–2017. They concluded that wheat and maize were primarily exported from countries with abundant water resources (e.g., the United States, Brazil, and Argentina). Rice was primarily exported from India and Pakistan, two of the most water-stressed countries, to water-stressed countries (e.g., Mexico and Egypt).

26.3.2 Virtual water trade: industrial products

Most often studies primarily focus on the virtual water present in food as food production consumes large amount of water. The impact of virtual water in industrial products has received less attention. This is despite the fact that industrial products have a substantially large virtual water footprint (Hassan *et al.*, 2017).

Hassan *et al.* (2017) conducted a study to analyse the virtual water footprint of industrial product exports from Malaysia. Malaysia was chosen for the study to understand how a water-abundant country can impact and improve water distribution globally. Also, Malaysia primarily trades industrial products over crop-based products. The study concluded that the export of products happens with water-abundant countries.

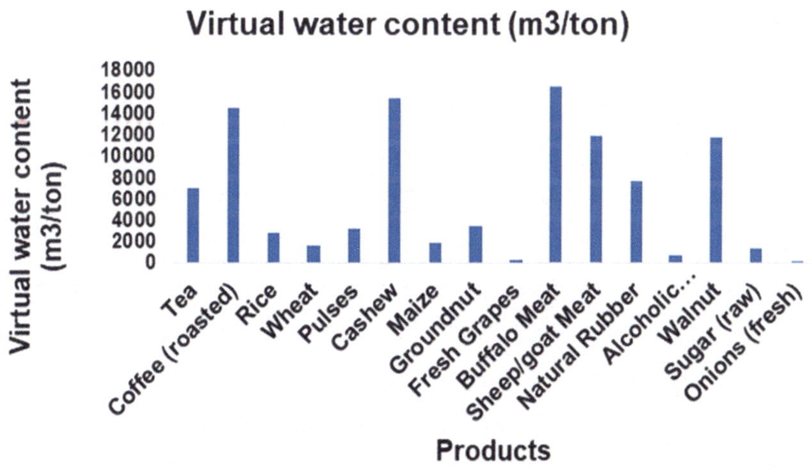

Figure 26.1 Virtual water content of various agricultural products. (*Source*: SreeVidhya & Elango, 2019).

26.4 VIRTUAL WATER EVALUATING METHODS

There are a variety of methods to account for virtual water. These methods can be further classified as top-down approaches and bottom-up approaches. Top-down approaches are indirect approaches based on production and trade data. Bottom-up approach is based on the direct data on consumption (Vanham & Bidoglio, 2015). Figure 26.2 summarizes the various methods used. Currently, there are two important methods used for computing virtual water. One method is the water footprint method and the other is the input–output analysis (IOA). The former method is widely used for estimating virtual water content in agricultural products. However, IOA includes the complete production chain, water input at local production, and import of water (Sun *et al.*, 2021).

26.4.1 Water footprint

The concept of 'water footprint' was developed as a tool to measure the virtual water consumption over a complete supply chain. It involves computing the fresh water used for production, the water involved in the delivery of the product to the customer, and the associated pollution (Hoekstra & Mekonnen, 2012). There are two methodologies by which water footprint assessments can be conducted. One is the integrated water resources management (IWRM) framework, the other is the life cycle analysis framework (LCA). However, these methods are quite complimentary in nature (Ray *et al.*, 2018).

The following method is largely used for measuring the water footprint of agricultural products. Typically, the virtual water content is equal to the water footprint of the crop. The water foot print of the agricultural product should be accounted for before calculating the virtual water trade volume (Yang *et al.*, 2013). Figure 26.3 describes the various components of water footprint. Therefore, the virtual water footprint of agricultural products is typically the sum of the blue water footprint and the green water footprint (Sun *et al.*, 2021).

26.4.1.1 Calculation

The calculation is based on the crop water requirement method (Allen *et al.*, 2006). CROPWAT model and the H08 global hydrological model are used for the calculation process. The most frequently used method, the CROPWAT8.0 (FAO Database), was designed by FAO. It is based on Penman–Monteith model (Allen, 2006). The crop water requirement is computed by estimating the accumulated crop evapotranspiration, ET_c (mm/day) (Sun *et al.*, 2021). Equation (26.1) depicts the complete calculation.

$$ET_c = K_c \times ET_0 \tag{26.1}$$

where K_c is the crop efficiency, ET_0 is the crop evapotranspiration.

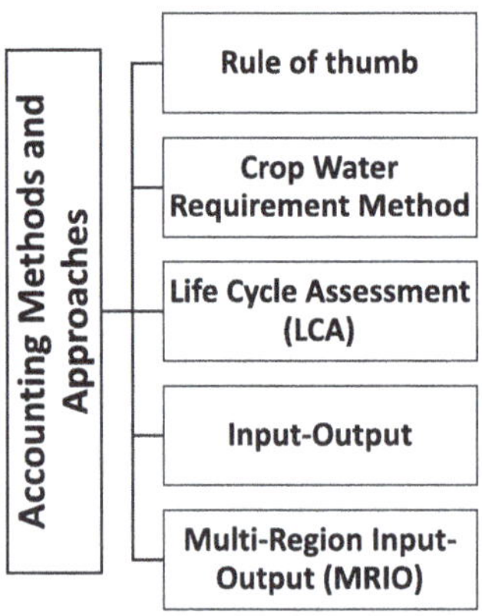

Figure 26.2 Different virtual water accounting methods. (*Source*: Adapted from Yang *et al.*, 2013).

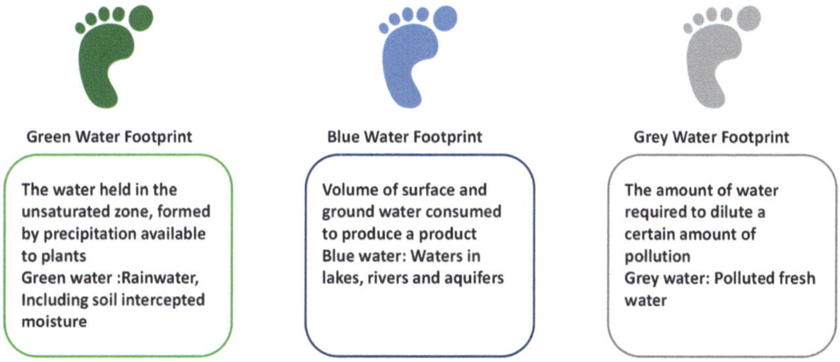

Figure 26.3 Different components of water footprint.

Through this the WF_{green} and WF_{blue} can be obtained. WF_{grey} can be computed by

$$WF_{grey} = (\propto \times AR)/((c_{max} - c_{nat}) \times Y) \tag{26.2}$$

AR is the amount of fertilizer used per hectare (kg/ha); α is the the leaching rate (the proportion of pollution entering water body to the total amount of chemicals used); c_{max} is the maximum allowable concentration (kg/m^3); c_{nat} is the natural concentration for the pollutant considered (kg/m^3) and Y is the the crop yield (ton/ha).

Therefore, the water footprint is

$$WF = WF_{green} + WF_{blue} + WF_{grey} \tag{26.3}$$

$$VWT = WF \times TV \tag{26.4}$$

VWT is the virtual water trade in m^3 and TV is the trade volume in kg.

The virtual water source was specified in the simulation using two categories - green water and blue water - consistent with the global hydrological cycle when using the H08 model to estimate virtual water flow. Blue water was further classified into three types: streamflow, medium-sized reservoirs, and non-renewable and non-local blue water. The model outputs and statistics on crops, livestock products, and international trade will be combined to produce a global total estimate as well as a breakdown of virtual water exports by nation and region for major crops and livestock products (Sun *et al.*, 2021).

26.4.2 Input–output model

This model is an economic mathematical model based on input–output theory. The tool uses linear algebra to define mathematical models. The input–output model takes into account both the intermediate consumption and the indirect consumption of water. It helps in discovering the various internal links between the various sectors of the economy and the production steps, therefore aiding in economic analysis, budgeting, and forecasting. Therefore, the results from an input–output table not only represent the water footprint of the sector, year under study but also the water resource consumption of a significant time period around it. This is because the calculations involve various coefficients which depict the interdependence between various industries (Sun *et al.*, 2021).

Unlike other methods, the IO analysis method can calculate virtual water in industries, whereas other methods only focus on agriculture. It can clearly quantify the allocation of water in trade. Furthermore, the results produced by this method are more intuitive and precise. Many studies around the world have used this method to conduct virtual water analysis (Dietzenbacher & Velázquez, 2007; Guan & Hubacek, 2007; Lenzen, 2009; Zhao *et al.*, 2010). Some of these studies attempted to explain the flow of water in the socio-economic system using a building water account which is similar to the water footprint method (Dietzenbacher & Velázquez, 2007). However, because the virtual water content of industrial products is closely related to the manufacturing process, it is difficult to calculate and is typically omitted in this calculation (Shi & Zhan, 2015).

26.4.2.1 Calculation

The model can be defined on the basis of the number of regions involved. Therefore, there can be single-region models as well as multi-region models. These models can help in computing the virtual water trade among the different regions, sectors, and industries. Furthermore, through these models, the virtual water flow path in the trade can also be tracked (Sun *et al.*, 2021).

The basic steps involved in an input–output model are (Figure 26.4)

(1) The water consumption is added into the traditional input–output matrix of n × n type. This gives a mixed input–output table of economic and water resource consumption.
(2) According to Leontief input–output model (Leontief, 1986), a model for water resources can be established.

Figure 26.4 is a brief representation of the steps involved in computing virtual water using input–output models.

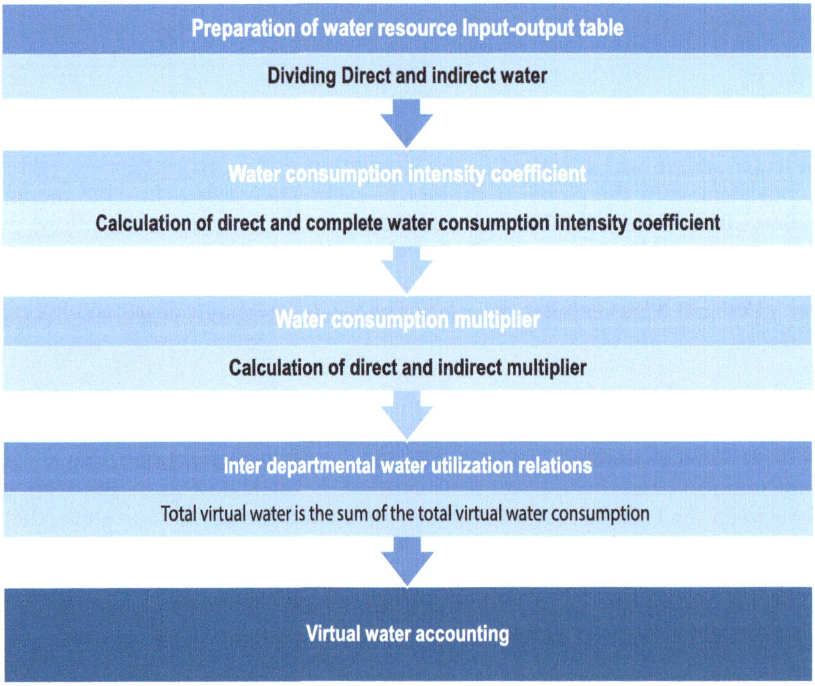

Figure 26.4 Brief representation of the steps involved in computing virtual water using input–output models. (*Source*: Adapted from Sun *et al.*, 2021).

26.5 VIRTUAL WATER AND INDIA

Among the emerging economies, India ranks as one of the highest global water consumers. In 1995, it was the highest water user. In 2008, India ranked second in terms of global water usage with 13% water usage share. According to UNESCO (2022), in 2017, Asia accounted for two-thirds of global ground water extraction. India ranked as one of the largest in Asia. In India, the agriculture sector has the highest share of water consumption. It is estimated to have a share of 65–70%, few states use around 90% of their water resources. Agriculture is also the primary source of livelihood for 70% of the population in India. Even though agriculture has a vast impact, the water use efficiency is only 30%.

India figures as an exporter of its scarce water resource from the virtual water trade flow assessments done on a global level. India being a vast nation, it is important to note that there is virtual water exchange happening in both national and sub-national levels. Therefore, India has two types of virtual water flows. One is on the national level due to international trade. The other flow is on the interstate level (Katyaini & Barua, 2017). Figure 26.5 gives the virtual water flow in India.

26.5.1 National-level trade

The period of 1960–2010 is a representative timeline for understanding virtual water flows. The total virtual water import for the period of 1960–1970 was 3% of the total water used in production. This reduced to negligible values during the period of 1980–2000. Currently, the import stands at 1%. On the contrary, the export is around 25×10^9 m^3 which estimates to the annual demand of 13 million people or 4% of the total water involved in production per year (Goswami & Nishad, 2015).

Rice is a major crop grown and consumed in India with production increasing threefold from 50 million tonnes to 172 million tonnes between 1961 and 2018. Similarly, rice consumption has increased in line with this trend. Rice exports, on the other hand, were very low until 1990 when they began to rise and reached around 17 million tonnes in 2018, accounting for about 10% of total production. In terms of rice export, India is one of the major virtual water exporters with virtual water exports totalling around 18 billion m^3 and being exported to 199 countries and territories (Nishad

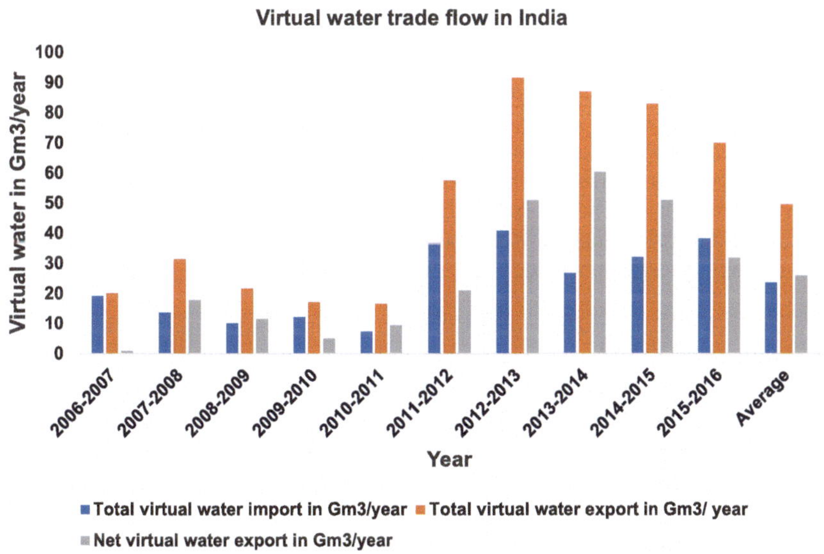

Figure 26.5 Virtual water flow in India for the period of 2006–2016. (*Source*: Adapted from data of SreeVidhya & Elango, 2019).

Table 26.1 Studies of virtual water trade flow of various agricultural products across various time periods.

Reference	Time Period	Commodity of Trade	Virtual Water Import (Bm^3/year)	Virtual Water Export (Bm^3/year)	Net Virtual Water (Bm^3/year)
Hoekstra and Hung (2002)	1995–1999	Crop and livestock	2.4	34.6	−32.2
Brindha (2017)	1999	Crop and livestock	31	8	23
Kumar and Jain (2007)	2001–2005	Crop and livestock	71.7	45.7	26
Gupta (2008)	2001–2006	Crop and livestock	41.2	51.6	−10.4
Goswami and Nishad (2015)	1961–2010	Crop and livestock	5.9	44	−38.1
Brindha (2017)	1986–2013	Crop and livestock	32.6	59	−26.4
SreeVidhya and Elango (2019)	2006–2016	Crop and livestock	23.72	49.69	
Katyaini *et al.* (2020)	1996–2014	Crops	21.24	89.23	

& Kumar, 2022). Table 26.1 lists out the studies of virtual water trade flow of various agricultural products across various time periods.

26.5.2 State-level trade

Katyaini and Barua (2017) conducted a study to understand the interstate virtual water flows. The study was conducted over two time periods, 1996–2005 and 2005–2014. The study resulted in establishing that north India has the highest water losses (219.8 TL/year), west has the highest water savings (11.1 TL/year), and South has the second highest savings (9.7 TL/year). The majority of virtual water inflows in the west and the south come from the north. It is important to note that virtual water flows are also directed from relatively water-rich east and north-east zones to water-scarce south and north zones.

Interstate VW-flows from 1996 to 2005 resulted in water savings in the west zone states of Gujarat, Maharashtra, and Rajasthan, as well as the south zone states of Andhra Pradesh, Karnataka, Kerala, and Tamil Nadu. VW-flows, on the other hand, resulted in water loss in Delhi, Haryana, Himachal Pradesh, Punjab, and Uttar Pradesh. While Maharashtra had the highest water savings (24.589 TL/year), Punjab had the highest water loss (3.724 TL/year), despite being a water-deficit state (Katyaini & Barua, 2016).

The virtual water flows from 2005 to 2014 resulted in Tamil Nadu having the highest water savings. It was followed by Maharashtra. Even in the second period, Punjab had the maximum water losses. This is precisely because Punjab has been practising unsustainable water management and has very few or no governmental policies for efficient water management. On the contrary, Tamil Nadu and Maharashtra are at the forefront of water resource management (Katyaini & Barua, 2017).

26.6 VIRTUAL WATER MANAGEMENT: POLICY RECOMMENDATIONS

A nation's overall sustainability is dependent on two major primary resources: cultivable land and water resources. While the former is essentially immobile, water can and is transported across countries via water embedded in agricultural items. Thus, a food or agricultural product trade network is accompanied by a network of virtual water trading. Goswami and Nishad (2015) have concluded that the magnitude of such a virtual trade of water can have an impact on a country's overall sustainability. In particular, a net export of water via agricultural export can slowly lead India to an irreversible loss of water sustainability.

For India, it is estimated that the water sustainability loss through virtual export is less than 300 years for water requirement and less than 500 years for available water. Increased food demand and climate change can further shorten these time scales (Goswami & Nishad, 2015; Valin *et al.*, 2014). The virtual water research at the subnational level has received little attention, despite its major role in water scarcity management (Hoff *et al.*, 2014). Katyaini and Barua (2017) and Mubako (2011) added that for a diverse country like India, the national average of virtual water flows should be investigated further at the subnational level. Addressing these concerns will be only effective when tough policy control is also introduced.

The idea of water scarcity was first considered in policy development in 1987 as part of National Water Policy (NWP). To further acknowledge the challenge of water scarcity, NWP in the years 2002 and 2012 suggested guidelines for sustainable water use (Katyaini *et al.*, 2020). Even though water footprint was included in the NWP 2012, states have yet to revise and implement their water policies in accordance with the NWP 2012. It is important to assess interstate VW-flows for water policies to adequately address water scarcity. Due to a lack of science-policy interface, incorporating the most recent scientific knowledge into water policy formulation remains a challenge. Another challenge is the difficulty in adapting the production system to water scarcity in order to make usage of water resources sustainable (Katyaini & Barua, 2017).

Katyaini *et al.* (2020) have identified the important parameters that should be addressed to create inclusive and sound policies for the Indian context. The policies should be aimed towards achieving water sustainability by improving the water use patterns. This measure will cater to both international and national virtual water trade flows. It is important to identify and involve the virtual water flows through inter-state trade. Prioritizing water allocation should be done for states with low water resources and high virtual water export. States like Punjab, Haryana, and Delhi on the one hand have acute water shortage and over exploitation of groundwater while, on the other, they are net virtual water exporters. As a result, there is a potential to inform state-specific strategies for reducing groundwater overexploitation and establishing subnational water allocation priorities.

Stakeholders' participation and literacy is vital in addressing virtual water flows. VW-flows research is relevant for enabling informed participation of stakeholders and increasing their water literacy because it draws on the links between water use, agricultural production, and inter-state movements. There is evidence from other studies, such as WF at farm level, on how farmers can best use the water and land resources to which they have access, and can find assistance in decisions on local crop choice (Hogeboom & Hoekstra, 2017). Furthermore, the sustainable food or agricultural policy must be based on zero trade deficit in virtual water.

26.7 SUMMARY

Global assessment of virtual water export reveals that India is a major water exporter country, exporting around 32 billion m^3 of water or 1.6% of total available water and contributing a 24% share of global virtual water export, while virtual water import is almost non-existent. India has become the world's leading virtual water exporter in terms of rice exports. The magnitude of virtual water trade may have an impact on water sustainability and thus food sustainability. Therefore, inclusive research on virtual water flows is a requirement for the current scenario in India. Along with research, science-based policies, policies with water scarcity assessment and virtual water calculation should be introduced.

REFERENCES

Aldaya M. M., Martínez-Santos P. and Llamas M. R. (2010). Incorporating the water footprint and virtual water into policy: reflections from the Mancha Occidental region, Spain. *Water Resources Management*, **24**(5), 941–958, https://doi.org/10.1007/s11269-009-9480-8

Allan J. A. (1998). Virtual water: a strategic resource global solutions to regional deficits. *Ground Water*, **36**(4), 545–546, https://doi.org/10.1111/J.1745-6584.1998.TB02825.X

Allen R. G. (2006). *FAO Irrigation and Drainage Paper Crop by. 56.*

Allen R. G., Pereira L. S., Raes D. and Smith M. (2006). Guidelines for computing crop water requirements. FAO Irrigation and Drainage Paper Crop by. *Remote Sensing of Environment*, **300**(56), 173. Retrieved from http://www.kimberly.uidaho.edu/water/fao56/

Brindha K. (2017). International virtual water flows from agricultural and livestock products of India. *Journal of Cleaner Production*, **161**, 922–930, https://doi.org/10.1016/j.jclepro.2017.06.005

Delbourg E. and Dinar S. (2020). The globalization of virtual water flows: explaining trade patterns of a scarce resource. *World Development*, **131**, 104917, https://doi.org/10.1016/j.worlddev.2020.104917

Dietzenbacher E. and Velázquez E. (2007). Analysing Andalusian virtual water trade in an input–output framework. *Regional Studies*, **41**(2), 185–196, https://doi.org/10.1080/00343400600929077

Goswami P. and Nishad S. N. (2015). Virtual water trade and time scales for loss of water sustainability: a comparative regional analysis. *Scientific Reports*, **5**, 1–11, https://doi.org/10.1038/srep09306

Guan D. and Hubacek K. (2007). Assessment of regional trade and virtual water flows in China. *Ecological Economics*, **61**(1), 159–170, https://doi.org/10.1016/j.ecolecon.2006.02.022

Gupta K. B. (2008). Water footprint of India and its implications for international trade in food products. *South Asia Economic Journal*, **9**(2), 419–433, https://doi.org/10.1177/139156140800900208

Harris F., Dalin C., Cuevas S., L N. R., Adhya T., Joy E. J. M., Scheelbeek P. F. D., Kayatz B., Nicholas O., Shankar B., Dangour A. D. and Green R. (2020). Trading water: virtual water flows through interstate cereal trade in India. *Environmental Research Letters*, **15**(12), 125005, https://doi.org/10.1088/1748-9326/abc37a

Hassan A., Saari M. Y. and Tengku Ismail T. H. (2017). Virtual water trade in industrial products: evidence from Malaysia. *Environment, Development and Sustainability*, **19**(3), 877–894, https://doi.org/10.1007/s10668-016-9770-2

Hoekstra A. Y. and Hung P. Q. (2002). Virtual water trade: a quantification of virtual water flows between nations in relation to international crop trade. *Value of Water Research Report Series*. NO. *11 IHE, 11*, https://doi.org/10.4324/9780203867785-15

Hoekstra A. Y. and Mekonnen M. M. (2012). The water footprint of humanity. *Proceedings of the National Academy of Sciences*, **109**(9), 3232–3237, https://doi.org/10.1073/pnas.1109936109

Hoff H., Döll P., Fader M., Gerten D., Hauser S. and Siebert S. (2014). Water footprints of cities – indicators for sustainable consumption and production. *Hydrology and Earth System Sciences*, **18**(1), 213–226, https://doi.org/10.5194/hess-18-213-2014

Hogeboom R. J. and Hoekstra A. Y. (2017). Water and land footprints and economic productivity as factors in local crop choice: the case of silk in Malawi. *Water*, **9**(10), 802, https://doi.org/10.3390/W9100802

Katyaini S. and Barua A. (2016). Water policy at science-policy interface – challenges and opportunities for India. *Water Policy*, **18**(2), 288–303, https://doi.org/10.2166/wp.2015.086

Katyaini S. and Barua A. (2017). Assessment of interstate virtual water flows embedded in agriculture to mitigate water scarcity in India (1996–2014). *Water Resources Research*, **53**(8), 7382–7400, https://doi.org/10.1002/2016WR020247

Katyaini S., Barua A. and Duarte R. (2020). Science-policy interface on water scarcity in India: giving 'visibility' to unsustainable virtual water flows (1996–2014). *Journal of Cleaner Production*, **275**, 124059, https://doi.org/10.1016/j.jclepro.2020.124059

Kumar V. and Jain S. K. (2007). Status of virtual water trade from India. *Current Science*, **93**(8), 1093–1099.

Lenzen M. (2009). Understanding virtual water flows: A multi-region input-output case study of Victoria. *Water Resources Research*, **45**, 9, https://doi.org/10.1029/2008WR007649

Leontief W. (1986). Input-Output Economics. Published by Oxford University Press, Inc., 200 Madison Avenue, New York, New York 10016.

Mekonnen M. M. and Hoekstra A. Y. (2012). A global assessment of the water footprint of farm animal products. *Ecosystems*, **15**(3), 401–415, https://doi.org/10.1007/S10021-011-9517-8

Mubako S. (2011). Frameworks for estimating virtual water flows among U.S. states. Dissertation, Southern Illinois University at Carbondale.

Nishad S. N. and Kumar N. (2022). Virtual water trade and its implications on water sustainability. Water Supply, **22**(2), 1704–1715, https://doi.org/10.2166/ws.2021.322

Pastor A. V., Palazzo A., Havlik P., Biemans H., Wada Y., Obersteiner M., Kabat P. and Ludwig F. (2019). The global nexus of food–trade–water sustaining environmental flows by 2050. *Nature Sustainability*, **2**(6), 499–507, https://doi.org/10.1038/s41893-019-0287-1

Ray C., McInnes D. and Sanderson M. (2018). Virtual water: its implications on agriculture and trade. *Water International*, **43**(6), 717–730, https://doi.org/10.1080/02508060.2018.1515564

Shi C. and Zhan J. (2015). An input-output table based analysis on the virtual water by sectors with the five northwest provinces in China. *Physics and Chemistry of the Earth*, **79–82**, 47–53, https://doi.org/10.1016/j.pce.2015.03.004

SreeVidhya K. S. and Elango L. (2019). Temporal variation in export and import of virtual water through popular crop and livestock products by India. *Groundwater for Sustainable Development*, **8**, 468–473, https://doi.org/10.1016/j.gsd.2019.01.002

Sun J. X., Yin Y. L., Sun S. K., Wang Y. B., Yu X. and Yan K. (2021). Review on research status of virtual water: The perspective of accounting methods, impact assessment and limitations. *Agricultural Water Management*, **243**, 106407, https://doi.org/10.1016/j.agwat.2020.106407

UNESCO (2022). Groundwater, Making the invisible visible groundwater, UN World Water Development Report 2022.

Valin H., Sands R. D., van der Mensbrugghe D., Nelson G. C., Ahammad H., Blanc E., Bodirsky B., Fujimori S., Hasegawa T., Havlik P., Heyhoe E., Kyle P., Mason-D'Croz D., Paltsev S., Rolinski S., Tabeau A., van Meijl H., von Lampe M. and Willenbockel D. (2014). The future of food demand: understanding differences in global economic models. *Agricultural Economics*, **45**(1), 51–67, https://doi.org/10.1111/agec.12089

Vallino E., Ridolfi L. and Laio F. (2021). Trade of economically and physically scarce virtual water in the global food network. *Scientific Reports*, **11**(1), 1–18, https://doi.org/10.1038/s41598-021-01514-w

Vanham D. and Bidoglio G. (2015). A review on the indicator water footprint for the EU28. *Ecological Indicators*, **26**, 61–75, https://doi.org/10.1016/j.ecolind.2012.10.021

Wu H., Jin R., Liu A., Jiang S. and Chai L. (2022). Savings and losses of scarce virtual water in the international trade of wheat, maize, and rice. *International Journal of Environmental Research and Public Health*, **19**(7), 4119, https://doi.org/10.3390/ijerph19074119

Yang H., Pfister S. and Bhaduri A. (2013). Accounting for a scarce resource: virtual water and water footprint in the global water system. *Current Opinion in Environmental Sustainability*, **5**(6), 599–606, https://doi.org/10.1016/j.cosust.2013.10.003

Zhan-Ming C. and Chen G. Q. (2013). Virtual water accounting for the globalized world economy: national water footprint and international virtual water trade. *Ecological Indicators*, **28**, 142–149, https://doi.org/10.1016/j.ecolind.2012.07.024

Zhao X., Yang H., Yang Z., Chen B. and Qin Y. (2010). Applying the input-output method to account for water footprint and virtual water trade in the Haihe River basin in China. *Environmental Science and Technology*, **44**(23), 9150–9156, https://doi.org/10.1021/es100886r

Index

T

U

V

W

Z

IWA Publishing's authorised EU representative for General Product Safety Regulations is Diane D'Arras, 15 rue Duret, 75116 Paris, France, e-mail: safety@iwap.co.uk.

Printed and bound by CPI Group (UK) Ltd, Croydon, CR0 4YY
12/05/2026
02108890-0001